WITHDRAWN
UTSA LIBRARIES

1374

Digital Multimedia Perception and Design

Gheorghita Ghinea, Brunel University, UK

Sherry Y. Chen, Brunel University, UK

IDEA GROUP PUBLISHING

Hershey • London • Melbourne • Singapore

Acquisitions Editor:	Michelle Potter
Development Editor:	Kristin Roth
Senior Managing Editor:	Amanda Appicello
Managing Editor:	Jennifer Neidig
Copy Editor:	Susanna Svidanovich
Typesetter:	Marko Primorac
Cover Design:	Lisa Tosheff
Printed at:	Yurchak Printing Inc.

Published in the United States of America by
    Idea Group Publishing (an imprint of Idea Group Inc.)
    701 E. Chocolate Avenue
    Hershey PA 17033
    Tel: 717-533-8845
    Fax: 717-533-8661
    E-mail: cust@idea-group.com
    Web site: http://www.idea-group.com

and in the United Kingdom by
    Idea Group Publishing (an imprint of Idea Group Inc.)
    3 Henrietta Street
    Covent Garden
    London WC2E 8LU
    Tel: 44 20 7240 0856
    Fax: 44 20 7379 0609
    Web site: http://www.eurospanonline.com

Copyright © 2006 by Idea Group Inc. All rights reserved. No part of this book may be reproduced, stored or distributed in any form or by any means, electronic or mechanical, including photocopying, without written permission from the publisher.

Product or company names used in this book are for identification purposes only. Inclusion of the names of the products or companies does not indicate a claim of ownership by IGI of the trademark or registered trademark.

Library of Congress Cataloging-in-Publication Data

Digital multimedia perception and design / Gheorghita Ghinea and Sherry Chen, editors.
    p. cm.
   Summary: "This book provides a well-rounded synopsis of the state-of-the-art in perceptual-based multimedia design"--Provided by publisher.
   ISBN 1-59140-860-1 (hardcover) -- ISBN 1-59140-861-X (softcover) -- ISBN 1-59140-862-8 (ebook)
   1. Multimedia systems. 2. System design. 3. Perception. I. Ghinea, Georghita. II. Chen, Sherry, 1961-
   QA76.575.G525 2006
   006.7--dc22
                                2006009287
British Cataloguing in Publication Data
A Cataloguing in Publication record for this book is available from the British Library.

All work contributed to this book is new, previously-unpublished material. The views expressed in this book are those of the authors, but not necessarily of the publisher.

Library
University of Texas
at San Antonio

Digital Multimedia Perception and Design

Table of Contents

Section I: Perceptual Modeling in Multimedia

Section IV: Multimedia Communication and Adaptation

Foreword

From Quality of Service to User Perception

Over the last 15 years, multimedia systems and communication have made great progress in investigating mechanisms, policies in our operating systems, distributed systems, networks to support different types of distributed multimedia applications and their Quality of Service (QoS) parameters such as throughput, end-to-end delay, loss rate, and others. However, as it was shown in many multimedia publications including books, research papers, and reports, satisfying QoS only is not enough to achieve the full user acceptance of developed multimedia services. QoS provisioning is a necessary but not a sufficient condition in the end-to-end acceptance of multimedia. Researchers in human-computer interfaces (HCI), machine learning, database, and other user-driven research domains indicated through many venues the need of transition from QoS to user perception, and the close connection between QoS and user perception. The results are there, but they are spread across different workshops, conferences, and journals of many disciplines such as systems, networks, vision, speech, and HCI. For example, looking back at the IFIP/IEEE International Workshop on Quality of Service (IWQoS), a leading working conference in QoS, only a few user perception papers relating QoS to user perception appeared over the last 10 years.

Hence, it is great to see the effort of this book, *Digital Multimedia Perception and Design*, which concentrates the various aspects of user perception in one place. The book provides a very broad view on multimedia perception and design, making user perception its focal point. The presented book nicely ties the QoS and user perception; gives a clean summarization of the mathematics of perception to support personalization of queries to multimedia databases; brings cognitive science into the picture to derive cognitive states of users and detect individual differences; stresses the need for visualization of multi-dimensional auditory information; explores affective computing to recognize user's happiness and wellbeing; includes user studies to understand the influence of multimedia senses such as smell, touch, sight, and sound, with respect to user perception in multimedia design; shows the differences in user perception if we take into account hand preferences between left- and right-handed users; explores theoretically and experimentally the usage of machine learning in multimedia design and perception; and investigates user perception when personalizing different types of multimedia content such as Web-based multimedia content, multimedia databases, video-on-demand, and others.

I expect that this book will be of great assistance to many researchers and developers involved in multimedia design and understanding of user perception at the multimedia human-computer interface. Furthermore, I am certain it will be very useful to many researchers and developers in the area of distributed multimedia systems and applications, since this book will assist in a better understanding of how to make the difficult transition from QoS to user perception. I congratulate the editors and the authors for their worthwhile and successful endeavor, and wish the readers, especially multimedia designers, a productive usage of this interesting material.

Klara Nahrstedt
University of Illinois at Urbana-Champaign, USA

Preface

Multimedia is an interdisciplinary, application-oriented technology that capitalizes on the multi-sensory nature of humans and the ability of computers to store, manipulate, and convey non-numerical information such as video, graphics, and audio in addition to numerical and textual information. Multimedia has the intrinsic goal of improving the effectiveness of computer-to-human interaction and, ultimately, of human-to-human communication.

While the Human Computer Interaction (HCI) community looks predominantly at the application layer and the telecommunications community at the lower end of the ISO OSI stack, little work has been published in bridging the gap between these two communities. Indeed, the human element is often neglected in Quality of Service (QoS) negotiation protocols. Not only does this have a negative and undesirable impact on the user's experience of multimedia, it also discards the potential for more economical resource allocation strategies With the proliferation of ubiquitous multimedia in predominantly bandwidth-constrained environments, more research is needed towards integrating and mapping perceptual considerations across the protocol stack and building truly end-to-end multimedia communication solutions.

Objectives and Structure of This Book

The book provides a comprehensive review of state-of-the-art in the area of perception-based multimedia design, covering both theories and applications in areas such as virtual reality, multi-sensory-based computing, ubiquitous communications, personalization, and adaptation according to user perceptual needs. The book covers a wide spectrum of up-to-date issues in the field, providing insight into analytical and architectural aspects of perception-based multimedia systems, as well as applications in real-world settings.

The book is structured in four main sections, reflecting the varied themes and interdisciplinarity associated with perceptual multimedia. Thus, the first section deals with perceptual modeling in multimedia, which is an important issue if one is to truly integrate human perceptual requirements into multimedia design. While multimedia is traditionally associated with the auditory and visual senses, there is no intrinsic reason why multimedia should not cover the other three senses of humans, and research done in this respect forms the focus of the second section of the book. The following section presents work dealing with a core issue at the heart of perceptual multimedia, namely the integration of human factors in the design and development of multimedia applications. Multimedia applications need to adapt at both ends: to the different characteristics and needs of the user, and to the changing requirements of the communication medium. This is the focus of the last section of the book, which following advances in networking technologies, presents research in multimedia communication and adaptation.

Perceptual Modeling in Multimedia

Understanding and modeling perceptual requirements is a challenging but very much exciting issue, due to the fickle nature of humans, their wants, and their needs. To address this issue, this section starts with **Chapter I**, by Cavallaro and Winkler, which presents work done in respect of emulating human vision. Their idea is to prioritize the visual data in order to improve the compression performance of video coders and the prediction performance of perceptual quality metrics. They propose an encoder and quality metric which incorporate visual attention and use a semantic segmentation stage, which takes into consideration some aspects of the cognitive behavior of people when watching multimedia video. Their semantic model corresponds to a specific human abstraction, which need not necessarily be characterized by perceptual uniformity. Since the semantics are defined through human abstraction, the definition of the semantic partition depends on the task to be performed, and they present the results of evaluating their proposed model in respect of segmenting moving objects and faces in multimedia video.

Chapter II summarizes the work on Mathematics of Perception performed by Chang's research team between 2000 and 2005. To support personalization, a search engine must comprehend users' query concepts (or perceptions), which are subjective and complicated to model. Traditionally, such query-concept comprehension has been performed through a process called "relevance feedback." Their work formulates relevance feedback as a machine-learning problem when used with a small, biased training dataset. His team has pioneered and developed a method of query-concept learning as the learning of a binary classifier to separate what a user wants from what she or he does not want, sorted out in a projected space. They have developed and published several algorithms to reduce data dimensions, to maximize the usefulness of selected training instances, to conduct learning on imbalanced datasets, to accurately account for perceptual similarity, to conduct indexing and learning in a non-metric, high-dimensional space, and to integrate perceptual features with keywords and contextual information.

In **Chapter III**, May sketches a semantic taxonomy of representational forms (or "sign types" in the terminology of semiotics) relevant for the compositional analysis, design, and indexing of multimodal multimedia, specifically in the domain of instructional design of learning objects, and contributes to current attempts to bridge the "semantic gap" between the technical and spatio-temporal description of media objects and the contextual perception and interpretation by users of their content.

Chapter IV rounds off this section. Gulliver proposes a model of distributed multimedia, to allow a more structured analysis of the current literature concerning video and audio information. The model segregates studies implementing quality variation and/or assessment into three discrete information abstractions (the network, media, and content levels) and from two perspectives (the technical and user perspectives). In so doing, Gulliver places current research in the context of a quality structure, and thus highlights the need for fuller incorporation of the user perspective in multimedia quality assessment.

Multimedia and the Five Senses

Gulliver's chapter is a good link to this section, as his model incorporates four of the human senses, olfactory (smell), tactile / haptic (touch), visual (sight), and auditory (sound), and naturally leads to work done in respect of making multimedia applications truly multi-sensual and multimodal. **Chapter V**, by Axelrod and Hone, explores affective computing, illustrating how innovative technologies are capable of emotional recognition and display. Research in this domain has emphasized solving the technical difficulties involved, through the design of ever more complex recognition algorithms. But fundamental questions about the use of such technology remain neglected. Can it really improve human-computer interaction? For which types of application is it suitable? How is it best implemented? What ethical considerations are there? This field is reviewed in this chapter, and the authors discuss the need for user-centered design in multimedia systems. They then conclude the chapter by describing and giving evidence from a study that explored some of the user issues in affective computing.

In **Chapter VI**, Bussell examines the use of haptic feedback in a multimedia simulation as a means of conveying information about physical science concepts. The case study presented investigates the effects of force feedback on children's conceptions of gravity, mass, and related concepts following experimentation with a force-feedback-enabled simulation. Guidelines for applying these technologies effectively for educational purposes are discussed. This chapter adds to the limited research on the application of haptic feedback for conceptual learning, and provides a basis for further research into the effects of computer-based haptic feedback on children's cognition.

In **Chapter VII**, Giannakis investigates the use of visual texture for the visualization of multi-dimensional auditory information. In his work, 20 subjects with a strong musical background performed a series of association tasks between high-level perceptual dimensions of visual texture and steady-state features of auditory timbre. The results indicated strong and intuitive mappings among (a) texture contrast and sharpness, (b) texture coarseness-granularity and compactness, and (c) texture periodicity and sensory dissonance. The findings contribute in setting the necessary groundwork for the application of empirically-derived auditory-visual mappings in multimedia environments.

Chapter VIII winds up this section. In their chapter, Crosby and Ikehara describe their research focused on deriving changing cognitive state information from the patterns of data acquired from the user with the goal of using this information to improve the presentation of multimedia computer information. Detecting individual differences via performance and psychometric tools is supplemented by using real-time physiological sensors. Described is an example computer task that demonstrates how cognitive load is manipulated. The different types of physiological and cognitive state measures are discussed, along with their advantages and disadvantages. Experimental results from eye tracking and the pressures applied to a computer mouse are described in greater detail. Finally, adaptive information filtering is discussed as a model for using the physiological information to improve computer performance. Study results provide support that we can create effective ways to adapt to a person's cognition in real time and thus facilitate real-world tasks.

Human Factors

Multimedia technology is widely applied to Web-based applications, which are faced by diverse users, in terms of their background, skills, and knowledge, and, we would argue, perceptual requirements. Therefore, human factors represent an essential issue in the design of perceptual multimedia systems, which is addressed in this section.

In **Chapter IX**, Ghinea and Chen investigate the effect of cognitive style on the perceived quality of distributed multimedia. They use two dimensions of cognitive style analysis – field dependent/independent and verbalizer/visualizer – and the quality of perception metric to characterize the human perceptual experience. This is a metric which takes into account multimedia's "infotainment" (combined informational and entertainment) nature, and comprises not only a human's subjective level of enjoyment with regards to multimedia content quality, but also his/her ability to analyze, synthesizes, and assimilate the informational content of such presentations. Their results show that multimedia content and dynamism are strong factors influencing perceptual quality.

Chapter X, by Kalyuga, provides an overview of theoretical frameworks and empirical evidence for the design of adaptive multimedia that is tailored to individual levels of user expertise to optimize cognitive resources available for learning. Recent studies indicate that multimedia design principles which benefit low-knowledge users may disadvantage more experienced ones due to the increase in the cognitive load required for the integration of presented information with the available knowledge base. The major implication for multimedia design is the need to tailor instructional formats to individual levels of expertise. The suggested adaptive procedure is based on empirically-established interactions between levels of user proficiency and formats of multimedia presentations (the expertise reversal effect), and on real-time monitoring of users' expertise using rapid cognitive diagnostic methods.

Chapter XI, by Tilinger and Sik-Lanyi, presents the differences between left- and right-handed persons in the aspect of computer-presented information and virtual realities. It introduces five test scenarios and their results addressing this question. They show that there are moderate differences between groups preferring different hands, and that the different needs of left- and right-handed people may play an important role in user-friendly interface and virtual environment design, since about a tenth of the population is left-handed. This could help to undo the difficulties which left-handed and ambidextrous individuals routinely encounter in their daily lives.

Multimedia Communication and Adaptation

To address the diverse needs of users and the underlying communication networks, adaptation is a necessity in distributed multimedia systems. The fluctuating bandwidth and time-varying delays of best-effort networks makes providing good quality streaming a challenge; when such quality has to take into account perceptual considerations, the problem is compounded. This section kicks off with **Chapter XII**, by Cranley and Murphy. Their work describes research that proposes that an optimal adaptation trajectory through the set of possible video encodings exists, and indicates how to adapt transmission in response to changes in network conditions in order to maximize user-perceived quality.

Chapter XIII, by Wikstrand, introduces a framework for understanding, developing, and controlling networked services and applications based on a division into three layers. While previous framework models mostly strive to be content- and usage-independent, his proposed model incorporates user considerations. The model is based on three layers: usage, application, and network. By setting targets for and measuring "utility" in appropriate layers of the model, it becomes possible to consider new ways to solve problems and of integrating perceptual requirements in the design of distributed multimedia systems.

Last, but by no means least, the final contribution of our book is **Chapter XIV**, by Germanakos and Mourlas. They investigate the new multi-channel constraints and opportunities emerged by mobile and wireless technologies, as well as the new user demanding requirements that arise. It further examines the relationship between the adaptation and personalization research considerations, and proposes a three-layer architecture for adaptation and personalization of Web-based multimedia content based on the "new" user profile, with visual, emotional, and cognitive processing parameters incorporated.

Summary

Perceptual multimedia design is an exciting but notoriously difficult research area — first, foremost, and last, because it attempts to integrate human requirements and considerations in multimedia communication systems and applications. This book is, to the best of our knowledge, the first book to reunite state-of-the-art research in this challenging area. It will be useful to professionals and academic researchers working in various areas, such as multimedia telecommunications, Web-based technologies, human-computer interaction, virtual reality, content infrastructure, customization, and personalization. However, this is a first small step. We hope that readers of our book will be imbued with passion and motivation to take this fascinating research area forward, capable of meeting the challenges of the 21st century.

Gheorghita Ghinea and Sherry Y. Chen

Acknowledgments

The successful completion of this book would not have been possible without the help of a number of people, to whom we wish to express our gratitude.

We are very grateful to the authors of this book for contributing their excellent contributions and to the reviewers for providing high-quality, constructive comments and helpful suggestions.

Moreover, we would like to thank all of the staff at Idea Group Inc, particularly Mehdi Khosrow-Pour, Jan Travers, and Kristin Roth for their invaluable assistance.

Last but certainly not least, this book would have been impossible without constant support, patience, encouragement, and love from our families.

Gheorghita Ghinea and Sherry Y. Chen

January 2006

Section I

Perceptual Modeling in Multimedia

Chapter I

Perceptual Semantics

Andrea Cavallaro, Queen Mary University of London, UK

Stefan Winkler,
National University of Singapore and Genista Corporation, Singapore

Abstract

The design of image and video compression or transmission systems is driven by the need for reducing the bandwidth and storage requirements of the content while maintaining its visual quality. Therefore, the objective is to define codecs that maximize perceived quality as well as automated metrics that reliably measure perceived quality. One of the common shortcomings of traditional video coders and quality metrics is the fact that they treat the entire scene uniformly, assuming that people look at every pixel of the image or video. In reality, we focus only on particular areas of the scene. In this chapter, we prioritize the visual data accordingly in order to improve the compression performance of video coders and the prediction performance of perceptual quality metrics. The proposed encoder and quality metric incorporate visual attention and use a semantic segmentation stage, which takes into account certain aspects of the cognitive behavior of people when watching a video. This semantic model corresponds to a specific human abstraction, which need not necessarily be characterized by perceptual uniformity. In particular, we concentrate on segmenting moving objects and faces, and we evaluate the perceptual impact on video coding and on quality evaluation.

Copyright © 2006, Idea Group Inc. Copying or distributing in print or electronic forms without written permission of Idea Group Inc. is prohibited.

Introduction

The development of new compression or transmission systems is driven by the need of reducing the bandwidth and storage requirements of images and video while increasing their perceived visual quality. Traditional compression schemes aim at minimizing the coding residual in terms of mean squared error (MSE) or peak signal-to-noise ratio (PSNR). This is optimal from a purely mathematical but not a perceptual point of view. Ultimately, perception is the more appropriate and more relevant benchmark. Therefore, the objective must be to define a codec that maximizes perceived visual quality such that it produces better quality at the same bit rate as a traditional encoder or the same visual quality at a lower bit rate (Cavallaro, 2005b).

In addition to achieving maximum perceived quality in the encoding process, an important concern for content providers is to guarantee a certain level of quality of service during content distribution and transmission. This requires reliable methods of quality assessment. Although subjective viewing experiments are a widely accepted method for obtaining meaningful quality ratings for a given set of test material, they are necessarily limited in scope and do not lend themselves to monitoring and control applications, where a large amount of content has to be evaluated in real-time or at least very quickly. Automatic quality metrics are desirable tools to facilitate this task. The objective here is to design metrics that predict perceived quality better than PSNR (Winkler, 2005a).

One of the common shortcomings of traditional video coders and quality metrics is the fact that they treat the entire scene uniformly, assuming that people look at every pixel of the image or video. In reality, we focus only on particular areas of the scene, which has important implications on the way the video should be analyzed and processed.

In this chapter, we take the above observations into account and attempt to emulate the human visual system. The idea is to prioritize the visual data in order to improve the compression performance of video coders and the prediction performance of perceptual quality metrics. The proposed encoder and quality metric incorporate visual attention and use a semantic segmentation stage (Figure 1). The semantic segmentation stage takes into account some aspects of the cognitive behavior of people when watching a video. To represent the semantic model of a specific cognitive task, we decompose each frame of the reference sequence into sets of mutually- exclusive and jointly-exhaustive segments. This semantic model corresponds to a specific human abstraction, which need not necessarily be characterized by perceptual uniformity. Since the semantics (i.e., the meaning) are defined through human abstraction, the definition of the semantic partition depends on the task to be performed. In particular, we will concentrate on segmenting moving objects and faces, and we will evaluate the perceptual impact on video coding and on quality evaluation.

The chapter is organized as follows: The section "Cognitive Behavior" discusses the factors influencing the cognitive behavior of people watching a video. The section "Semantic Segmentation" introduces the segmentation stage that generates a semantic partition to be used in video coding and quality evaluation. In "Perceptual Semantics for Video Coding" and "Perceptual Semantics for Video Quality Assessment", we describe

Copyright © 2006, Idea Group Inc. Copying or distributing in print or electronic forms without written permission of Idea Group Inc. is prohibited.

Figure 1. Flow diagram of the encoder and the quality metric that incorporate factors influencing visual attention and use a semantic segmentation stage

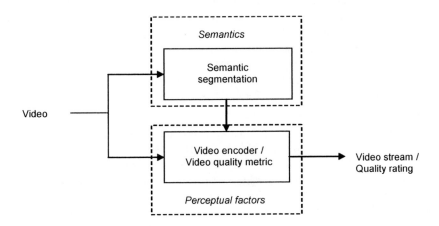

how the cognitive behavior can be incorporated into a video coder and a quality metric, respectively. Moreover, the compression performance of the proposed encoder and the prediction performance of the proposed metrics are discussed. In the final section, we draw some conclusions and describe the directions of our current work.

Cognitive Behavior

Visual Attention

When watching images or video, we focus on particular areas of the scene. We do not scan a scene in raster fashion; instead, our visual attention tends to jump from one point to another. These so-called saccades are driven by a fixation mechanism, which directs our eyes towards objects of interest (Wandell, 1995). Saccades are high-speed eye movements that occur at a rate of 2-3 Hz. We are unaware of these movements because the visual image is suppressed during saccades.

Studies have shown that the direction of gaze is not completely idiosyncratic to individual viewers. Instead, a significant number of viewers will focus on the same regions of a scene (Endo, 1994; Stelmach, 1994). Yarbus (1967) demonstrated in his classic experiments that the saccadic patterns depend on the visual scene as well as the cognitive task to be performed. In other words, "we do not see, we look" (Bajcsy, 1988, p. 1). We focus our visual attention according to task at hand and the scene content.

The reason why this behavior should be taken into account in video encoding and quality assessment is that our visual acuity is not uniform across the entire visual field. In

Copyright © 2006, Idea Group Inc. Copying or distributing in print or electronic forms without written permission of Idea Group Inc. is prohibited.

general, visual acuity is highest only in a relatively small cone around the optical axis (the direction of gaze) and decreases with distance from the center. This is partly due to the deterioration of the optical quality of the eye towards the periphery and partly due to the layout of the retina (Banks, Sekuler, & Anderson, 1991). The central region of the retina around the optical axis is called the fovea. It contains the highest density of cones, which are the photoreceptors responsible for vision under usual image or video viewing conditions. Outside of the fovea, the cone density decreases rapidly. This explains why vision is sharp only around the present focus of attention, and the perception of the peripheral field of vision is blurred. In other words, contrast sensitivity is reduced with increasing eccentricity (Robson & Graham, 1981).[1] This effect is also frequency-dependent, as shown in Figure 2.

While the processes governing perception are not completely understood, many different factors contributing to visual attention have been identified (Wolfe, 1998). These include simple stimulus properties such as contrast (regions with high contrast attract attention), size (large objects are more interesting than small ones), orientation, shape (edge-like features or objects with corners are preferred over smooth shapes), hue and intensity (an object with very different color from the background or specific bright colors stands out), or flicker and motion (see below). In general, an object or feature that stands out from its surroundings in terms of any of the earlier-mentioned factors is more likely to attract our attention. Additionally, the location (most viewers focus on the center of a scene), foreground/background relationship, and context (the task, personal interest, or motivation) influence visual attention. Finally, moving objects and people, in particular their faces (eyes, mouth) and hands, draw our attention.

Figure 2. Loss of spatial contrast sensitivity as a function of eccentricity and stimulus frequency (measured in cycles per degree [cpd]), based on a model by Geisler and Perry (1998)

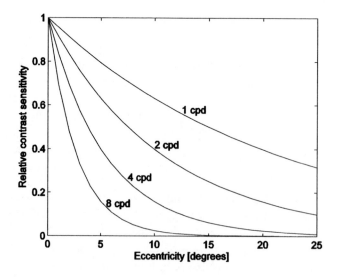

Copyright © 2006, Idea Group Inc. Copying or distributing in print or electronic forms without written permission of Idea Group Inc. is prohibited.

Computational Models of Attention

While most vision models and quality metrics are limited to lower-level aspects of vision, the cognitive behavior of people when watching video cannot be ignored. Cognitive behavior may differ greatly between individuals and situations, which makes it very difficult to generalize. Furthermore, little is known about the relative importance given to the factors listed above by the human visual system. Nonetheless, based on the above studies and factors contributing to visual attention, a variety of computational models of attention have been developed (Itti & Koch, 2001; Wolfe, 1998). Maeder, Diederich, and Niebur (1996) proposed constructing an importance map for a sequence as a prediction for the focus of attention, taking into account perceptual factors such as edge strength, texture energy, contrast, color variation, and homogeneity. Osberger and Rohaly (2001) developed a segmentation-based model with motion estimation and skin detection, taking into account color, contrast, size, shape, location, and foreground. It was calibrated and tested with eye movement data gathered from an experiment using a large database of still and video scenes. Recently, Navalpakkam and Itti (2005) demonstrated a model that uses task-specific keywords to find relevant objects in a scene and extracts their low-level features to build a visual map of task-relevance.

In this chapter, we focus on two important high-level aspects, namely faces and moving objects.

Attraction of Faces and Moving Objects

People and especially faces are among the objects attracting the most attention.. If there are faces of people in a scene, we will look at them immediately. Furthermore, because of our familiarity with people's faces, we are very sensitive to distortions or artifacts occurring in them. The importance of faces is underlined by a study of image appeal in consumer photography (Savakis, 2000). People in the picture and their facial expressions are among the most important criteria for image selection.

In a similar manner, viewers may also track specific moving objects in a scene. In fact, motion (in particular, in the periphery) tends to attract the viewers' attention. The spatial acuity of the human visual system depends on the velocity of the image on the retina: As the retinal image velocity increases, spatial acuity decreases. The visual system addresses this problem by tracking moving objects with smooth-pursuit eye movements, which minimizes retinal image velocity and keeps the object of interest on the fovea. Smooth pursuit works well even for high velocities, but it is impeded by large accelerations and unpredictable motion (Eckert, 1993). On the other hand, tracking a particular movement will reduce the spatial acuity for the background and objects moving in different directions or at different velocities. An appropriate adjustment of the spatio-temporal contrast sensitivity function (CSF) as outlined by Daly (1998) to account for some of these sensitivity changes can be considered as a first step in modeling such phenomena.

Copyright © 2006, Idea Group Inc. Copying or distributing in print or electronic forms without written permission of Idea Group Inc. is prohibited.

Based on the observations of this section, the proposed video coder and perceptual quality metric take into account both low-level and high-level aspects of vision. To achieve this, a segmentation stage is added to the video coder and to the quality metric to find regions of interest. The segmentation output then guides a pre-processing step in coding and a pooling process in video quality evaluation by giving more weight to the regions with semantically higher importance.

Semantic Segmentation

The high-level contribution to the cognitive behavior of people when watching a video is taken into account by means of semantic segmentation. To represent the semantic model of a specific cognitive task, we decompose each frame of the sequence into sets of mutually-exclusive and jointly-exhaustive segments. In general, the topology of this semantic partition cannot be expressed using homogeneity criteria, because the elements of such a partition do not necessarily possess invariant properties. As a consequence, some knowledge of the objects we want to segment is required. We will consider two cases of such *a priori* information, namely segmentation of faces and segmentation of moving objects. The final partition will be composed of foreground areas and background areas of each image.

Color segmentation and feature classification can be exploited to segment faces of people. A number of relatively robust algorithms for face segmentation are based on the fact that human skin colors are confined to a narrow region in the chrominance (C_b, C_r) plane (Gu, 1999), when the global illumination of the scene does not change significantly. Otherwise, methods based on tracking the evolution of the skin-color distribution at each frame based on translation, scaling, and rotation of skin color patches in the color space can be used (Sigal, 2004). One of the limits of approaches based on color segmentation is that the resulting partition may include faces as well as body parts. To overcome this problem, a combination of color segmentation with facial feature extraction (Hsu, 2002) can be used. Other approaches use feature classifiers only. Viola and Jones (2004) proposed a face detector based on a cascade of simple classifiers and on the integral image. This detector is a multi-stage classification that works as follows. First, features similar to Haar basis functions are extracted from the gray-level integral image. Next, a learning step called AdaBoost is used to select a small number of relevant features. This pruning process selects weak classifiers that depend on one feature only. Finally, the resulting classifiers are combined in a cascade structure. With such an approach, a face cannot be reliably detected when it appears small or not frontal. However, in such a situation the face in general does not attract viewers' attention as much as a frontal or large face. Therefore, this limitation will not affect the proposed approach in a significant way. Figure 3 shows an example of face detection for the test sequence Susie.

To segment moving objects, motion information is used as semantics. The motion of an object is usually different from the motion of background and other surrounding objects. For this reason, many extraction methods make use of motion information in video sequences to segment objects (Cavallaro, 2004a). Change detection is a typical tool used

Copyright © 2006, Idea Group Inc. Copying or distributing in print or electronic forms without written permission of Idea Group Inc. is prohibited.

Figure 3. (a) Example of face detection result on the test sequence Susie and (b) corresponding segmentation mask (white: foreground, black: background)

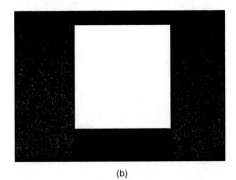

(a)   (b)

to tackle the problem of object segmentation based on motion. Different change detection techniques can be employed for moving camera and static camera conditions. If the camera moves, change detection aims at recognizing coherent and incoherent moving areas. The former correspond to background areas, the latter to video objects. If the camera is static, the goal of change detection is to recognize moving objects (foreground) and the static background. The semantic segmentation discussed here addresses the static camera problem and is applicable in the case of a moving camera after global motion compensation.

The change detector decides whether in each pixel position the foreground signal corresponding to an object is present. This decision is taken by thresholding the frame difference between the current frame and a frame representing the background. The frame representing the background is dynamically generated based on temporal information (Cavallaro, 2001). The thresholding aims at discarding the effect of the camera noise after frame differencing. A locally adaptive threshold, $T(i,j)$, is used that models the noise statistics and applies a significance test. To this end, we want to determine the probability that frame difference at a given position (i,j) is due to noise, and not to other causes. Let us suppose that there is no moving object in the frame difference. We refer to this hypothesis as the null hypothesis, H_0. Let $g(i,j)$ be the sum of the absolute values of the frame difference in an observation window of q pixels around (i,j). Moreover, let us assume that the camera noise is additive and follows a Gaussian distribution with variance σ. Given H_0, the conditional probability density function (pdf) of the frame difference follows a χ_q^2 distribution with q degrees of freedom defined by:

$$f(g(i,j) \mid H_o) = \frac{1}{2^{q/2} \sigma^q \Gamma(q/2)} g(i,j)^{(q-2)/2} e^{-g(i,j)^2/2\sigma^2} .$$

Copyright © 2006, Idea Group Inc. Copying or distributing in print or electronic forms without written permission of Idea Group Inc. is prohibited.

Figure 4. Example of moving object segmentation. (a) Original images; (b) moving object segmentation mask (white: foreground, black: background)

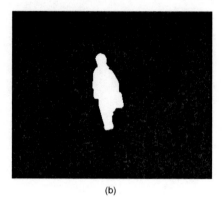

(a)   (b)

$\Gamma(\cdot)$ is the Gamma function, which is defined as $\Gamma(x+1) = x\Gamma(x)$, and $\Gamma(1/2) = \sqrt{\pi}$. To obtain a good trade-off between robustness to noise and accuracy in the detection, we choose $q=25$ (5-by-5 window centered in *(i,j)*). It is now possible to derive the significance test as:

$$P\{g(i,j) \geq \tau(i,j) \mid H_0\} = \frac{\Gamma(q/2, g(i,j)^2 / 2\sigma^2)}{\Gamma(q/2)}.$$

When this probability is smaller than a certain significance level, α, we consider that H_0 is not satisfied at the pixel position *(i,j)*. Therefore we label that pixel as belonging to a moving object. The significance level α is a stable parameter that does not need manual tuning over a sequence or for different sequences. Experimental results indicate that valid values fall in the range from 10^{-2} to 10^{-6}.

An example of a moving object segmentation result is shown in Figure 4.

The video partitions generated with semantic segmentation are then used in the video encoding process and in the quality evaluation metric as described in the following sections.

Perceptual Semantics for Video Coding

A video encoder that exploits semantic segmentation facilitates the improvement of video compression efficiency in terms of bandwidth requirements as well as visual quality at low bit rates. Using semantic decomposition, the encoder may adapt its

Copyright © 2006, Idea Group Inc. Copying or distributing in print or electronic forms without written permission of Idea Group Inc. is prohibited.

Figure 5. Using semantic segmentation for perceptual video coding

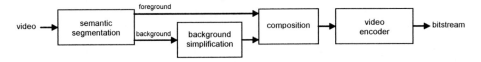

behavior to code relevant and non-relevant portions of a frame differently. This adaptation can be achieved with a traditional frame-based encoder by semantically pre-filtering the video prior to coding (Figure 5).

Video Coders

The semantic pre-filtering is obtained by exploiting semantics in a traditional frame-based encoding framework, such as MPEG-1. The use of the decomposition of the scene into meaningful objects prior to encoding, referred here as semantic pre-filtering, helps support low-bandwidth transmission. The areas belonging to the foreground class are used as the region of interest. The areas not included in the region of interest are lowered in importance by using a low-pass filter. The latter solution simplifies the information in the background, while still retaining essential contextual information. The simplification of the background allows for a reduction of information to be coded. The use of a simplified background so as to enhance the relevant objects aims at taking advantage of the task-oriented behavior of the human visual system for improving compression ratios. The work reported by Bradley (2003) demonstrates that an overall increase in image quality can be obtained when the increase in quality of the relevant areas of an

Figure 6. Simplification of contextual information: (a) Sample frame from the test sequence Hall Monitor; (b) simplification of the whole frame using a low-pass filter; (c) selective low-pass filtering based on semantic segmentation results allows one to simplify the information in the background while still retaining essential contextual information

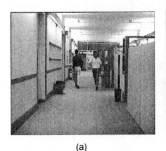

(a)   (b)   (c)

Copyright © 2006, Idea Group Inc. Copying or distributing in print or electronic forms without written permission of Idea Group Inc. is prohibited.

image more than compensates for the decrease in quality of the image background. An example of this solution, which aims at mimicking the blur occurring in the retina due to foveation (Itti, 2004) using high-level semantic cues, is reported in Figure 6.

Another way to take into account less relevant portions of an image before coding is to take advantage of the specifics of the coding algorithm. In the case of block-based coding, each background macro-block can be replaced by its DC value only. While this approach also has the effect of frequency reduction and loss of detail, it may lead to unacceptably strong blocking artifacts in the video.

An alternative approach is using object-based encoding. In such a case, the encoder needs to support the coding of individual video objects, such as for MPEG-4, object-based mode (Sikora, 1997). With this solution, each video object is assigned to a distinct object class, according to its importance in the scene. The encoding quality can be set depending on the object class: the higher the relevance, the higher the encoding quality. One advantage of this approach is the possibility of controlling the sequencing of objects. Video objects may be encoded with different degrees of compression, thus allowing better granularity for the areas in the video that are of more interest to the viewer. Moreover, objects may be decoded in their order of priority, and the relevant content can be viewed without having to reconstruct the entire image. Finally, sprite coding could be used when an image representing the background is sent to the receiver once and then objects are encoded and composed with an appropriate portion of the background at the receiver side. However, these solutions require that objects are tracked after segmentation (Cavallaro, 2005a), thus increasing the complexity of the approach.

Results

The results presented in this section illustrate the impact of semantic pre-filtering on the encoding performance of a frame-based coder. Sample results are shown from the MPEG-4 test sequence Hall Monitor and from the MPEG-7 test sequence Highway. Both sequences are in CIF format at 25 Hz. The background is simplified using a Gaussian 9x9 low-pass filter with $\mu=0$ and $\sigma=2$, where μ and σ are the mean and standard deviation of the filter, respectively. The TMPGEnc 2.521.58.169 encoder using constant bit-rate (CBR) rate control was used for the encoding.

Figure 7 shows a sample frame from each test sequence coded with MPEG-1 at 150 kbit/s with and without semantic pre-filtering. Figure 8 shows magnified excerpts of both test sequences coded with MPEG–1 at 150 kbit/s. Figure 8(a, b) shows a blue truck entering the scene at the beginning of the Highway sequence. Coding artifacts are less disturbing on the object in Figure 8(b) than in Figure 8(a). Moreover, the front-left wheel of the truck is only visible with semantic pre-filtering, Figure 8(b). Similar observations can be made for Figure 8(c, d), which shows the person that carries a monitor in the Hall Monitor sequence. The amount of coding artifacts is notably reduced by semantic pre-filtering as shown in Figure 8(d). In particular, the person's facial features and clothes are clearly visible in Figure 8(d), whereas they are corrupted by coding artifacts in Figure 8(c).

Copyright © 2006, Idea Group Inc. Copying or distributing in print or electronic forms without written permission of Idea Group Inc. is prohibited.

Figure 7. Comparison between standard MPEG–1 encoding and perceptually pre-filtered MPEG–1 encoding at 150 kbit/s. Frame 190 of Hall Monitor: (a) MPEG–1 encoded; (b) MPEG–1 encoded after perceptual pre-filtering; Frame 44 of Highway (c) MPEG–1 encoded; (d) MPEG–1 encoded after perceptual pre-filtering. It is possible to notice that coding artifacts are less disturbing on objects of interest in (b) and (d) than in (a) and (c), respectively.

(a)

(b)

(c)

(d)

Perceptual Semantics for Video Quality Assessment

A quality measure based on semantic segmentation is useful for end-to-end communication architectures aiming at including perceptual requirements for an improved delivery of multimedia information. Predicting subjective ratings using an automatic visual quality metric with higher accuracy than peak signal-to-noise ratio (PSNR) has

Copyright © 2006, Idea Group Inc. Copying or distributing in print or electronic forms without written permission of Idea Group Inc. is prohibited.

Figure 8. Details of sequences encoded at 150 kbit/s: (a) Frame 44 of Highway MPEG–1 encoded; (b) Frame 44 of Highway MPEG–1 encoded after perceptual pre-filtering (c) Frame 280 of Hall Monitor MPEG–1 encoded; (d) Frame 280 of Hall Monitor MPEG–1 encoded after perceptual pre-filtering; notice that the front left wheel of the truck is visible with semantic pre-filtering only (b). The person's facial features and clothes are clearly visible in (d), whereas they are corrupted by coding artifacts in (c).

(a)   (b)   (c)   (d)

been the topic of much research in recent years. However, none of today's metrics quite achieves the reliability of subjective experiments.

Two approaches for perceptual quality metric design can be distinguished (Winkler, 2005b): One class of metrics implements a general model of low-level visual processing in the retina and the early visual cortex. Metrics in this class typically require access to the reference video for difference analysis. The other class of metrics looks for specific features in the image, for example, compression artifacts arising from a certain type of codec, and estimates their annoyance.

To demonstrate the use of perceptual semantics with both classes of metrics, we describe a specific implementation from each class here. The first is a full-reference perceptual distortion metric (PDM) based on a vision model, and the second is a no-reference video quality metric based on the analysis of common artifacts. These are then combined with perceptual semantics to achieve a better prediction performance (Cavallaro & Winkler, 2004).

Full-Reference Quality Metric

The full-reference PDM is based on a contrast gain control model of the human visual system that incorporates spatial and temporal aspects of vision as well as color

Copyright © 2006, Idea Group Inc. Copying or distributing in print or electronic forms without written permission of Idea Group Inc. is prohibited.

Figure 9. Block diagram of the perceptual distortion metric (PDM)

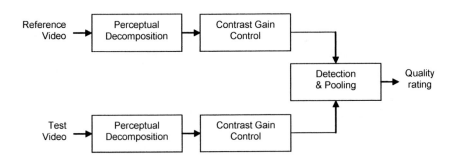

perception (Winkler, 1999). A block diagram of this metric is shown in Figure 9. The metric requires both the test sequence and the corresponding reference sequence as inputs.

The input videos are first converted from YUV or RGB to an opponent-color space, which is closer to the perceptual representation of color information. Each of the resulting three components is then subjected to a spatio-temporal decomposition, which is implemented as a filter bank. This emulates the different mechanisms in the visual system, which separate the input signals according to their color, frequency, orientation, and other characteristics. The sensitivity to the information in every channel is different, so the channels are weighted according to contrast sensitivity data.

The subsequent contrast gain control stage models pattern masking, which is one of the most critical components of video quality assessment, because the visibility of distortions is highly dependent on the local background. Masking is strongest between stimuli located in the same perceptual channel, and becomes weaker for stimuli with differing characteristics. The perceptual decomposition allows us to take many of these intra- and inter-channel masking effects into account. Within the process of contrast gain control, masking occurs through the inhibitory effect of the various perceptual channels.

At the output of the contrast gain control stage, the differences between the reference and the test video are computed for each channel and then combined ("pooled") into a distortion measure or quality rating according to the rules of probability summation.

No-Reference Quality Metric

The no-reference quality metric estimates visual quality based on the analysis of common coding and transmission artifacts found in the video (Winkler & Campos, 2003). These artifacts include blockiness, blur, and jerkiness. A block diagram of this metric is shown in Figure 10. It does not need any information about the reference sequence.[2] The use of a no-reference metric is particularly interesting here because semantic segmentation does not require a reference video either.

Copyright © 2006, Idea Group Inc. Copying or distributing in print or electronic forms without written permission of Idea Group Inc. is prohibited.

Figure 10. Block diagram of no-reference quality metric

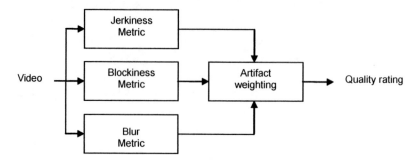

The blockiness metric looks for typical block patterns in a frame. These block patterns are an artifact common to DCT block-based image and video compression methods such as JPEG or MPEG. Blocks often form regular horizontal or vertical structures in an image, which appear as characteristic peaks in the spectrum of spatial difference signals. If the blocks are not regular, each block has to be identified individually based on its texture and boundaries. The spectral power peaks and the boundary contrasts are used to compute a measure of overall blockiness.

Blur is a perceptual measure of the loss of fine detail in the image. It is due to the attenuation of high frequencies at some stage of the recording or encoding process. The blur metric is based on the assumption that edges, which often represent object contours, are generally sharp. Compression has a smearing effect on these edges. The blur metric thus looks for significant edges in an image and measures their width. This local edge smear is then averaged over the entire image for a global estimate of blur.

The jerkiness metric is a temporal metric dedicated to measuring the motion rendition in the video. Jerkiness is a perceptual measure of frozen pictures or motion that does not look smooth. The primary causes of jerkiness are network congestion and/or packet loss. It can also be introduced by the encoder dropping or repeating frames in an effort to achieve the given bit-rate constraints. The jerkiness metric takes into account the video frame rate as well as the amount of motion activity in the video.

Both of these metrics can make local quality measurements in small sub-regions over a few frames in every video. The process of combining these low-level contributions into an overall quality rating is guided by the result of the semantic segmentation stage as shown in Figure 11. The metrics described here attempt to emulate the human visual system to prioritize the visual data in order to improve the prediction performance of a perceptual distortion metric. The resulting semantic video quality metric thus incorporates both low-level and high-level aspects of vision. Low-level aspects are inherent to the metric and include color perception, contrast sensitivity, masking and artifact visibility. High-level aspects take into account the cognitive behavior of an observer when watching a video through semantic segmentation.

Copyright © 2006, Idea Group Inc. Copying or distributing in print or electronic forms without written permission of Idea Group Inc. is prohibited.

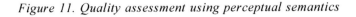

Figure 11. Quality assessment using perceptual semantics

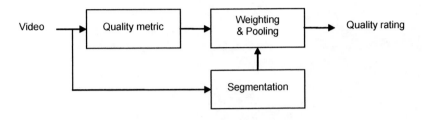

Evaluation and Results

To evaluate the improvement of the prediction performance due to face segmentation, we compare the ratings of the regular full-frame metrics with those of the segmentation-supported metrics for different data sets. We focus our evaluation on the special case when faces attract the attention of an observer. We quantify the influence of variations of quality of the relevant parts of the content on perceived video quality.

We used test material and subjective ratings from three different subjective testing databases:

1. VQEG Phase I database (VQEG, 2000): This database comprises mainly TV material with 16 test conditions. Three relevant scenes were selected from this database to evaluate the full-reference PDM.

2. PC video database (Winkler, 2001): This database was created with CIF-size video and various DirectShow codecs at bit-rates of 1-2 Mb/s, for a total of eight test conditions. We picked two scenes from this database to evaluate the full-reference PDM.

3. Internet streaming database (Winkler & Campos, 2003): This database contains clips encoded with MPEG-4, Real Media and Windows Media at 256 and 512 kb/s, as well as some packet loss (seven conditions in total). Four scenes from this database were used. Due to the test conditions here, these sequences cannot be properly aligned with the reference. Therefore, we use this set for the evaluation of our no-reference metric.

The scenes we selected from these databases contain faces at various scales and with various amounts of head and camera movements. Some examples are shown in Figure 12.

The results of the evaluation for our three data sets are shown in Figure 13. Segmentation generally leads to a better agreement between the metric's predictions and the subjective

Copyright © 2006, Idea Group Inc. Copying or distributing in print or electronic forms without written permission of Idea Group Inc. is prohibited.

Figure 12. Sample frames from selected test sequences

Figure 13. Prediction performance with and without segmentation; correlations are shown for the metrics applied uniformly across the full frame (gray bars), with an emphasis on the areas resulting from face segmentation (black bars), and the complementary emphasis (white bars)

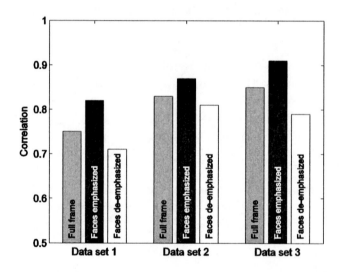

Copyright © 2006, Idea Group Inc. Copying or distributing in print or electronic forms without written permission of Idea Group Inc. is prohibited.

ratings. The trend is the same for all three data sets, which indicates that face segmentation is useful for augmenting the predictions of quality metrics. Moreover, giving lower weights to faces leads to a reduced prediction performance. This observation also supports the conclusion that semantic segmentation helps predicting perceived quality. As expected, the improvement is most noticeable for the scenes where faces cover a substantial part of the frame. Segmentation is least beneficial for sequences in which the faces are quite small and the distortions in the background introduced by some test conditions are more annoying to viewers than in other regions (as is the case with data set 2).

Summary

We analyzed factors influencing visual attention and how they can be incorporated in a video encoder and in a perceptual distortion metric using segmentation. We looked at the special cases of moving object and face segmentation, and we evaluated the performance improvement of a frame-based encoder and video quality metrics combined with a segmentation stage.

In video coding, the advantages of segmentation support have been demonstrated at low-bit rates with test sequences containing moving objects. In quality evaluation, the advantages of segmentation support have been demonstrated with test sequences showing human faces, resulting in better agreement of the predictions of a perceptual quality metric with subjective ratings. Even if the segmentation support is adding complexity to the overall system, the algorithms used for the detection of faces and moving objects are running in real-time on standard personal computers. Moreover, the proposed perceptual video coding method does not add any overhead to the bitstream, thus allowing interoperability with existing decoders.

Current work includes the extension of the technique and results to the evaluation of multimedia information, including graphics, audio, and video for applications such as augmented reality and immersive spaces.

References

Bajcsy, R. (1988). Active perception. *Proceedings of the IEEE, 76*(8), 996-1005.

Banks, M. S., Sekuler, A. B., & Anderson, S. J. (1991). Peripheral spatial vision: Limits imposed by optics, photoreceptors, and receptor pooling. *Journal of the Optical Society of America, A 8*, 1775-1787.

Bradley, A. P., & Stentiford, F. W. M. (2003). Visual attention for region of interest coding in JPEG-2000. *Journal of Visual Communication and Image Representation, 14*, 232-250.

Copyright © 2006, Idea Group Inc. Copying or distributing in print or electronic forms without written permission of Idea Group Inc. is prohibited.

Cavallaro, A., & Ebrahimi, T. (2001). Video object extraction based on adaptive background and statistical change detection. *Proceedings of SPIE Visual Communications and Image Processing, Vol. 4310* (pp. 465-475). San Jose, CA: SPIE.

Cavallaro, A., & Ebrahimi, T. (2004). Interaction between high-level and low-level image analysis for semantic video object extraction. *Journal on Applied Signal Processing, 2004*(6), 786-797.

Cavallaro, A., Steiger, O., & Ebrahimi, T. (2005). Tracking video objects in cluttered background. *IEEE Trans. Circuits and Systems for Video Technology, 15*(4), 575-584.

Cavallaro, A., & Steiger, O., & Ebrahimi, T. (in press). Semantic video analysis for adaptive content delivery and automatic description. *IEEE Trans. Circuits and Systems for Video Technology, 15*(19), 1200-1209.

Cavallaro, A., & Winkler, S. (2004). Segmentation-driven perceptual quality metrics. *Proceedings of the International Conference on Image Processing* (pp. 3543-3546). Singapore: IEEE.

Daly, S. (1998). Engineering observations from spatio-velocity and spatio-temporal visual models. *Proceedings of SPIE Human Vision and Electronic Imaging: Vol. 3299* (pp. 180-191). San Jose, CA: SPIE.

Eckert, M. P., & Buchsbaum, G. (1993). The significance of eye movements and image acceleration for coding television image sequences. In A. B. Watson (Ed.), *Digital images and human vision* (pp. 89-98). Cambridge, MA: MIT Press.

Endo, C., Asada, T., Haneishi, H., & Miyake, Y. (1994). Analysis of the eye movements and its applications to image evaluation. *Proceedings of the Color Imaging Conference* (pp. 153-155). Scottsdale, AZ: IS&T/SID.

Geisler, W. S., & Perry, J. S. (1998). A real-time foveated multi-resolution system for low-bandwidth video communication. *Proceedings of SPIE Human Vision and Electronic Imaging: Vol. 3299* (pp. 294-305).

Gu, L., & Bone, D. (1999). Skin color region detection in MPEG video sequences. *Proceedings of the International Conference on Image Analysis and Processing* (pp. 898-903). Venice, Italy: IAPR.

Hsu, L., Abdel-Mottaleb, M., & Jain, A. K. (2002). Face detection in color images. *IEEE Trans. Pattern Analysis and Machine Intelligence, 24*(5), 696-706.

Itti, L. (2004). Automatic foveation for video compression using a neurobiological model of visual attention. *IEEE Trans. Image Processing, 13*(10), 1304-1318.

Itti, L., & Koch, C. (2001). Computational modeling of visual attention. *Nature Reviews Neuroscience, 2*(3), 194-203.

Maeder, A., Diederich, J., & Niebur, E. (1996). Limiting human perception for image sequences. *Proceedings of SPIE Human Vision and Electronic Imaging: Vol. 2657* (pp. 330-337). San Jose, CA: SPIE.

Navalpakkam, V., & Itti, L. (2005). Modeling the influence of task on attention. *Vision Research, 45*, 205-231.

Copyright © 2006, Idea Group Inc. Copying or distributing in print or electronic forms without written permission of Idea Group Inc. is prohibited.

Osberger, W., & Rohaly, A. M. (2001). Automatic detection of regions of interest in complex video sequences. *Proceedings of SPIE Human Vision and Electronic Imaging: Vol. 4299* (pp. 361-372). San Jose, CA: SPIE.

Robson, J. G., & Graham, N. (1981). Probability summation and regional variation in contrast sensitivity across the visual field. *Vision Research, 21*, 409-418.

Savakis, A. E., Etz, S. P., & Loui, A. C. (2000). Evaluation of image appeal in consumer photography. *Proceedings of SPIE Human Vision and Electronic Imaging: Vol. 3959* (pp. 111-120). San Jose, CA: SPIE.

Sigal, L., Sclaroff, S., & Athitsos, V. (2004). Skin color-based video segmentation under time-varying illumination. *IEEE Trans. Pattern Analysis and Machine Intelligence, 26*(7), 862-877.

Sikora, T. (1997). The MPEG-4 video standard verification model. *IEEE Trans. Circuits and Systems for Video Technology, 7*(1), 19-31.

Stelmach, L. B., & Tam, W. J. (1994). Processing image sequences based on eye movements. *Proceedings of SPIE Human Vision, Visual Processing, and Digital Display: Vol. 2179* (pp. 90-98). San Jose, CA: SPIE.

Viola, P., & Jones, M. (2004). Robust real-time object detection. *International Journal of Computer Vision, 57*(2), 137-154.

VQEG. (2000). Final report from the Video Quality Experts Group on the validation of objective models of video quality assessment. Retrieved from http://www.vqeg.org/

Wandell, B. (1995). *Foundations of vision.* Sunderland, MA: Sinauer Associates.

Winkler, S. (1999). A perceptual distortion metric for digital color video. *Proceedings of SPIE Human Vision and Electronic Imaging: Vol. 3644* (pp. 175-184). San Jose, CA: SPIE.

Winkler, S. (2001). Visual fidelity and perceived quality: Towards comprehensive metrics. *Proceedings of SPIE Human Vision and Electronic Imaging: Vol. 4299* (pp. 114-125). San Jose, CA: SPIE.

Winkler, S. (2005a). *Digital video quality—Vision models and metrics.* Chichester, UK: John Wiley & Sons.

Winkler, S. (2005b). Video quality metrics—A review. In H. R. Wu & K. R. Rao (Eds.), *Digital video image quality and perceptual coding* (pp. 155-179). Boca Raton, FL: CRC Press.

Winkler, S., & Campos, R. (2003). Video quality evaluation for Internet streaming applications. *Proceedings of SPIE Human Vision and Electronic Imaging: Vol. 5007* (pp. 104-115). San Jose, CA: SPIE.

Wolfe, J. M. (1998). Visual search: A review. In H. Pashler (Ed.), *Attention* (pp. 13-74). London: University College London Press.

Wright, M. J., & Johnston, A. (1983). Spatio-temporal contrast sensitivity and visual field locus. *Vision Research, 23*(10), 983-.989.

Yarbus, A. (1967). *Eye movements and vision.* New York: Plenum Press.

Copyright © 2006, Idea Group Inc. Copying or distributing in print or electronic forms without written permission of Idea Group Inc. is prohibited.

Endnotes

[1]   Note that this applies only to spatial contrast sensitivity. The temporal contrast sensitivity does not vary across the visual field (Wright & Johnston, 1983).

[2]   The no-reference metric used here is part of Genista's quality measurements solutions; see http://www.genista.com for more information.

Copyright © 2006, Idea Group Inc. Copying or distributing in print or electronic forms without written permission of Idea Group Inc. is prohibited.

Chapter II

The Mathematics of Perception:

Statistical Models and Algorithms for Image Annotation and Retrieval

Edward Y. Chang, University of California, USA

Abstract

This chapter summarizes the work on Mathematics of Perception performed by my research team between 2000 and 2005. To support personalization, a search engine must comprehend users' query concepts (or perceptions), which are subjective and complicated to model. Traditionally, such query-concept comprehension has been performed through a process called "relevance feedback." Our work formulates relevance feedback as a machine-learning problem when used with a small, biased training dataset. The problem arises because traditional machine learning algorithms cannot effectively learn a target concept when the training dataset is small and biased. My team has pioneered in developing a method of query-concept learning as the learning of a binary classifier to separate what a user wants from what she or he does not want, sorted out in a projected space. We have developed and published several algorithms to reduce data dimensions, to maximize the usefulness of selected training instances, to conduct learning on unbalanced datasets, to accurately account for perceptual similarity, to conduct indexing and learning in a non-metric, high-dimensional space, and to integrate perceptual features with keywords and contextual

Copyright © 2006, Idea Group Inc. Copying or distributing in print or electronic forms without written permission of Idea Group Inc. is prohibited.

information. The technology of mathematics of perception encompasses an array of algorithms, and has been licensed by major companies for solving their image annotation, retrieval, and filtering problems.

Introduction

The principal design goal of a visual information retrieval system is to return data (images or video clips) that accurately match users' queries (for example, a search for pictures of a deer). To achieve this design goal, the system must first comprehend a user's query concept (i.e., a user's perception) thoroughly, and then find data in the low-level input space (formed by a set of perceptual features) that match the concept accurately (Chang, Li, Wu, & Goh, 2003). Statistical learning techniques can assist achieving the design goal via two complementary avenues: *semantic annotation* and *query-concept learning*.

Semantic annotation provides visual data with semantic labels to support keyword-based searches (for example, landscape, sunset, animals, and so forth). Several researchers have proposed semi-automatic annotation methods to propagate keywords from a small set of annotated images to the other images (e.g., Barnard & Forsyth, 2000; Chang, Goh, Sychay, & Wu, 2003; Goh, Chang, & Cheng, 2001; Lavrenko, Manmatha, & Jeon, 2003). Although semantic annotation can provide some relevant query results, annotation is often subjective and narrowly construed. When it is, query performance may be compromised. To thoroughly understand a query concept, with all of its semantics and subjectivity, a system must obtain the target concept from the user directly via query-concept learning (Chang & Li, 2003; Tong & Chang, 2001). Semantic annotation can assist, but not replace, query-concept learning.

Both semantic annotation and query-concept learning can be cast into the form of a supervised learning problem (Hastie, Tibshirani, & Friedman, 2001; Mitchell, 1997), which consists of three steps. First, a representative set of perceptual features (e.g., color, texture, and shape) is extracted from each training instance. Second, each training feature-vector x_i is assigned semantic labels g_i. Third, a classifier $f(.)$ is trained by a supervised learning algorithm, based on the labeled instances, to predict the class labels of a query instance x_q. Given a query instance x_q represented by its low-level features, the semantic labels g_q of x_q can be predicted by $g_q = f(x_q)$. In essence, these steps learn a mapping between the perceptual features and a human perceived concept or concepts.

At first it might seem that traditional supervised learning methods such as neural networks, decision trees, and Support Vector Machines could be directly applied to perform semantic annotation and query-concept learning. Unfortunately, such traditional learning algorithms are not adequate to deal with the technical challenges posed by these two tasks. To illustrate, let D denote the number of low-level features, N the number of training instances, N^+ the number of positive training instances, and N^- the number of negative training instances ($N = N^+ + N^-$). Two major technical challenges arise:

1.   Scarcity of training data. The features-to-semantics mapping problem often comes up against the $D > N$ challenge. For instance, in the query-concept learning

Copyright © 2006, Idea Group Inc. Copying or distributing in print or electronic forms without written permission of Idea Group Inc. is prohibited.

scenario, the number of low-level features that characterize an image (D) is greater than the number of images a user would be willing to label (N) during a relevance feedback session. As pointed out by Donoho (2000), the theories underlying "classical" data analysis are based on the assumptions that D < N, and N approaches infinity. But when D > N, the basic methodology which was used in the classical situation is not similarly applicable.

2. Imbalance of training classes. The target class in the training pool is typically outnumbered by the non-target classes ($N^- \gg N^+$). For instance, in a k-class classification problem where each class has about the same number of training instances, the target class is outnumbered by the non-target classes by a ratio of k-1:1. The class boundary of imbalanced training classes tends to skew toward the target class when k is large. This skew makes class prediction less reliable.

We use an example to explain the above technical challenges. Figures 1-3 show an example query using a *Perception-Based Image Retrieval* (PBIR) prototype (Chang, Cheng, Lai, Wu, Chang, & Wu, 2001) developed at UC Santa Barbara on a *300,000*-image dataset. This dataset had been manually annotated by professionals. The prototype extracted perceptual features (described in the next section) from the images, using the PBIR learning methods to combine keywords and perceptual features in a synergistic way to support retrieval. The figures demonstrate how a query concept is learned in an iterative process by the PBIR search engine to improve search results. The user interface shows two frames. The frame on the left-hand side is the feedback frame, on which the user marks images relevant to his or her query concept. On the right-hand side, the search engine returns what it interprets as matching this far from the image database.

Figure 1. Cat query initial screen

Copyright © 2006, Idea Group Inc. Copying or distributing in print or electronic forms without written permission of Idea Group Inc. is prohibited.

To query "cat," we first enter the keyword *cat* in the query box to get the first screen of results in Figure 1. The right-side frame shows a couple of images containing domestic cats, but several images containing tigers or lions. This is because many tiger/lion images were annotated with "wild cat" or "cat." To disambiguate the concept, we click on a couple of domestic cat images on the feedback frame (left side, in gray borders). The search engine refines the class boundary accordingly, and then returns the second screen in Figure 2. In this figure, we can see that the images in the result frame (right side) have been much improved. All returned images contain a domestic cat or two. After we perform another couple of rounds of feedback to make some further refinements, we obtain the more satisfactory results shown in Figure 3.

This example illustrates three critical points. First, keywords alone cannot retrieve images effectively because words may have varied meanings or senses. This is called the *word-aliasing* problem. Second, the number of labeled instances that can be collected from a user is limited. Through three feedback iterations, we can gather just $16 \times 3 = 48$ training instances, whereas the feature dimension of this dataset is *144* (Chang, Li, & Li, 2000). Since most users would not be willing to give more than three iterations of feedback, we encounter the problem of scarcity of training data. Third, the negatives outnumber the relevant or positive instances being clicked on. This is known as the problem of imbalanced training data.

In the remaining sections of this article, we present our proposed statistical methods for dealing with the above challenges for mapping perceptual features to perception. We depict the perceptual features we use to characterize images. We then present. the training-data scarcity problem in more detail and outlines three proposed remedies: *active learning, recursive subspace co-training,* and *adaptive dimensionality reduc-*

Figure 2. Cat query after iteration #2

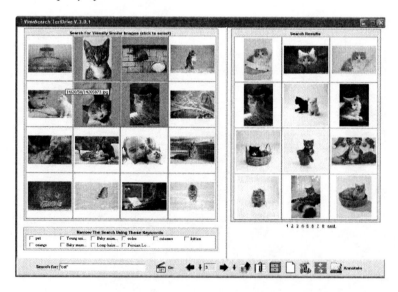

Copyright © 2006, Idea Group Inc. Copying or distributing in print or electronic forms without written permission of Idea Group Inc. is prohibited.

Figure 3. Cat query after iteration #3

tion. The section titled "Imbalance of Training Classes" discusses the problem of training-class imbalance and outlines two remedies, *class-boundary alignment* and *quasi-bagging*. Finally, we sketch EXTENT, a proposed architecture for combining context and content (perceptual features) to annotate images, and CDAL, an active learning algorithm that considers both keywords and perceptual features.

Perceptual Features

The first step to model perception is to extract features from images that can represent how the human visual process perceives an image (Chang, Li, & Li, 2000). Our vision tends to pay less attention to the border pixels, and it distributes effort unevenly by paying closer attention to ambiguous regions. Our visual system adjusts to the environment and adapts to the visual goals. For instance, we may not be able to tell if a figure is a real person or a statue at first glance. If we pay closer attention to the figure's surface (texture), we may be able to identify it as a statue or person. We thus believe that an image search engine must also be adaptive to the goals of a search task. We can think of our visual system as being divided into two parts: Our eyes (the front-end) perceive images, and our brain (the back-end that is equipped with a knowledge database and an inference engine) recognizes images. The front-end collects visual data for the back-end to allow high-level processing. The back-end instructs the front-end to zoom, pan, and collect

Copyright © 2006, Idea Group Inc. Copying or distributing in print or electronic forms without written permission of Idea Group Inc. is prohibited.

visual data with different features. (A feature can be regarded as a particular way of perceiving an image.) The front-end responds flexibly in perceiving visual data by selecting, ordering, and weighting visual features differently. These features can be categorized into *holistic* and *local* perceptual features.

Holistic perceptual feature: For each image, we first extract global features, which include *color, texture, shape*, and so forth (Rui, Huang, & Chang, 1999). We divide pixels on an image into color sets. A color set is the collection of pixels in an image that have the same color. We divide the visible color space into eleven culture colors (Tong & Chang, 2001). An image thus consists of at most eleven color sets, where a color set is characterized by three attributes of the image: its color, moments, and centroid. These attributes are defined as follows:

- **Color:** The culture color of the pixel collection. Each culture color spans a range of wavelengths. We keep the mean and variance of the color set on the three axes: hue, saturation, and intensity. We also record the number of pixels belonging to the color set.

- **Moments:** The moment invariants (Hu, 1962) of the color set. Moment invariants are properties of connected regions in binary images that are invariant to translation, rotation, and scaling.

- **Centroid:** The center of gravity of the color set in x and y.

We also extract texture features in three orientations (vertical, horizontal, and diagonal) and three resolutions (coarse, medium, and fine). Both color and texture are characterized in multiple resolutions. Shapes are treated as attributes of color and texture and characterized by moments.

Local perceptual features: Scale Invariant Feature Transform (SIFT) has become a very popular local feature representation in recent years (Lowe, 1999, 2004). SIFT extracts local, distinctive, invariant features. SIFT consists of four major stages: 1) scale-space extreme detection, 2) keypoint localization, 3) orientation assignment, and 4) keypoint descriptor. Features generated by SIFT have been shown to be robust in matching across an affined distortion, change in 3D viewpoint, addition of noise, and change of illumination.

The key of our features is that they represent images in multiple scales, orientations, and aspects. These features will then be combined in a non-linear way to formulate a query concept. Our multi-resolution representation not only provides a rich representation, it can also significantly conserve computational cost (see Chang & Li, 2003, for details).

Scarcity of Training Data

Some of the first image retrieval systems required a sample picture from the user to begin a search (e.g., Flickner et al., 1995; Gupta & Jain, 1997; Manjunath & Ma, 1996; Smith &

Copyright © 2006, Idea Group Inc. Copying or distributing in print or electronic forms without written permission of Idea Group Inc. is prohibited.

Chang, 1996). Yet, the limitations of that procedure quickly become obvious. A user, for example, might not have a picture of "panda" when he or she begins to seek one. Even if a query image with a panda is available, the system cannot precisely tell what the user seeks. The user might seek for an animal panda, a group of pandas, or a panda toy. Relevance feedback[1] provides the means for understanding the user's query concept better, by letting him or her give feedback regarding which images come closer to the query concept. But users might be unwilling to take the time for numerous iterations. If one begins with a Google text search, for example, it would be time-consuming to examine even just a small number of documents listed by the search engine. Fortunately for image retrieval, the time it takes to decide whether an image is relevant to a query concept is much shorter. Even so, we cannot always rely on users to provide a large number of relevant/irrelevant instances to use for training. Thus, a typical learning task for the classifier might be to infer a target concept with fewer than one hundred training instances, although the number of low-level features (the dimension of the *input space*) is usually much higher ($D > N$). (The distinction between input space and *feature space* is important. Input space refers to the space where the feature vectors reside. Feature space refers to the space onto which the feature vectors are projected via kernel methods (Aizerman, Braverman, & Rozonoer, 1964).)

Semantic annotation (attaching text labels to images) suffers from training-data scarcity in a slightly different way. Although it might be technically feasible to label a large number of training images, doing so would not be economically practical. Given a fixed amount of budget and time to label N training instances, we would like the labeled pool to provide maximum information to enable a classifier to propagate annotation with maximal accuracy (Chang, Goh, Sychay, & Wu, 2003).

Both query-concept learning and annotation are problems related to supervised learning. Supervised learning embodies two approaches: *generative* and *discriminant*. The generative approach estimates the probability density function for each class (or concept), and then uses the density function to predict an unseen instance. The discriminant approach predicts the posterior probability of an unseen instance directly. In cognitive science, the most popular probabilistic model employed for the task of classification is Bayesian Networks (BNs) (e.g., Tenenbaum 1999; Tenenbaum & Griffiths, 2001). BNs are easy to interpret, if the model of a concept can be learned. However, it is well known that learning a BN model with even just a moderate number of features is an NP-hard problem (Friedman & Koller, 2000). In the $D > N$ scenario that image perception comes up against, learning a BN is simply impossible. (Stated in an analogical way, one cannot solve D variables with $N < D$ equations).

The discriminant approach, on the other hand, can at least offer a solution even when $D > N$. In particular, the dual formulation of *Support Vector Machines* (SVMs) (Vapnik, 1995) consists of N free variables to solve, independent of the value of D. Even though the solution might overfit the data when $N > D$, it is the best that one can achieve under the circumstance. Practical experience tells us that if training instances are well chosen and representative, the concern of overfitting can be substantially mitigated.

To address the training-data scarcity ($D > N$) challenge for mapping features to perception, we consider three approaches. First, given N training instances from which to select, we choose the N most informative (most useful) instances for learning the target concept. Second, using the selected N instances, we infer additional training

Copyright © 2006, Idea Group Inc. Copying or distributing in print or electronic forms without written permission of Idea Group Inc. is prohibited.

instances for increasing N. Third, we would employ dimensionality reduction methods for reducing D.

Approach 1: Making N Instances Most Useful

The first challenge of *query-concept learning* is to find some relevant objects so that the concept boundary (between images that are relevant and those that are irrelevant) can be fuzzily identified. For this discussion, we assume that keywords are not available to assist query processing. We would like to see how far statistical methods could go to learn a concept from image content (content refers to perceptual features like color, texture, and shape). Finding a relevant object can be difficult if only a small fraction of the dataset satisfies the target concept. For instance, suppose the number of desired objects in a one-million-image dataset is *1,000* (i.e., *0.1%* of the dataset). If we randomly select *20* objects per round for users to identify relevant objects, the probability of finding a positive sample after five rounds of random sampling is just *10%*—clearly not acceptable.

We can improve the odds with an intelligent sampling method like the *Maximizing Expected Generalization Algorithm* (MEGA) (Chang & Li, 2003), which finds relevant instances quickly, to initialize *query-concept learning*. To explain without getting into mathematical details, MEGA uses given negative labeled instances to remove more negative instances. In addition, MEGA uses positive instances to eliminate irrelevant features. More specifically, MEGA models a query concept in k-CNF, which can express almost all practical query concepts; and MEGA uses k-DNF to model the sampling space. For instance, a 2-CNF of three features a, b, and c can be written as $(a \wedge b \wedge c \wedge (a \vee b) \wedge (a \vee b) \wedge (a \vee b))$; and a 2-DNF as $(a \vee b \vee c) \vee (a \wedge b) \vee (a \wedge b) \vee (a \wedge b))$. A positively-labeled instance can eliminate terms in k-CNF, and hence makes it more general. A negatively-labeled instance can eliminate terms in k-DNF, and hence makes it more restrictive to remove additional negative instances. The two complementary steps—concept refinement and sample selection—increase the probability of finding relevant instances, even if only negative instances are available initially. To continue with the one-million-image dataset example, if we can reduce the search space from one million to thirty thousand by reducing negative instances, we simultaneously increase the odds of finding a positive instance from *0.1%* to *3.3%*. Thus, we increase from *10%* to *97%* the probability of finding one positive instance after five rounds of *20* random samples.

Once some relevant and some irrelevant instances have been identified, we can employ SVMActive (Tong & Chang, 2001) to refine the class boundary. Intuitively, SVMActive works by combining the following three ideas:

1.   SVMActive regards the task of learning a target concept as one of learning an SVM binary classifier. An SVM captures the query concept by separating the relevant images from the irrelevant images with a hyperplane in a projected space (see Figure 4), usually a very high-dimensional one. The projected points on one side of the hyperplane are considered relevant to the query concept and the rest irrelevant.

Copyright © 2006, Idea Group Inc. Copying or distributing in print or electronic forms without written permission of Idea Group Inc. is prohibited.

Figure 4. SVM hyperplane, a simple illustrative example

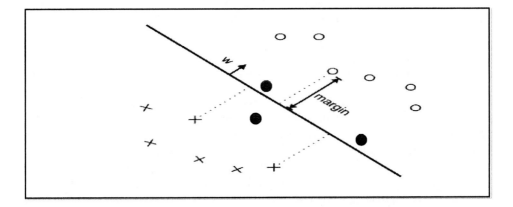

2. SVMActive learns the classifier quickly via active learning. The active part of SVMActive selects the most informative labeled instances with which to train the SVM classifier. This step ensures fast convergence to the query concept in a small number of feedback rounds. The most informative instances are the ones in the margin nearest the hyperplane (the filled circles in Figure 4). These instances are most uncertain about their positive or negative membership, whereas the points farther away from the hyperplane on the positive (negative) side are more certain to be positive (negative).

3. Once the classifier is trained, SVMActive returns the top-k most relevant images. These are the k images farthest from the hyperplane on the query concept side.

In short, we can use MEGA to find initial relevant images, and then switch to SVMActive for refining the binary classifier and ranking returned images. This method can effectively learn a query concept when the *complexity* of a concept is not too high. We will define *concept complexity* later in the chapter, and present remedies for searching concepts of high complexity.

Remarks

Tibshirani remarks that when N < D, the employed statistical model should be simple (or the number of free variables should be small) so that the resulting solution would not overfit the data. The nearest shrunken centroids scheme, proposed by Tibshirani and Hastie (2003), shows promising results on analyzing DNA micro-array data, where N < D. However, perceptual features interact with one another in high orders to depict an image concept (Chang & Li, 2003). Therefore, employing a kernel function to work with

Copyright © 2006, Idea Group Inc. Copying or distributing in print or electronic forms without written permission of Idea Group Inc. is prohibited.

SVMs seems to be the logical choice. As previously mentioned, the dual formulation of SVMs consists of only N free variables (N α parameters), independent of D. Carefully selecting training data through active learning can alleviate concerns about overfitting. For limitations of active learning, see Goh, Chang, and Lai (2004).

Approach 2: Increasing N

The second approach is to increase the number of training instances from the given N instances (Lai, Chang, & Cheng, 2002). As counter-intuitive as this may sound, the scarcity of negative training instances is more problematic than the scarcity of positive training instances for our learning algorithms. This is because describing a concept such as "tigers" may be challenging, but describing a negated concept (in this case the "non-tiger" concept) can require potentially infinite information. (Only a finite number of images are needed to adequately portray the tiger concept, but the number of non-tiger images is infinite.) Negative training instances are significantly easier to come by, but at the same time, the number of negative samples needed to depict a negated concept can be enormous.   Thus, we need a substantial number of negative training instances to establish the class boundary accurately between negative and positive instances.

Let us revisit the SVMActive algorithm. At each round of sampling, it selects objects that are in the margin and also close to the dividing hyperplane. If the margin is wide (because of the lack of support at the negative side), the probability is low that a positive training instance can be found within that wide margin. Our goal is to find enough positive training instances to depict the target concept. However, the difficulty in gathering sufficient negative training instances slows down progress for finding positive training instances. The learning process is analogous to mining gold in a stream. The most productive way to harvest gold (relevant instances) is to quickly filter out water, mud, and sand (negative-labeled instances) (Lai, Chang, & Cheng, 2002).

For solving the problem of insufficient training data, *transductive learning* has been proposed to work with different learning algorithms for various applications (Joachims, 1999). The basic idea of transduction is to leverage the unlabeled data near the labeled data. Suppose the nearest neighbors of negative training instances are always negative. We can then use the neighboring instances to substantially increase the negative-labeled pool. However, transduction must be performed with care, since the mislabeling of data caused by wrong "guesses" can degrade the results. The performance of transduction is, unfortunately, consistently inconsistent. Its performance could be both application-dependent and dataset-dependent. For image retrieval, transduction may not be suitable for the following two reasons:

1.   Nearest-neighbor sampling. Most image query-concepts are not convex, nor do they reside continuously in the input space. For instance, flowers of different colors tend to spread out in the color-feature dimension. Thus, to describe either a flower or a non-flower concept, we might need to *explore* the input space more aggressively. That is, we want to look for other possible flower (non-flower) images away from the neighborhood of the already known flower (non-flower) images.

Copyright © 2006, Idea Group Inc. Copying or distributing in print or electronic forms without written permission of Idea Group Inc. is prohibited.

Transductive learning, however, *exploits* only in the nearest neighborhood of the existing labeled instances. Suppose a picture of a lake (one of many non-flower instances) is labeled negative. Using the pictures neighboring this lake picture, which are likely to contain lakes or something visually similar, may not be very productive in refining the description of the negated concept (non-flower).

2.   Noise factor. Mislabeled data due to noise can reduce the prediction accuracy of the classifier. In the usual case where relevant images are rare, performing trans- duction runs a very high risk of introducing false relevant instances, which thereby reduces class-prediction accuracy.

To avoid the aforementioned problems, Lai, Chang, and Cheng (2002) proposed employ- ing *co-training* to increase N in a more meaningful and accurate way. The problem of using a large unlabeled sample pool to boost performance of a learning algorithm is considered within the framework of co-training (Blum & Mitchell, 1998). A broad definition of co-training is to provide each instance with multiple distinct views. Distinct views can be provided by different learning algorithms — that is, by using MEGA and SVMActive at the same time. Information sharing between distinct views can increase especially the negative-labeled pool. Here, we propose another co-training method, which provides each training instance with distinct views, via subspace learning, to boost the pool of negative-labeled instances. As in query-concept learning scenario, where the percentage of positive-labeled instances is rare, boosting the positive pool directly may not be productive. Thus, we attempt to boost the negative pool to increase the probability of finding positive instances. This method recursively conducts sub- space co-training at each feedback iteration in the following steps:

1.   Divide the input space into G subspaces.

2.   Conduct parallel training in these G subspaces using labeled training dataset L.

3.   Use the G resulting learners to label the unlabeled pool and yield a new set of labeled instances L'.

4.   Replace L with L'.

5.   Go back to Step 1 until no more labeled instances can be inferred (that is, until L' $=\varnothing$).

Approach 3: Reducing D

When D is large, the training instances in the high-dimensional input space become very sparse (even when N > D). The sparsely populated space causes two theoretical obstacles for conducting statistical inferences. First, given a query instance, its nearest neighbors tend to be equally distanced when data are uniformly distributed (Beyer, Goldstein, Ramakrishnan, & Shaft, 1999). (Since most image datasets are not uniformly distributed in the input space, this concern may not be applicable to our case.) Second, the nearest neighbors are not "local" to the query point. This second mathematical problem makes class prediction difficult.

Copyright © 2006, Idea Group Inc. Copying or distributing in print or electronic forms without written permission of Idea Group Inc. is prohibited.

Many methods have been proposed for performing dimensionality reduction, such as principal component analysis (PCA) and independent component analysis (ICA). Recently, several dimensionality-reduction algorithms have been proposed to find nonlinear manifolds embedded in a high-dimensional space. Among the proposed methods, Isomap (Tenenbaum, Silva, & Langford, 2000), local linear embedding (LLE) (Bengio, Paiement, Vincent, Delalleau, Roux, & Quimet, 2004), and kernel PCA (KPCA) (Ham, Lee, Mika, & Scholkopf, 2004) have been applied to tasks of image clustering and image retrieval (He, Ma, & Zhang, 2004). However, the scenarios in which manifold learning appears to be effective seem rather contrived. For instance, the widely used Swiss-roll example (Bengio et al., 2004) is a three-dimensional structure on which data are densely populated. Several examples of face and object images presented in the literatures change their poses only slightly from one image to another, so manifolds can easily be discovered. Moreover, none of the demonstrated scenarios have considered noise as a factor to seriously challenge manifold learning.

The major drawback of these dimension-reduction methods in the context of query-concept learning is that they reduce dimensionality in a universal mode with respect to the entire dataset, and with respect to all users. Reducing dimensionality in such a one-size-fits-all way entirely disregards individuals' subjectivity. Our work (Li & Chang, 2003; Li, Chang, & Wu, 2002) shows that a perceptual distance function is not only *partial* in dimensionality, but also *dynamic* in the subspace where the similarity between two images is measured. We thus propose using *Dynamic Partial Function* (DPF) to perform adaptive dimensionality reduction.

Let us briefly summarize why DPF works. Some important work in the cognitive psychology community has provided ample evidence to show that not all features need to be considered in measuring similarity between a pair of objects (images) (Goldstein, 1999; Tversky, 1977). Most recently, Goldstone (1994) showed that similarity perception is the *process* that determines the respects (features or attributes) for measuring similarity. More precisely, a similarity function for measuring a pair of objects is formulated only *after* the objects are compared, not *before* the comparison is made; and the respects for the comparison are *dynamically* activated in this formulation process. Let us use a simple example to explain. Suppose we are asked to name two places that are similar to England. Among several possibilities, "Scotland and New England" could be a reasonable answer. However, the respects in which England is similar to Scotland differ from those in which England is similar to New England. If we use the shared attributes of England and Scotland to compare England and New England, the latter pair might not seem similar, and vice versa. Thus, a distance function using a fixed set of respects (such as the traditional weighted Minkowski function) cannot capture objects that are similar in different sets of respects. (See Li & Chang (2003) for further information about how DPF dynamically select a subset of features in a pair-dependent way.)

A seamless way to integrate this partial and dynamic dimensionality reduction method into query-concept learning is to introduce a new DPF kernel function as:

$$K(\mathbf{x}, \mathbf{x'}) = \exp\left(\Sigma j \in \Delta m \; |\mathbf{x}j - \mathbf{x'}j|^2 / 2\,\sigma^2\right),$$

Copyright © 2006, Idea Group Inc. Copying or distributing in print or electronic forms without written permission of Idea Group Inc. is prohibited.

where Dm is the set that contains m out of D dimensions that have the smallest feature differences between instances **x** and **x'**, the j suffix denotes the jth feature of image **x** and **x'**.

Imbalance of Training Classes

A subtle but serious problem that hinders the process of classification is the skewed class boundary caused by imbalanced training data. To illustrate this problem, we use a 2D checkerboard example. Figure 5 shows a checkerboard of 200´200 square divided into four quadrants. The top-left and bottom-right quadrants are occupied by negative (majority) instances, but the top-right and bottom-left quadrants contain only positive (minority) instances. The lines between the classes are the "ideal" boundary that separates the two

Figure 5. Two-class checker board setup

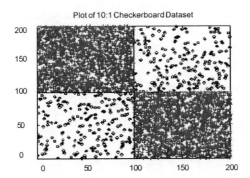

Figure 6. Biased boundary

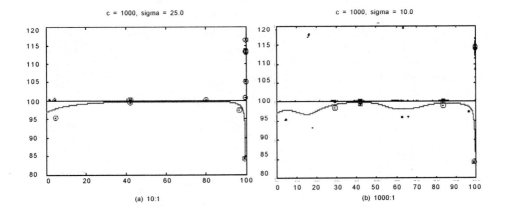

Copyright © 2006, Idea Group Inc. Copying or distributing in print or electronic forms without written permission of Idea Group Inc. is prohibited.

classes. In the rest of this section, we will use *positive* when referring to minority instances, and negative when referring to majority instances.

Using SVMs as the classifier, we plot in Figures 6(a) and (b) the boundary distortion between the two left quadrants of the checkerboard under two different negative/positive training-data ratios. Figure 6(a) shows the SVM class boundary when the ratio of the number of negative instances (in the quadrant above) to the number of positive instances (in the quadrant below) is *10:1*. Figure 6(b) shows the boundary when the ratio increases to *10,000:1*. The boundary in Figure 6(b) is much more skewed toward the positive quadrant than the boundary in Figure 6(a), which results in a higher incidence of false negatives.

Both query-concept learning and semantic annotation may suffer from the imbalanced training-class problem, since the target class is always overwhelmingly outnumbered by the other classes (that is, the relevant class is often rare and outnumbered by the irrelevant class). We consider two approaches for remedying the imbalanced training-data problem for SVMs: the algorithmic approach and the data-processing approach. The algorithmic approach focuses on modifying the kernel function or the kernel matrix, and the data-processing approach focuses on processing the training data.

Approach 1: Algorithmic Remedy

Examining the class prediction function of SVMs

$$\text{sgn}(f(\mathbf{x})) = \Sigma i\, y_i\, \alpha_i\, K(\mathbf{x}, \mathbf{x}_i) + b,\, i = 1, \ldots n,$$

we can identify three parameters that affect the decision outcome: b, αi, and K. In Wu and Gang (2003, 2004), we show that the only effective algorithmic method for tackling the problem of imbalanced training classes is through modifying the kernel K or the kernel matrix[2] $K_{i,j}$ adaptively to the training data distribution. We propose using adaptive conformal transformation to dynamically enlarge (or scale) the spatial resolution of minority and majority areas around the separating boundary surface in the feature space, such that the SVM margin could be increased while the boundary becomes less skewed. Conformal transformation modifies the kernel function by changing the Riemannian geometrical structure induced by the mapping function. For data that do not have a vector-space representation (such as video sequence data), we apply transformation directly on the kernel matrix. Our experimental results on both UCI and real-world image/video datasets show the proposed algorithm to be very effective in correcting the skewed boundary. For further details, see Wu and Gang (2003, 2004).

Copyright © 2006, Idea Group Inc. Copying or distributing in print or electronic forms without written permission of Idea Group Inc. is prohibited.

Approach 2: Data Processing Remedy

The *bagging* process subsamples training data into a number of *bags*, trains each bag, and aggregates the decisions of the bags to make final class predictions. We propose *quasi-bagging*, which performs subsampling somewhat differently. For each bag, quasi-bagging subsamples only the majority class, but takes the entire minority class. By subsampling just the majority class, we improve class balance in each training bag. At the same time, using a large number of bags reduces the variance caused by the subsampling process. See Chang, Li, Wu, and Goh (2003) for details.

Synergistic Context and Content Fusion

So far, we have addressed the problem of query-concept learning: how to find the right combination of perceptual features to depict a concept. We have not discussed how keywords can be tagged on images to provide text-based searches. Using perceptual features alone, certain semantics cannot be depicted (perceptual features cannot reliably tell e.g., where, when, or who). Keyword annotation can be complementary to perceptual features. In this section, we broaden the scope of our discussion by bringing keyword annotation into the "picture." We first discuss how keywords can be semi-automatically provided for images; we then discuss how content and context (defined shortly) can be effectively combined to improve image organization and retrieval quality.

Applying Keywords to Images

Traditional keyword annotation has been performed either by people manually, or by statistical methods that propagate annotation from annotated images to unannotated ones via image content. Although progress has been made in semi-automatic annotation, the achieved quality is far from satisfactory. At least a couple of obstacles make image annotation difficult. First, images are taken under varying environmental conditions (e.g., lighting, movements, etc.) and with varied resolutions. The same Eiffel tower pictures taken by two different models of cameras, from two different angles, and at different times of day seem different enough to be mislabeled. Thus, keyword propagation via image content can be unreliable. Second, although some progress has been made (e.g., Tu, Chen, Yuille, & Zhu, 2005), reliable object segmentation is not attainable in the foreseeable future (Barnard & Forsyth, 2000). When we are able to segment objects with high accuracy, recognizing objects will be easier.

Fortunately, recent hardware advances are beginning to assist. With advances in lenses on camera phones, many pundits believe that camera phones will replace digital and film cameras within a few years. A camera with the ability to record time, location, and voice, and to provide camera parameters (which can tell the environmental characteristics at the photo-taking time) offers renewal of optimism for achieving high-quality automatic annotation.

Copyright © 2006, Idea Group Inc. Copying or distributing in print or electronic forms without written permission of Idea Group Inc. is prohibited.

Semantic labels can be roughly divided into two categories: *wh* labels and *non-wh* labels. Wh-semantics include: time (*when*), people (*who*), location (*where*), landmarks (*what*), and event (inferred from *when, who, where,* and *what*). Providing the *when* and *where* information is trivial. Already cameras can provide time, and we can easily infer location from GPS or CellID. However, determining the *what* and *who* requires contextual information in addition to time, location, and photo content. More precisely, contextual information provided by cameras includes *time, location, camera parameters, user profile, and voice.* Content of images consists of perceptual features, which can be categorized into *holistic features* (color, shape, and texture characteristics of an image), and *local features* (edges and salient points of regions or objects in an image). Besides context and content, another important source of information (which has been largely ignored) is the relationships between semantic labels (which we refer to as *semantic ontology*). To explain the importance of having a semantic ontology, let us consider an example with two semantic labels: *outdoor* and *sunset.* When considering contextual information alone, we may be able to infer the outdoor label from camera parameters: focal length and lighting condition (Boutell & Luo, 2005). We can infer *sunset* from time and location (Naaman, Paepcke, & Garcia-Molina, 2003). Notice that inferring outdoor and sunset do not rely on any common contextual modality. However, we can say that a sunset photo is outdoor with certainty (but not the other way). By considering semantic relationships between labels, photo annotation can take advantage of contextual information in a "transitive" way.

To fuse *context, content,* and *semantic ontology* in a synergistic way, Chang (2005) proposed EXTENT, an inferencing framework to generate semantic labels for photos. EXTENT uses an influence diagram (Wu, Chang, & Tseng, 2005) to conduct semantic inferencing. The variables on the diagram can either be *decision variables* (i.e., causes), or *chance variables* or (i.e., effects). For image annotation, decision variables include time, location, user profile, and camera parameters. Chance variables are semantic labels. However, some variables may play both roles. For instance, time can affect some camera parameters (such as exposure time and flash on/off), and hence these camera parameters are both decision and chance variables. Finally, the influence diagram connects decision variables with chance variables with arcs weighted by *causal strength* (Novick & Cheng, 2004).

To construct an influence diagram, we rely on both domain knowledge and data. In general, learning such a probabilistic graphical model from data is an NP hard problem (Friedman & Koller, 2000). Fortunately, for image annotation, we have abundant prior knowledge about the relationships between context, content, and semantic labels, and we can use them to substantially reduce the hypothesis space to search for the right model. For instance, time, location, and user profile, are independent of each other (no arc exists between them in the diagram). Camera parameters such as exposure time and flash on/off depend on time, but are independent of other modalities (hence some fixed arcs can be determined). The semantic ontology provides us the relationships between words. The only causal relationships that we must learn from data are those between context/content and semantic labels (and their causal strengths).

Let us use a model generated by our algorithm to illustrate the power of our fusion model.

Figure 7 shows the learned diagram for two semantic labels: outdoor and sunset. In addition to the arcs that show the causal relationships between the variables, and

Copyright © 2006, Idea Group Inc. Copying or distributing in print or electronic forms without written permission of Idea Group Inc. is prohibited.

Figure 7. Causal diagram

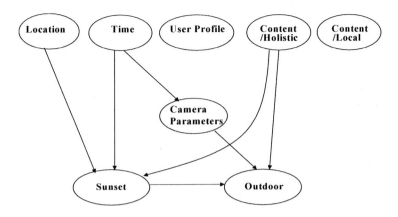

between the variables and the labels, we also see that the relationship between outdoor and sunset can also be encoded.

Case Study: Landmark Recognition

We present a case study on landmark recognition using the influence diagram. Recognizing objects (landmarks or people) in a photo remains a very challenging computer-vision research problem. However, with available contextual information, such as time, location, and a person's social network, recognizing objects among a much-limited set of candidates is not such a daunting task. For example, given the photo-taking location as downtown Santa Barbara, the interesting landmarks in the photo are limited. Given a birthday party of a person, the attendees are most likely to be his/her friends and family. With candidate objects being limited, matching becomes easier, and matching time becomes shorter, so we can afford to employ more expensive matching techniques for improved accuracy.

Some recent works annotate digital images using the metadata of spatial and temporal context. Nevertheless, these methods suffer from several drawbacks. In Naaman, Paepcke, and Garcia-Molina (2003) as well as Davis, King, Good, and Sarvas (2004), GPS devices are used to infer the location where a photo is taken. However, the objects in a photo may be far away from the camera, not at the GPS coordinate. For instance, one can take a picture of the Bay Bridge from many different locations in the cities of Berkeley, Oakland, or San Francisco. The GPS information of the camera cannot definitely infer landmarks in a photo. Second, a landmark in a photo might be occluded and hence not deemed important. For instance, a ship might occlude a bridge; a person might occlude a building, and so forth. Also, a person can be inside or outside of a landmark. A robust system must analyze

Copyright © 2006, Idea Group Inc. Copying or distributing in print or electronic forms without written permission of Idea Group Inc. is prohibited.

content to identify the landmarks precisely. The work of Naaman, Paepcke, and Garcia-Molina (2003) uses temporal context to find time-of-day, weather, and season information. These additional metadata provide only a few drops in the bucket to quench thirst, and the proposed method cannot scale up for annotating vast quantities of other semantics.

For landmark recognition, we use two modalities: *location* and *local features*. We select these two modalities to construct our influence diagram based on our domain knowledge. Although other metadata might also play a role in determining a landmark, we keep this case study simple.

For local features, we follow the SIFT-line of work proposed by Lowe (1999). Our testbed (referred to hereafter as the Stanford dataset) was obtained from Naaman (2003). The dataset was constructed by collecting photographs taken by visitors to the Stanford Visitor Center. All photographs were taken on the Stanford campus and are annotated with GPS information. From this dataset, we used a subset containing about *1,000* images, and added *60* pictures taken by the author of this article. To evaluate the EXTENT system, we selected photos with GPS coordinates around the Memorial Church and Hoover Tower (two important landmarks on the Stanford campus). All images were rescaled to *320′240*, before SIFT feature extraction was performed.

Each image in the dataset was individually processed and matched with sample images that contain candidate landmarks. We used a distance threshold to separate likely matches from non-matches. If the distance (computed using the SIFT features) between the query image and a landmark sample came within the threshold, the landmark would be a possible match. If no possible match was found (after comparing with all landmark samples), we concluded that the image contained no landmark. Otherwise, the best match was used to annotate the image.

We made three queries on the dataset: "Memory Church," "Hoover Tower," and "Rodin Garden" (Figure 8). Each query has *10* images containing the landmark taken at different time, from different angles, in the dataset. For each query, we used one of the ten images

Figure 8. Landmark queries

Copyright © 2006, Idea Group Inc. Copying or distributing in print or electronic forms without written permission of Idea Group Inc. is prohibited.

as the query, and the other nine were mingled in the *1,000*-image dataset. We performed this leave-one-out query for each of the ten images, and for three landmarks. Using GPS alone gives us just *33.3%* accuracy. This is because we have three landmarks in front of the Stanford Quad, and the information about the Quad cannot give us a specific landmark, and we have to resort to a random guess. Using just SIFT features also produces a disappointing result of *60%*. This is because too many Stanford buildings exhibit the same textures and shapes of the Hoover Tower. There are also a couple of locations displaying Rodin sculptures. Finally, combining GPS and SIFT gives us *95%* accuracy, a much better result, thanks to the synergistic fusion of the two modalities.

Keyword Assisted Concept-Dependent Active Learning

With keywords being available, the remaining question is how to use them effectively to assist image retrieval. One straightforward way is to turn image retrieval into text retrieval. However, subjective annotation and multiple senses of words prevent this approach from being invariably effective. Furthermore, even if the semantics of an image matches a query keyword, the user may prefer certain visual aspects that words cannot describe. Therefore, perceptual features must also be considered in order to offer a complete solution.

We present a *concept-dependent* active learning algorithm (Goh, Chang, & Lai, 2004), which uses keywords and perceptual features in a synergistic way. Concept-dependent active learning (CDAL) aims to solve one key problem: concept aliasing. Revisit the active learning algorithm presented in the second section. One major shortcoming of the method is that when a query-concept's *complexity* is high, active learning may not be able to find sufficient images relevant to the concept. This is because the concept can be cluttered with others in the perceptual space. More specifically, we formally define *concept complexity* as the level of difficulty in learning a target concept. To model concept complexity, we use two quantitative measures (Lai, Goh, & Chang, 2004): *scarcity* and *isolation*.

Scarcity measures how well represented a concept is in the retrieval system. We use *hit-rate*, defined as the percentage of data matching the concept, to indicate scarcity. As we assume that each keyword is equivalent to a concept, the hit-rate of a keyword is the number of images being annotated with that keyword. This parameter is dataset-dependent; while a concept such as outdoor is very general and may produce a high hit-rate, other general concepts such as eagles may be scarce simply because the system does not contain many matching images. Similarly, a very specific concept such as laboratory coat could have a high hit-rate solely because the system has many such images.

Isolation characterizes a concept's degree of separation from the other concepts. We measure two types of isolation: *input space* isolation and *keyword* isolation. The input space isolation is considered low (or poor) when the concept is commingled with others in the perceptual space.

When the keywords used to describe a concept has several meanings or senses, the keyword isolation is considered poor; very precise keywords provide good isolation. An

Copyright © 2006, Idea Group Inc. Copying or distributing in print or electronic forms without written permission of Idea Group Inc. is prohibited.

Figure 9. CDAL states and state transitions

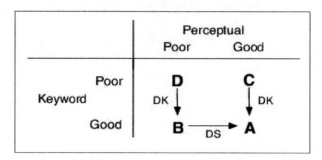

example of a poorly isolated keyword is "feline," which is often used to annotate images of tigers, lions and domestic cats. If the user seeds a query with "feline," it is difficult to tell which sense of the word the user has in mind. A well-isolated seed keyword like "Eiffel Tower" has less ambiguity; we can be quite sure that the user is thinking about the famous Parisian landmark.

We describe how scarcity and isolation can be quantified in Goh, Chang, and Lai (2004) and in Lai, Goh, and Chang (2004). Once the complexity of a concept can be measured, we can adjust SVMActive in a concept-dependent way. The key of CDAL is to reduce the complexity of the concept, so that the learnability of the target concept is improved.

CDAL consists of a state-transition table and two major algorithms: *disambiguate keywords* (DK) and *disambiguate input-space* (DS). First, CDAL addresses the scarcity problem by using keywords to seed a query. Thus, we eliminate the need to search the entire dataset for a positive image. The user can type in a keyword (or keywords) to describe the target concept. Images that are annotated with that keyword are added to the initial unlabeled pool U. If the number of images with matching keywords is small, we can perform query expansion using a thesaurus to obtain related words (synonyms) that have matching images in our dataset. For each related semantic, we select matching images and add them to U.

Using U to conduct learning rather than using the entire dataset improves the *hit-rate*. At this juncture, the learnability state can be in any of the four states depicted in Figure 9. The rows indicate whether the specified keyword enjoys good isolation or suffer from poor isolation. The columns show that the query concept may be well or poorly isolated in the input space formed by the perceptual features. The goal of CDAL is to move the learnability state of the target concept to the ideal state A, where both keywords and perceptual features enjoy good isolation. In the remainder of this section, we sketch how CDAL works. For further details, see Goh, Chang, and Lai (2004).

State C - Keyword Disambiguation

In state C, the keyword isolation is poor (due to aliasing). Thus, we first need to disambiguate the keywords by presenting images of various semantics to the user for selection. Once the semantic is understood, the learnability makes a transition to state A.

Copyright © 2006, Idea Group Inc. Copying or distributing in print or electronic forms without written permission of Idea Group Inc. is prohibited.

Algorithm DK helps state C to arrive at A. The algorithm first performs a lookup in the association table H (constructed through association mining on keywords of images) for the query keyword w. If two item-sets are found, the algorithm randomly selects from U images that are annotated by the co-occurred word set. The resulting sample set S is presented to the user to solicit feedback. Once feedback has been received, CDAL removes from U images labeled with the words associated with the irrelevant images. Notice a subtle but important difference between our query-expansion approach and the traditional method that uses a semantic ontology to find synonyms. For the task of image annotation (discussed in the fourth section), a semantic ontology is helpful. For image retrieval, however, semantic ontology may be counter-productive. This is because images labeled by a keyword's synonyms might not exist in the dataset. Using co-occurrence, mined by association rules on the dataset, provides information about the dataset, and hence is useful for disambiguating keyword semantics in a dataset-dependent way.

State B - Input-space Disambiguation

State B has poor perceptual isolation but good keyword isolation. If the number of matching images yielded by the good keyword is small, we can perform query expansion to enlarge the unlabeled sample pool. For these poorly isolated concepts, we attempt to improve the perceptual isolation of the sample pool through input-space disambiguation. CDAL employs DS to achieve this improvement. The algorithm removes from U the instances that can interfere with the learnability of the target concept.

Given the current labeled pool of images, DS first generates two word lists: P from positive-labeled images and N from negative-labeled images. In the third step, DS checks the annotation of each image in the unlabeled pool for matches to words in P and N. If there are more negative keyword matches, CDAL removes that image from U. Once perceptual isolation is enhanced, state B makes the transition to A.

State D - Keyword & Space Disambiguation

State D suffers from both poor semantic and poor input-space isolation. To remedy the problem, CDAL first improves the keyword isolation using the same DK strategy as in state C. Thereafter, the state is moved to state B, and the perceptual isolation is improved by applying state B's method for disambiguation in the input space. Figure 8 shows the path that D takes to reach the ideal state A.

Summary

We have discussed in this article several statistical models and algorithms for addressing challenges of image perception. Essentially, these models and algorithms attempt to model human perception of images, and thereby achieve more intuitive and effective

Copyright © 2006, Idea Group Inc. Copying or distributing in print or electronic forms without written permission of Idea Group Inc. is prohibited.

image annotation and retrieval. We enumerated three major technical challenges: scarcity of training instances, imbalanced training data, and fusion of context and content. These challenges run parallel to the challenges human perception comes up against. For scarcity of training instances, most of us learned in our childhood to recognize, for example, an animal by seeing just a few example pictures. For imbalanced training data, information relevant to each concept we attempt to learn plays a minority role in our knowledge base and experience. For context and content fusion, we seem to be at ease to correctly interpret an expression in different contexts. We showed the mathematical difficulties in dealing with these challenges, and presented our solutions. With our integrated approach, combining keyword with perceptual features and combining context with content, we have built systems that can perform at a satisfactory level and can be used in practical scenarios.

We are furthering our work in several exciting directions. Our recent papers (Chang, Hoi, Wang, Ma, & Lyu, 2005; Wu, Chang, & Panda, 2005) can provide an overview on our future research topics and preliminary results.

Acknowledgments

I would like to thank contributions of my PhD students and VIMA colleagues: Beitao Li, Simon Tong, Kingshy Goh, Gang Wu, Yi Wu, Yi-Leh Wu, Tony Wu, Tim Cheng, Wei-Cheng Lai, and Cheng-Wei Chang. I would also like to thank the funding supported by two NSF grants: NSF Career IIS-0133802 and NSF ITR IIS-0219885.

References

Aizerman, M. A., Braverman, E. M., & Rozonoer, L. I. (1964). Theoretical foundations of the potential function method in pattern recognition learning. *Automation and Remote Control, 25*, 821-837.

Barnard, K., & Forsyth, D. (2000). Learning the semantics of words and pictures. *International Conference on Computer Vision: Vol. 2* (pp. 408-415).

Bengio, Y., Paiement, J. F., Vincent, P., Delalleau, O., Roux, N. L., & Quimet, M. (2004). Out-of sample extensions for LLE, isomap, MDS, eigenmaps, and spectral clustering. *Neural Information and Processing System (NIPS), 16,* 177-184.

Beyer, K., Goldstein, J., Ramakrishnan, R., & Shaft, U. (1999). When is nearest neighbor meaningful. *Proceedings of ICDT* (pp. 217-235).

Blum, A., & Mitchell, T. (1998). Combining labeled and unlabeled data with co-training. *Proceedings of the 11th Annual Conference on Computational Learning Theory (COLT)* (pp. 92-100).

Copyright © 2006, Idea Group Inc. Copying or distributing in print or electronic forms without written permission of Idea Group Inc. is prohibited.

Boutell, M., & Luo, J. (2005). Beyond pixels: Exploiting camera metadata for photo classification. *Pattern Recognition, Special Issue on Image Understanding for Digital Photos, 38*(6).

Chang, E. Y. (2005). *EXTENT: Combining context, content, and semantic ontology for image annotation.* ACM SIGMOD CVDB Workshop.

Chang, E. Y., Cheng, K. T., Lai, W. C., Wu, C. T., Chang, C. W., & Wu, Y. L. (2001). PBIR: A system that learns subjective image query concepts. *ACM International Conference on Multimedia* (pp. 611-614).

Chang, E. Y., Goh, K., Sychay, G., & Wu, G. (2003). CBSA: Content-based soft annotation for multimodal image retrieval using bayes point machines. *IEEE Transactions on Circuits and Systems for Video Technology Special Issue on Conceptual and Dynamical Aspects of Multimedia Content Description, 13*(1), 26-38.

Chang, E. Y., Hoi, C. H., Wang, X., Ma, W. Y., & Lyu, M. (2005). *A unified machine learning paradigm for large-scale personalized information management* (invited paper). The 5th Emerging Information Technology Conference.

Chang, E. Y., & Li, B. (2003). MEGA—The maximizing expected generalization algorithm for learning complex query concepts. *ACM Transactions on Information Systems (TOIS), 21*(4), 347-382.

Chang, E. Y., Li, B., & Li, C. (2000). Toward perception-based image retrieval. *IEEE Workshop on Content-Based Access of Image and Video Libraries* (pp. 101-105).

Chang, E. Y., Li, B., Wu, G., & Goh, K. S. (2003). Statistical learning for effective visual information retrieval. *IEEE International Conference on Image Processing (ICIP)* (pp. 609-612).

Davis, M., King, S., Good, N., & Sarvas, R. (2004). *From context to content: Leveraging context to infer media metadata.* ACM International Conference on Multimedia.

Donoho, D. L. (2000). *High-dimensional data analysis: The curses and blessings of dimensionality.* Aide-memoire presented at the American Math. Society Lecture on Math Challenges of the 21st Century.

Flickner, M. et al. (1995). Query by image and video content: The QBIC system. *IEEE Computer Magazine, 28*(9), 23-32.

Friedman, N., & Koller, D. (2000). *Learning Bayesian networks from data* (tutorial). NIPS.

Goh, K., Chang, E. Y., & Cheng, K. T. (2001). SVM binary classifier ensembles for multi-class image classification. *ACM International Conference on Information and Knowledge Management (CIKM)* (pp. 395-402).

Goh, K., Chang, E. Y., & Lai, W. C. (2004). Concept-dependent multimodal active learning for image retrieval. *ACM International Conference on Multimedia* (pp. 564-571).

Goldstein, B. E. Goldstein (1999). *Sensation and perception* (5th ed.). Pacific Grove: Brooks/Cole.

Goldstone, R. L. (1994). Similarity, interactive activation, and mapping. *Journal of Experimental Psychology: Learning, Memory, and Cognition, 20*, 3-28.

Gupta, A., & Jain, R. (1997). Visual information retrieval. *Comm. of the ACM, 40*(5), 69-79.

Copyright © 2006, Idea Group Inc. Copying or distributing in print or electronic forms without written permission of Idea Group Inc. is prohibited.

Ham, J., Lee, D. D., Mika, S., & Scholkopf, B. (2004). *A kernel view of dimensionality reduction of manifolds.* International Conference on Machine Learning (ICML).

Hastie, T., Tibshirani, R., & Friedman, J. (2001). *The elements of statistical learning.* New York: Springer.

He, X., Ma, W. Y., & Zhang, H. J. (2004). Learning an image manifold for retrieval. *Proceedings of ACM Multimedia Conference* (pp. 17-23).

Heckerman, D., & Shachter, E. (1994). *Decision-theoretic foundations for causal reasoning* (Microsoft Tech. Rep. No. MSR-TR-94-11).

Hu, M. K. (1962). *Visual pattern recognition by moment invariants. IRE Trans. Information Theory, IT-8* (pp. 179-187).

Joachims, T. (1999). Transductive inference for text classification using support vector machines. *Proceedings of International Conference on Machine Learning* (pp. 200-209).

Lai, W. C., Chang, E. Y., & Cheng, K. T. (2002). *Hybrid learning schemes for multimedia information retrieval.* IEEE Pacific-Rim Conference on Multimedia.

Lai, W. C., Goh, K., & Chang, E. Y. (2004). On scalability of active learning for formulating query concepts. *Workshop on Computer Vision Meets Databases (CVDB) in cooperation with ACM SIGMOD* (pp. 11-18).

Lavrenko, V., Manmatha, R., & Jeon, J. (2003). A model for learning the semantics of pictures. *Proceedings of the Seventeenth Annual Conference on Neural Information Processing Systems (NIPS),* Vancouver, British Columbia (Vol. 16, pp. 553-560).

Li, B., & Chang, E. Y. (2003). Discovery of a perceptual distance function for measuring image similarity. *ACM Multimedia Systems Journal Special Issue on Content-Based Image Retrieval, 8*(6), 512-522.

Li, B., Chang, E. Y., & Wu, C. T. (2002). *DPF - A perceptual distance function for image retrieval.* IEEE International Conference on Image Processing (ICIP).

Lowe, D. G. (1999). *Object recognition from local scale-invariant features.* International Conference on Computer Vision.

Lowe, D. G. (2004). Distinctive image features from scale-invariant keypoints. *International Journal of Computer Vision, 60*(2), 91-110.

Lu, Y., Hu, C., Zhu, X., Zhang, H. J., & Yang, Q. (2000). A unified semantics and feature based image retrieval technique using relevance feedback. *Proceedings of ACM International Multimedia Conference* (pp. 31-37).

Manjunath, B. S., & Ma, W. Y. (1996). Texture features for browsing and retrieval of image data. *IEEE Transactions on Pattern Analysis and Machine Intelligence, 18*(8) 837-842.

Mitchell, T. (1997). *Machine learning.* New York: McGraw-Hill.

Naaman, M., Paepcke, A., & Garcia-Molina, H. (2003). *From where to what: Metadata sharing for digital photographs with geographic coordinates.* International Conference on Cooperative Information Systems (CoopIS).

Copyright © 2006, Idea Group Inc. Copying or distributing in print or electronic forms without written permission of Idea Group Inc. is prohibited.

Novick, L., & Cheng, P. (2004). Assessing interactive causal influence. *Psychological Review, 111*(2), 455-485.

Rocchio, J. (1971). Relevance feedback in information retrieval. In G. Salton (Ed.), *The SMART retrieval system in automatic document processing* (pp. 313-323). Englewood Cliffs, NJ: Prentice Hall.

Rui, Y., Huang, T. S., & Chang, S.-F. (1999). Image retrieval: Current techniques, promising directions, and open issues. *Journal of Visual Communication and Image Representation, 10*(1), 39-62.

Rui, Y., Huang, T. S., & Mehrotra, S. (1997). Content-based image retrieval with relevance feedback in MARS. *Proceedings of IEEE International Conference on Image Processing* (pp. 815-818).

Smith, J. R., & Chang, S. F. (1996). VisualSEEk: A fully automated content-based image query system. *Proceedings of ACM Multimedia Conference* (pp. 87-98).

Tenenbaum, J. B. (1999). *A Bayesian framework for concept learning.* PhD thesis, Massachusetts Institute of Technology.

Tenenbaum, J. B., & Griffiths, T. L. (2001). Generalization, similarity, and Bayesian inference. *Behavioral and Brain Sciences, 24*, 629-641.

Tenenbaum, J. B., Silva, V. de, & Langford, J. C. (2000). A global geometric framework for nonlinear dimensionality reduction. *Science, 290*(5500), 2319-2323.

Tibshirani, R., Hastie, T., Narasimhan, B., & Chu, G. (2003). Class prediction by nearest shrunken centroids, with applications to DNA microarrays. *Statistical Science, 18*(1), 104-117.

Tong, S., & Chang, E. Y. (2001). Support vector machine active learning for image retrieval. *Proceedings of ACM International Conference on Multimedia* (pp. 107-118).

Tu, Z. W., Chen, X. R., Yuille, A. L., & Zhu, S. C. (2005). Image parsing: Unifying segmentation, detection, and recognition. *International Journal of Computer Vision, 63*(2), 113-140.

Tversky, A. (1977). Features of similarity. *Psychological Review, 84*, 327-352.

Vapnik, V. (1995). *The nature of statistical learning theory.* New York: Springer.

Wu, G., & Chang, E. Y. (2003). Adaptive feature-space conformal transformation for imbalanced-data learning. *The Twentieth International Conference on Machine Learning (ICML)* (pp. 816-823).

Wu, G., & Chang, E. Y. (2004). Aligning boundary in kernel space for learning imbalanced dataset. *IEEE International Conference on Data Mining (ICDM)* (pp. 265-272).

Wu, G., Chang, E. Y., & Panda, N. (2005). Formulating context-dependent similarity functions. *Proceedings of ACM International Conference on Multimedia (MM)* (pp. 725-734).

Wu, Y., Chang, E. Y., & Tseng, B. (2005). Multimodal metadata fusion using causal strength. *Proceedings of ACM International Conference on Multimedia* (pp. 872-881).

Copyright © 2006, Idea Group Inc. Copying or distributing in print or electronic forms without written permission of Idea Group Inc. is prohibited.

Endnotes

[1]   Relevance feedback was first proposed in the information retrieval community for text (Rocchio 1971), and was introduced to the image retrieval community by the works of Rui, Huang, and Mehrotra (1997) and Lu, Hu, Zhu, Zhang, and Yang (2000).

[2]   Given a kernel function K and a set of instances $S = \{x_i, y_i\}$, $i=1,\ldots,n$, the kernel matrix (Gram matrix) is the matrix of all possible inner-products of pairs from S, $K_{i,j} = K(x_i, x_j)$.

Copyright © 2006, Idea Group Inc. Copying or distributing in print or electronic forms without written permission of Idea Group Inc. is prohibited.

Chapter III

Feature-Based Multimedia Semantics:
Representational Forms for Instructional Multimedia Design

Michael May, LearningLab DTU, Technical University of Denmark, Denmark

Abstract

The aim of this chapter is to sketch a semantic taxonomy of representational forms (or "sign types" in the terminology of semiotics) relevant for the compositional analysis, design, and indexing of multimodal multimedia, specifically, in the domain of instructional design of learning objects, and to contribute to current attempts to bridge the "semantic gap" between the technical and spatio-temporal description of media objects and the contextual perception and interpretation by users of their content.

Introduction

In the present chapter, it is suggested that descriptions of *graphical and multimedia content*, as it is known from multimedia databases, multimedia programming, and graphic design, should be extended to include *taxonomic information about the representa-*

Copyright © 2006, Idea Group Inc. Copying or distributing in print or electronic forms without written permission of Idea Group Inc. is prohibited.

tional forms (or "sign types" in the terminology of semiotics) combined in multimedia presentations.

From the point of view of design, there is a fundamental *rhetorical* problem in deciding which *combination of media types and representational forms* ("sign types", "modalities of expression") would be adequate for a particular multimedia presentation within a particular context-of-use. This rhetorical problem is necessarily present in all forms of information, interface, and interaction design, although it is not the whole story of designing usable and adequate presentations, since there are many criteria of usability as well as adequacy of presentations (i.e., presentations that are relevant, intelligible, efficient, consistent, aesthetic, etc.) and many levels of design that have to be considered in the design or automated generation of presentations (functional specification, spatial layout, temporal organisation, forms of interaction, forms of cognitive support for users, contextual relations to activities and situations of use, etc.). In this chapter, we will only address the rhetorical problem of "media and modality" and specifically the problem of taxonomy, that is, the need for some kind of classification of media types and representational forms to support a conceptual understanding of the design space of available types and its operational specification in particular design contexts (where the example given here is multimedia content specification for instructional design).

Taxonomies of Media and Representational Forms

Taxonomies in Graphic Design

One might think that taxonomic issues in graphic design have been settled for a long time, since the representation and layout of simple graphical objects like two-dimensional documents with "text" and "illustrations" appear simple compared to the full complexity of computer-based animation, video, gesture, and haptics. This appearance of simplicity is an illusion, however, and the problems of representation and layout of multimodal documents have not yet been fully solved (Bateman, Delin, & Henschel, 2002; Bateman, Kamps, Kleinz, & Reichenberger, 2001). It is important, from the point of view presented in the present chapter, to note that *multimedia* communication as well as *multimodality* (of representations) does not originate with modern computer technology or electronic audio-visual technologies like television. We know multimodal representations from the printed graphical media in the form of combinations of language, images, and diagrams, and before the invention of printing, from combinations of writing and drawing within the graphical media. The fundamental origin of multimedia communication and multimodality, however, is coextensive with the origin of language, since the embodied language of human speech combining multiple media (speech, gestures) and multiple forms of representation (natural language discourse, intonation patterns, schematic "conversational gestures", facial expression, etc.) is an advanced form of multimodal communication (Quek, McNeil, Bryll, Duncan, Ma, Kirbas, McCullough, & Ansari, 2002)

Copyright © 2006, Idea Group Inc. Copying or distributing in print or electronic forms without written permission of Idea Group Inc. is prohibited.

that in some ways can play the role of a natural ideal for designed technically-mediated communication.

The taxonomic problem of identifying and combining different types of media and representational forms (or "modalities") have been addressed in design theories based on the printed graphical media, but it was intensified with the advent of computer-based multimedia systems. It will be useful to start the present analysis of problems of taxonomy with a few early attempts to classify graphical objects.

With a few exceptions, the focus on taxonomic issues within *graphical design* has been on representational forms derived from examples of "business graphics", and it has been restricted to static graphical media objects. In the *scientific visualisation* community, on the other hand, there has also been some interest in taxonomic issues, but with a focus on images, maps, and graphs used for data visualisation within simulations and other forms of dynamic graphics. Early work in scientific visualisation was driven by presentations of examples of new applications and graphic algorithms, whereas the conceptual foundation of the domain was neglected. Recent work in the domain of *information visualisation*, common to business graphics and scientific visualisation, has attempted to present the area as based on knowledge of perception (Ware, 2000). This is an important step forward, but what is still missing is an understanding of the syntactic and semantic features of graphical expressions, and how these features are expressed in other media (acoustic, haptic, gestic).

With the increasing importance of the Internet and Web-based presentations, a new quest for taxonomies has arisen in studies of automated presentation agents and multimodal document design. It appears that, even from the point of view of document design restricted to static two-dimensional documents including text, pictures, and diagrams, there is still no consensus on a systematic description of multimodal documents (Bateman, Delin, & Henschel, 2002).

A taxonomy of graphics in design was proposed by Michael Twyman (1979). "Graphical languages" is here analysed along two dimensions: *configuration* and *mode of symbolisation*. The focus of Twyman was on static graphics and the role of eye movements in reading graphics. Configuration thus refers to different forms of linear, branched, matrix-like, or non-linear reading of graphical forms, whereas the "mode of symbolisation" refers to the verbal/numerical, verbal and pictorial, pure pictorial, or schematic organisation of the content. Ordinary text is usually configured typographically to be read in a linear fashion, whereas diagrams with branching nodes have a branching linear configuration, and pictures are examined in a non-linear fashion. Although we are presented with a classification of the "modes of symbolisation", there is no analysis of the semantic content as such. The classification is inconsistent in considering combined forms such as "verbal and pictorial" to be a class next to simple forms such as "pictorial" from which it must be derived. Twyman refers to the distinction between symbolic and iconic in semiotics, but does not find it relevant for the classification, and he does not utilise his own conception of "graphical languages". Even though the concept of a language implies syntactic and semantic structures, he only considers "pictorial" and "verbal" forms as aggregated and unanalysed wholes.

Another example of classification problems with "graphics" is seen in the work of Lohse, Walker, Biolsi, and Rueter (1991) and Lohse, Biolsi, Walker, and Rueter (1994). The methodology is here quite different; through a cluster analysis of individual subjective

Copyright © 2006, Idea Group Inc. Copying or distributing in print or electronic forms without written permission of Idea Group Inc. is prohibited.

classifications of a series of examples of graphics, they arrive at the following distinctions: graphs and tables, network charts, diagrams, maps, and icons. The authors try to interpret these classes along two dimensions, but they are given no real analysis. The five classes seem to range from symbolic (graphs and tables) to more "analogue" forms (maps and diagrams) in one dimension, and from explicit and "expanded" forms of expression (network charts) to more implicit and "condensed" forms of expression (icons) in another.

The work of Lohse, Walker, et al. (1991) and Lohse, Biolsi, et al. (1994) should be understood as referring to "folk taxonomies" derived from clustering of perceived differences among representations. This kind of taxonomy could be important for studies of individual styles of reasoning and self-generated external representations (Cox, 1999), but less important for systematic models of design choices in multimedia design. "Perceived differences" will reflect peoples' attitudes towards representations rather than their intrinsic characteristics, and folk taxonomies are not consistent classifications: Sometimes people will group representations together because they have the same use context, sometimes because they are expressed in the same media, and sometimes because they "look alike" (but disregarding their semantic and syntactic properties).

Web-based techniques for presentation based on HTMLand XML have enforced a clearer separation of document content from its layout as well as from different types of meta-data associated with documents, and this is reflected in recent theories of the structure and layout of multimodal documents. According to Bateman, Delin, and Henschel (2002) and Bateman, Kamps, Kleinz, and Reichenberger (2001), it will be necessary to distinguish five types of document structure: the *content structure* of the information communicated in a document, the *rhetorical structure* of the relations between content elements, the *layout structure* of the presented document, the *navigation structure* of the supported ways in which the document can be accessed, and the *linguistic structure* of the language used to present the document. The question of media types is *implicit* in this typology since only graphical media documents are considered in these typographic theories, whereas the question of representational forms (or "sign types") is hidden in what is called "the linguistic structure". For our purpose, the linguistic structure should be generalised to the *semiotic structure* of multimodal multimedia documents or "objects".

Media Types

In early approaches to multimedia the concept of "multimodality" was constructed from an engineering point of view as referring to combinations of technologies. According to Gourdol, Nigay, Salber, and Coutaz (1992) and Nigay and Coutaz (1993), for example, a modality is associated with a physical hardware device. A multimodal system is accordingly defined as a particular configuration of hardware, that is, as a system that supports many "modalities" for input and output. The taxonomy that can be constructed from this point of view will be restricted to dimensions of the technical control of hardware systems such as the support given by these modalities with respect to temporal execution (sequential or parallel) and the combined or independent nature of commands with regard

Copyright © 2006, Idea Group Inc. Copying or distributing in print or electronic forms without written permission of Idea Group Inc. is prohibited.

to the modalities. This taxonomy of "systems" does not consider the representational issues involved in multimedia, but from the perspective of users of applied multimedia systems, the important design decisions cannot be based on a characterisation of the physical devices and media, since it is precisely the representational and cognitive properties of the application that will determine what kind of support users can expect in their situated activity involving the application. An example is instructional design of multimedia applications for higher learning, where Recker, Ram, Shikano, and Stasko (1995) have stated the importance of what they call the "cognitive media types": Design of hypermedia and multimedia for learning should not be based directly on properties of the physical media, but on properties of the information presentation that are "cognitively relevant" to the learning objectives.

An important step forward towards a classification of media types was taken with the object- oriented approach to multimedia programming, but even with the focus on a conceptual foundation for multimedia (Gibbs & Tsichritzis, 1995), the software abstractions constructed to represent media types were mainly a reflection of the data types and digital file formats available at the time. The concept of media type introduced the idea of a template for distinguishing the representational from the operational aspects. With a template like

> Media type <name>
>> Representation
>>> <aspects of representation>
>> Operations
>>> <categories of operation>

Figure 1. Media classes as suggested by Gibbs and Tsichritzis (1995); the diagram has been redrawn to make the assumption of the non-temporal media classes explicit in the diagram

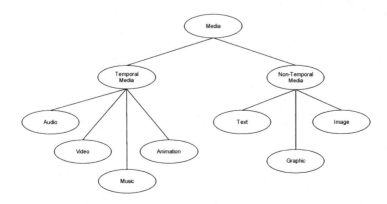

Copyright © 2006, Idea Group Inc. Copying or distributing in print or electronic forms without written permission of Idea Group Inc. is prohibited.

Gibbs and Tsichritzis (1995) identified a set of important media types as well as their associated forms of representation and supported operations. The actual classification of media types represented by media classes was, however, both incomplete and inconsistent.

The suggested media types are incomplete from a systematic point of view, where we would want to include haptic and gestic media types. Another problem is the inherent inconsistencies of the classification scheme. The distinction between temporal and non-temporal media appears clear cut, but on a closer analysis all so-called non-temporal media have temporal aspects. Text can indeed be statically displayed, but it could also be presented repetitively (blinking), sequentially (one word or noun phrase at a time displayed in a linear fashion as in the "ticker tape display"), or dynamically as in auto-scrolling text on a Web page. The same argument can be made for images and object-oriented graphics, that is, they are not inherently static, but can be presented in different temporal forms. The media type called "animation" now appears to be a derivation of repetitive, sequential, or dynamic versions of other media types. It could be an improve-

Figure 2. A systematic classification of presentation types as derived from media types and four temporal types, as they can be determined before the further articulation of representational forms (and before considering spatial and temporal layout issues)

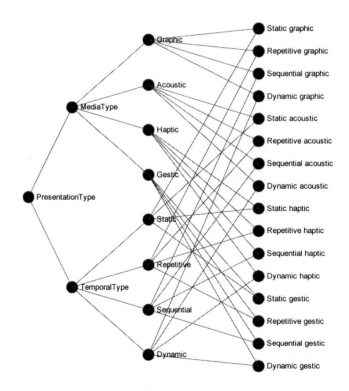

Copyright © 2006, Idea Group Inc. Copying or distributing in print or electronic forms without written permission of Idea Group Inc. is prohibited.

ment to differentiate the temporal properties of presentations from the specification of their media type (just as the interactivity of presentations is seen as a separate issue), and a classification scheme like the one constructed in Figure 2 could be defended.

From the classification presented in Figure 2, a series of questions will have to be answered, though. How can, for instance, the acoustic media be ascribed a static form? Although the acoustic media is inherently temporal at a physical level, it does make sense from a perceptual point of view to define a static acoustic media type with reference to sound (in the form of noise or un-modulated tones) that is constant within the duration of the presentation. After considering limit cases like static sound, the same temporal distinctions can be applied to all media resulting in a classification scheme (Figure 2) based on the systematic formation of all logical combinations. The temporal distinctions are static, repetitive, sequential, and dynamic (Bernsen, 1993, 1994).

With regard to the difference between "animation" and "video" suggested by the traditional classification scheme (Figure 1), this distinction breaks down when we consider modern techniques of animated computer graphics and virtual reality in the 3D computer games or in mixed-media film, where animated computer-generated characters and human actors can appear within the same media object. Although animated graphics and video have different technical foundations and origins, there is no longer any fundamental difference between them as systems of communication. The referential difference between "virtual" graphics and "real" photographic presentations have important cognitive and epistemic ramifications, but it does not determine the media type (i.e., graphic/photographic and animation/video are not distinctions in media type).

Finally "animation" is a mixed concept for another important reason: It includes the traditional graphical forms of communication utilised in animated drawings (sequential graphics) as well as the embodied movement-based communication realised in facial, hand, and full-body gestures. As channels of communication, we need to distinguish the *gestic media* ("gestures") from the graphical media, although presentations in both channels are visually perceived.

There is a more serious problem with the temporal specification of media types, though: Although temporal specifications in some cases seem to be fundamental to the internal properties essential to media types (as in the case of static graphics versus animation and video), this does not resolve the issue of how this internal temporality is articulated within the temporal organisation of the presentation as a whole. In accordance with the different levels of structure introduced for multimodal documents, we need to separate different issues of temporality in multimedia objects.

Temporal aspects of multimedia objects can appear as important for the type specification of the chosen presentation media for a particular representational content, for instance, whether a graph or a diagram within an instructional document is animated or not. With present-day techniques for multimedia multimodal presentations such as Java applets for interactive Internet documents or dynamically-constructed graphs within a Maple worksheet for doing mathematical calculations, the difference between static, sequentially-animated, or fully-dynamic graphs or diagrams turns out to be of minor importance at the level of media types. Static, animated, or dynamic versions of the same object will often be available as part of the temporal layout of the instructional document or otherwise put directly under the interactive control of the user. Consequently,

Copyright © 2006, Idea Group Inc. Copying or distributing in print or electronic forms without written permission of Idea Group Inc. is prohibited.

distinctions like static, repetitive, sequential, and dynamic should be located at the level of temporal layout and interactivity and not at the level of specification of media types (as an internal property of the type).

Gibbs and Tsichritzis (1995) are aware of another classification problem that arise with the combination of "audio" and "video": We would usually expect "video" to have some sound overlay, but since both audio and video can exist in a "pure" form, we should classify them according to these simple non-combined types and consider audio-visual media types as combinations of simple types (like video, audio, speech, and music).

A more difficult problem of classification, however, is associated with the inconsistent analysis of representational forms supported by different media. Speech is not considered as a type except as collapsed into the audio type, partly because it, in contrast to music represented in the midi encoding format, did not have a separate encoding at the time (cf. the recent VoiceXML format), and partly because there is no thorough analysis of the relation between media types and the representational forms they support.

In early approaches to multimedia content, many obvious relations that would be relevant for content management and for flexible display of multimedia content were almost "hidden" by the classification scheme. A good example is natural language, where the close relation between *text* and *speech* is lost if graphical and acoustic language presentations are only considered as presentations in two different media, and not also as representations within a single *representational form* (natural language). Similarly, the classification scheme hides the fact, well known to any graphical designer, that text is in fact a kind of graphical presentation that inherits many of the operations supported by other forms of object-oriented graphics: operations on shape, size, colour, shading, texture, and orientation (cf. the design space of fonts). These inconsistencies can be resolved, if we abstract the concept of media from specific technologies, programming languages, and their data types, in favour of a perception-based approach to different media, where media types are derived from human sensory abilities in combination with elaborated channels of communication. From this point of view, we only have four different physical presentation media and channels of communication subsumed under computerized control today: the graphic media that derives from visual perception, the acoustic media that derives from auditory perception, the haptic media that derives from tactile and kinaesthetic perception, and the gestic (or "gestural") media that derives from visual perception. The gestic and the graphic should be considered as separate channels of communication because the gestic media is based on the temporal dimension of movement in itself, rather than on its potential for producing graphic traces of movement (as in drawing on some interface).

Sign Types: Core Semantics and Emergent Semantics

From the point of view presented here, many problems of classification arise from the confusion of the physical media of presentation and communication with the representational forms ("sign types"). Natural language is a representational form that can present linguistic content across different media in the form of graphical language (text), in the form of acoustic language (speech), in the form of haptic language (Braille-embossed text), or in the form of encoded gestures (sign languages). A useful starting point for

Copyright © 2006, Idea Group Inc. Copying or distributing in print or electronic forms without written permission of Idea Group Inc. is prohibited.

understanding representational forms is, in fact, to consider them as representational "modalities", not to be confused with sensory modalities, that are invariant across different media (Stenning, Inder, & Neilson, 1995). In their analysis of media and representational forms (which they call "modalities"), Stenning, Inder, and Neilson focused on the invariant differences: The difference between diagrams and language, for instance, is invariant across the tactile and the graphical media. Where the media type is derived from how an object of presentation is communicated and perceived, the representational form (sign type) is derived from how it is interpreted. (the intended interpretation within a context). Braille text as well as graphical text is interpreted as language, whereas graphic and tactile diagrams are interpreted as diagrams. A semiotic and cognitive theory of information presentation is needed to explain this semantic invariance of the intended interpretation, by which representational forms are constituted. It can be claimed, however, that language is a unique representational system in providing this invariance across physical forms, that is, across the phonological and gestic system of expression and across different graphical writing systems, due to the differential nature of linguistic entities. Everything in language is differences, as it was proclaimed by de Saussure.

According to Stenning and Inder (1995), graphical representation systems are different because they rely on a less abstract mapping of relations between the represented domain and relations in the representation (within a presentation media). This should explain why there are graphic and tactile diagrams but apparently no acoustic equivalent, because the transient nature of acoustic media object does not allow diagrammatic relations to be presented and inspected. It is possible, however, to modify the idea of invariant properties as constitutive of the representational forms. According to the modified view, we will stipulate that only a reduced set of core properties are invariant across media and this core is what constitutes the representational form. On top of these core properties, each media affords and provides cognitive support for an extended set of properties and operations that are meaningful within the representational form, although they are not supported in all media.

From this point of view we could in fact have acoustic diagrams, but only in the very limited sense made available for the core properties of diagrams that are supported across all media. Acoustic diagrams would, therefore, be highly schematic with few details, but they could be constructed from external representations corresponding to the embodied image schemata described in cognitive semantics (Johnson, 1987; Lakoff, 1987). Acoustic diagrams would be based on image schematic asymmetries like left/right, up/down, centre/periphery, front/back with simple objects, relations, and events plotted against this background. An example could be the representation in sound of the relative position and movement of enemy aircrafts represented in the cockpit of a fighter aircraft through directional sound: The image schematic background for the acoustic diagram would in many cases be implicit (cf. the discussion of implicitness and expressiveness in Mackinlay & Genesereth, 1985), that is, it would not be explicitly represented but provided through the interpretation by the pilot according to the phenomena of distributed cognition (Zhang, 1996). Another example could be the use of directional 3D sound to indicate location of escape routes on passenger ships, where the traditional static graphic emergency signs could be misleading (in cases of fire blocking some escape routes) and difficult to see (due to smoke) (May, 2004).

Copyright © 2006, Idea Group Inc. Copying or distributing in print or electronic forms without written permission of Idea Group Inc. is prohibited.

It follows from this conceptualisation of media types and representational forms, that a distinction has to be made between the core semantics of the representational forms (sign types) and the extended semantic properties of the well-known prototypes associated with each representational form. The acoustic maps mentioned above are good examples. They are indeed maps, a diagrammatic representational form, in the restricted sense of the core semantics of the unimodal map, according to which a map is interpreted as a diagram supporting the representation of spatial localization of objects and/or events relative to a reference object. Prototypical maps as we know them from everyday life or professional work (city maps, road maps, sea charts, etc.) are graphical objects with a higher complexity resulting from (a) emergent properties of the expression of maps within the graphical media (such as support for graphically-based inferences about metrical properties like size and distance) and resulting from (b) properties inherited from other representational forms combined with the map object. Most maps are in fact complex multimodal objects because they combine useful features of many unimodal representational forms in order to achieve the usefulness of the prototypical map object. A unimodal city map without overlay of symbols in the form of street names, for instance, would not be very useful. In the case of city maps the combination of types can be understood as a spatially coordinated layering of different forms: a network chart (metro lines) on top of the basic map object (the street map), and symbols (street and metro names) as yet another layer. This conceptual layering of different representational forms is utilised constructively in geographical information systems (GIS), where each layer can be addressed and modified separately.

Towards a Semantics of Multimedia: The Semiotic Turn

A semiotic turn in the analysis of multimedia has been initiated by the Amsterdam multimedia group at CWI. Nack and Hardman (2002) refer to the work of Chatman on film theory (Chatman, 1978) in order to set up the fundamental conceptual aspects of multimedia: the form and substance of multimedia content and the form and substance of multimedia expression. It should be noted that these distinctions of form/substance and content/expression does not originate with film theory, but with the constitution of structural linguistics according to Ferdinand de Saussure (1993) and Louis Hjelmslev (1961). Saussure and Hjelmslev both stipulated a place for a future science of semiotics (or "semiology") derived from the linguistic study of language, but with a focus on the syntactic and semantic properties of other "sign systems". Building on Saussure's distinction between the plane of expression (the "signifiers") and the plane of content (the "signified") in language and on his foundation of linguistics on the differential forms in language as a system, as opposed to its physical "substance" (acoustic, graphic) and its psychological "substance" (ideas, concepts, etc.), Hjelmslev specified these four planes in language (Hjelmslev, 1961).

The four planes can be used to specify different levels of description for multimedia units as shown. A similar description is given by Nack and Hardman (2002), although with a different interpretation of the levels. A more complex semiotic framework for the analysis of multimedia based on structural linguistics in its further development by A. J. Greimas was given by Stockinger (1993).

Copyright © 2006, Idea Group Inc. Copying or distributing in print or electronic forms without written permission of Idea Group Inc. is prohibited.

Figure 3. The four planes in language as a system according to Saussure/Hjelmslev tradition

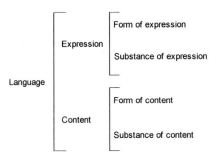

Note that the standard developed for multimedia content description interfaces for audio-visual media, better known as the ISO MPEG-7 standard, specifies multimedia units at the level of their substance of expression only, and the MPEG standard thus leaves the problem of mapping high-level semantics of media objects unsolved (Nack & Hardman, 2002). Even on its own level as a standard for the substance of expression, there has been some criticism of MPEG-7 for not being extensible and modular enough in its description definition language: It is criticised for its closed set of descriptors, for its non-modular language, and for not being sufficiently object-oriented in its data model (Troncy & Carrive, 2004).

On the level of the form of expression, our focus here is on the basic representational forms rather than on higher-level semantic structures of narration and interaction, and it is comparable to the three-dimensional TOMUS model presented in Purchase and

Figure 4. Four fundamental levels of description of multimedia units (not considering the pragmatic and social aspects of the use of multimedia in a specific context-of-use)

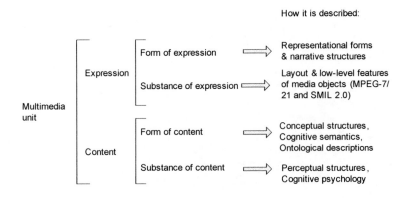

Copyright © 2006, Idea Group Inc. Copying or distributing in print or electronic forms without written permission of Idea Group Inc. is prohibited.

Naumann (2001). The TOMUS model is based on a differentiation between sign types, syntactic structures, and sensory modalities, the latter corresponding to *media* in the present study.

Some taxonomies of multimedia reproduce the difficulties encountered by taxonomies of graphics. An example is the ambiguous classification of media types into text, sound, graphics, motion, and multimedia as suggested by Heller & Martin (1995) and Heller, Martin, Haneef, and Gievska-Krliu (2001). "Motion" is here used to refer to video and animation, which seems like a confusion of form and substance of dynamic graphics. The haptic and gestic media are not considered. The relation between text and graphics is left unanalysed, and the inconsistency in not considering text as a form of graphics is not discovered. The model, however, introduces a second dimension of expression of media types covering three nominal types going from "elaboration" over "representation" to "abstraction". In the case of graphics, this expression dimension is exemplified by the difference between photographs and images at the "elaboration" end, blueprints and schematics as the intermediate form of "representation", and icons at the "abstraction" end. The expression dimension of Heller and Martin (1995) is close to the dimension of the sign in the TOMUS model of Purchase and Naumann (2001), with its subdivision of sign types into concrete-iconic, abstract-iconic, and symbolic.

The iconic-symbolic dimension of sign types have been derived from the semiotic analysis of signs in the work of C. S. Peirce. Where Saussure and later Hjelmslev conceived "semiology" as an extension of linguistics that would study the quasi-linguistic properties of other "sign systems", Peirce developed his more general conception of a "semiotic" from his analysis of logic. For Peirce the sign is not a dyadic relation

Figure 5. The TOMUS model of sign types (i.e., representational forms), syntax, and (sensory) modalities (i.e., media) according to Purchase and Naumann (2001)

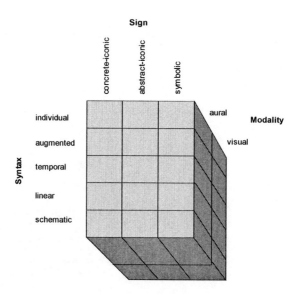

Copyright © 2006, Idea Group Inc. Copying or distributing in print or electronic forms without written permission of Idea Group Inc. is prohibited.

between expression and content, but a triadic relation between a physical representation (in his terminology, the "representamen"), and an interpretation (the "interpretant") of this representation as referring to an object in some respect. Within this triadic relation, it is the representation-object aspect which is categorised as being iconic, indexical, or symbolic by Peirce. The causal relation implied by the indexical category can be considered as a separate issue from the dimension of iconic – symbolic. A subdivision of the iconic – symbolic dimension was also used as a foundation of the sign typology presented in (May, 1993, 2001; May & Andersen, 2001), with the main categories being image, map, graph, conceptual diagram, language, and symbol. The image and the map category here corresponds to the concrete-iconic sign type in the taxonomy of Purchase and Naumann (2001), the diagrammatic forms of representation – graph and conceptual diagram – correspond to the abstract-iconic sign type, and language and symbol to the symbolic type. The conception of the "diagrammatic" in C. S. Peirce is somewhat broader (Figure 6), since it would include maps in the concrete-iconic end and natural as well as mathematical languages in the symbolic end.

With regard to media, the TOMUS model only distinguishes the aural and the visual senses supporting the acoustic and the graphic channel of communication. It is, however, an important feature of the model that all combinations of sensory modalities (or rather media) and sign types (representational forms) can be considered in a systematic way. We will not go into the discussion of the syntactic dimension of the model here.

The taxonomy suggested in May (2001) and May and Andersen (2001) relies on a set of fundamental principles, which makes it well suited to support a more formal approach. An initial version of this taxonomy was developed in May (1993) within the GRACE Esprit Basic Research project in collaboration with N. O. Bernsen, who developed his own version (Bernsen, 1993, 1994). It is suggested that:

- There is a limited number of media types derived from sensory modalities and forms of communication relevant for computer-based interaction;
- There is a limited number of possible un-combined representational forms ("unimodal sign types");

Figure 6. Conceptual diagram to indicate the relation between the sign types used in the TOMUS model compared with the taxonomy of representational forms

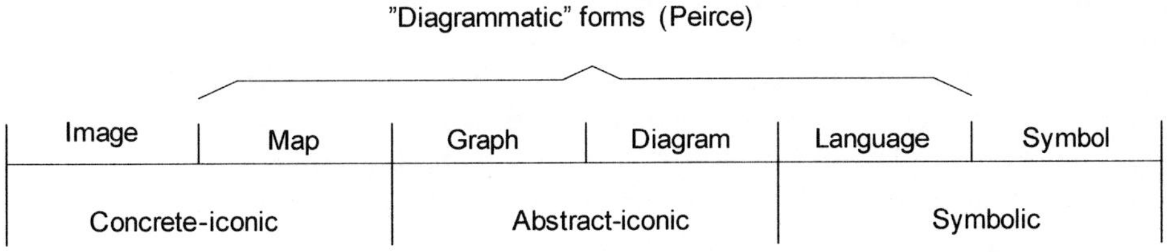

Copyright © 2006, Idea Group Inc. Copying or distributing in print or electronic forms without written permission of Idea Group Inc. is prohibited.

- It is possible to give a feature-based description of their semantic properties;
- Some of these properties are invariant across media types whereas other properties and operational possibilities are emergent with the expression of signs within a media;
- Invariant properties (the "core" semantics) are inherited to syntactic combinations of representational forms whereas their emergent properties are constrained by the specific combinations of media types and representational forms.

It follows that we cannot expect to find any kind of consistent hierarchical classification of prototypical multimedia objects or even graphical objects, because they will have to include complex objects with overlapping properties. The consistent alternative is to develop a feature-based classification, where we track the inheritance and combination of significant properties of sign types and media types. This will, of course, require that we can define the core semantics of the suggested representational forms and describe them in a way that will be relevant to the actual interpretation and use of complex media objects, where different media and representational forms are combined.

Feature-Based Multimedia Semantics

Feature-Structures and Multimedia Content

An early suggestion of a feature-based semantics of graphics was given by Alan Manning in a paper on "technical graphics" (Manning, 1989). Manning stipulated four types of technical graphics abstracting from their terminological variants and their apparent similarities and differences: chart, diagram, graph, and table. According to his analysis, these types can be distinguished by logical combination of two features: display of one unit (written as $-u$) or several units ($+u$) and the representation of one property ($-p$) or several properties ($+p$). Each graphical type in this simple semantic system can thus be described by a small set of feature structures:

chart: [-p, -u]   graph: [-p, +u]

diagram: [+p, -u]   table: [+p, +u]

A simple pie chart only displays one unit (some totality) and it only represents one property (fractions of the total), whereas a graph like a bar graph has several units (represented by individual bars) and again only one property (the amount represented by the height of each bar). A diagram, however, only represents one unit (some object), but displays several properties of the represented object, whereas a table can represent many units and display several properties of each represented object. Our interest here is in the formal approach using feature structures rather than the actual classification.

Copyright © 2006, Idea Group Inc. Copying or distributing in print or electronic forms without written permission of Idea Group Inc. is prohibited.

From a formal point of view, the feature structure approach corresponds to setting up a lattice of the logical combinations of features. This approach to taxonomy of conceptual structures has been formalised in *Formal Concept Analysis* (Ganter & Wille, 1999), where features ("attributes") and the concepts they specify (the "objects") are related in a matrix called a formal context. A formal context C:= (G, M, I) is defined as two sets G (from German "Gegenstände", Objects) and M (from German "Merkmale", attributes) with a relation I between G and M. The elements of G are the objects and the elements of M are the attributes or features of the context. From the formal context, all possible combinations of formal concepts can be generated. A formal concept is a pair (a, b) where a belongs to the set of objects G and b belongs to the set of attributes M. In the small example derived from Manning (1989), an example of a formal concept would be the specification of charts as ({Chart}, {-p,-u}). The full list of formal concepts for the context is shown in Figure 7 below together with the Hasse diagram of the corresponding lattice. To construct a lattice, we have to add two points: the "top" element corresponding to the empty set of objects and the union of all attributes and the "bottom" element corresponding to the full list of objects and the empty set of attributes.

Let us return for a moment to the analysis of the core semantics of a representational form (across all media) as opposed to its emergent semantics articulated within the graphical media. The diagrams used to locate city subway stations are sometimes referred to as "maps", although they are in fact network charts with features quite distinct from the features of maps. Subway charts will often be juxtaposed with timetables, which belong to yet another type of representation, because the combination of tables with information about the time schedule of trains and the network diagram (representing the connectivity and sequence of train stations) gives effective support for navigational decisions. Both representations are, however, virtually useless without the addition of a third representational form: symbolic representations for the denotation of different train stations. The

Figure 7. Formal concepts and lattice generated for the graphical types (objects) and the features (attributes) given by the example discussed earlier; the lattice drawing has been made using the Java-based Concept Explorer program "ConExp 1.2".

```
({}, {-p,-u,+p,+u})  = Top
({Chart}, {-p,-u})
({Graph}, {-p,+u})
({Diagram}, {-u,+p})
({Chart,Diagram}, {-u})
({Table}, {+p,+u})
({Chart,Graph}, {-p})
({Diagram,Table}, {+p})
({Graph,Table}, {+u})
({Chart,Diagram,Graph,Table}, {})  = Bottom
```

Copyright © 2006, Idea Group Inc. Copying or distributing in print or electronic forms without written permission of Idea Group Inc. is prohibited.

representational form of symbols is also needed to establish a coherent reference between the *annotation* to the network chart and the content of the table structure. The annotated network charts can furthermore be projected on an underlying map object giving rise to the familiar but complex prototype known as a "city map". City maps are complex multimodal representations that are not fully coherent in their inherent semantics, because the supported operations and the semantic properties of network charts do not fully agree with those of maps. Since network charts focus on the representation of topological properties like the connectivity of nodes and connecting lines in the network and not on its metric properties, the connecting lines can be deformed and distances between nodes can be changed: As long as the connectivity of the network is not changed, it will have the same interpretation (i.e., it will represent the same network). These operations are, however, not supported in a map and there is, therefore, a potential conflict of interpretation in maps with network overlay. In city maps the nodes representing subways stations have to be fixed to their correct location on the map, whereas the shape of connecting railway lines can be idealised and thus not true to the underlying map. This is acceptable because the geometric properties of railway lines are unimportant to passengers. On the other hand, city maps could support false inferences about these properties as seen on the map.

The core semantics of a representational form will necessarily be unimodal even though most concrete practical examples of external representation will be multimodal. We can artificially construct a unimodal map by "reverse engineering" of a city subway diagram like the one shown below to the left (an early version of the Copenhagen subway system), but it is virtually useless because the labelling of stations have been removed.

We can utilise the Formal Concept Analysis (FCA) for graphical types and generalise it to cover the full range of multimodal multimedia objects. FCA have already been extended to cover constructive aspects of diagram design (Kamps, 1999) and multimodal document layout (Bateman, Kamps, Kleinz, & Reichenberger, 2001), but what has not yet been fully recognised is the potential benefits of extending the description of objects with semiotic metadata about the representational forms articulated and combined within different media. This type of information would not only be relevant for the constructive generation of multimodal multimedia documents, but also for the design of flexible distributed documents, for instance, in supporting changes in representational form in moving Web-based documents to PDA interfaces. The issue of flexibility here is not just a question of how to deal with layout constraints imposed by different devices, but also a question of designing support for *adaptation* or *tailoring* through transformations of representational *scales* (Johansen & May, in press), media types as well as representational forms (sign types).

Consider again maps and network charts as two different representational forms. We can express what this difference means, in terms of the semantic features and supported operations associated with each form within different media, by setting up the formal context for these simple "unimodal" objects. The following example is just given as a sketch of what such an analysis could look like, and this simplified formal context only includes five objects and four attributes (semantic features and supported operations):

Copyright © 2006, Idea Group Inc. Copying or distributing in print or electronic forms without written permission of Idea Group Inc. is prohibited.

Figure 8. Artificially reconstructed "unimodal" network chart (derived from a subway map) (to the left) and the (re-) construction by annotation and projection of a multimodal map object by spatially coordinated layering of different representational spaces (to the right)

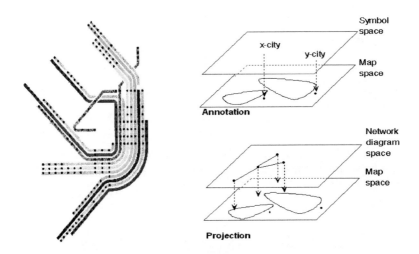

- **Connectivity:** Does the type require connectivity to be preserved?
- **Perceptual logic:** Does the type support "direct perception" of logical relations in a spatial structure (cf. Lakoff, 1990; May, 1999)?
- **Deformation:** Does the type support and allow deformation?
- **Object localization:** Does the type support localzation of objects relative to a background object?

It is implied by this analysis that the well-known graphical maps, although they are prototypical maps, do not exemplify the *core semantics* of maps. It is in fact the apparently peripheral case of the acoustic map (A-map in Figure 10) that exemplifies the *core* semantics of maps, because it is the attributes of acoustic maps (in the example here reduced to the feature "Object localization") that is shared by all maps. Graphical maps, of course, also have this feature, but in addition they have other emergent features derived from the graphical media (Connectivity and Perceptual logic).

We can visualise this through the inheritance hierarchy embedded within the lattice as shown in Figure 11. The core feature of maps (object localization) as exhibited by acoustic maps is inherited to haptic maps and graphical maps.

Lattice structures are well-suited to express feature structures, because they can express inheritance relations as well as systematic combinations of types. We can even use them to generate possible combinations of representational forms in cases where we might not know a prototypical example in advance, that is, we can use lattices to explore the design

Copyright © 2006, Idea Group Inc. Copying or distributing in print or electronic forms without written permission of Idea Group Inc. is prohibited.

Figure 9. Table indicating a part of the formal context specifying the map and the network chart types articulated in different media

	Connectivity	Perceptual logic	Deformation	Object localization
Graphical map (G-map)	+	+		+
Haptic map (H-map)	+			+
Acoustic map (A-map)				+
Graphical net (G-net)	+	+	+	
Haptic net (H-net)	+		+	

Figure 10. A Hasse-diagram of the lattice corresponding to the formal concepts specifying maps and network charts according to the simplified formal context sketched in Figure 9

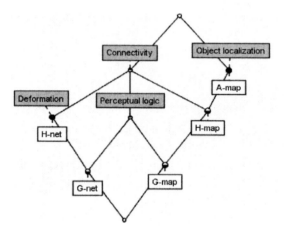

space of all possible type combinations and their expression in different media —given a full set of features necessary to distinguish the formal concepts involved.

The concept of a lattice is based on the concept of an ordering of types through the binary relation ≤. Defining this relation on a set, we obtain a partially ordered set (sometimes just called a partial order), if for all elements of the set, the following is true:

$x \leq x$ (Reflexivity)

$x \leq y \ \& \ y \leq x$ implies $x = y$ (Anti-symmetry)

$x \leq y \ \& \ y \leq z$ implies $x \leq z$ (Transitivity)

Copyright © 2006, Idea Group Inc. Copying or distributing in print or electronic forms without written permission of Idea Group Inc. is prohibited.

Figure 11. The lattice of Figure 10 used to show the inheritance of the "core feature" of maps (object localisation) from the acoustic map type; the object localisation feature is shared by all maps.

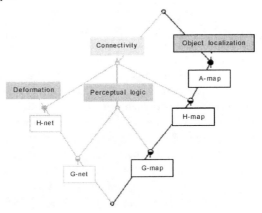

Figure 12. The lattice of Figure 10 used to show the specification of graphical nets (G-net) as the lattice "meet" of Deformation (a core semantic feature of network charts) and Perceptual logic (an emergent feature of the graphical media); by inheritance graphical nets are also characterized by Connectivity (a shared feature of the graphical and haptic media).

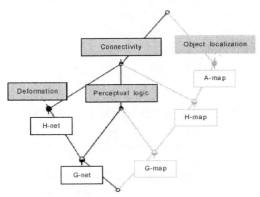

If every pair in the set has both a "supremum" (a least upper bound) and an "infimum" (a greatest lower bound) within the set, we have a lattice ordering $<L, \leq>$. From any lattice ordering $<L, \leq>$ we can construct a lattice algebra $<L, \vee, \wedge>$ with two binary operations called join ($\vee$) and meet ($\wedge$) by defining x y = sup{x,y} and x y = inf {x,y}. This equivalence of the order theoretic and the algebraic definition of a lattice is important, because it means that we can construct operations on a lattice given the ordering of its elements (Davey & Priestley, 1990). This is what is utilised in formal concept analysis (Ganter & Wille, 1999).

Copyright © 2006, Idea Group Inc. Copying or distributing in print or electronic forms without written permission of Idea Group Inc. is prohibited.

Feature-Based Taxonomy for Instructional Design

Related Approaches to Multimedia Semantics

The extension of feature-based approaches to multimedia semantics has followed different tracks, since the first attempts to specify features of graphical objects. One track has followed the work on automated presentation of information (Mackinlay, 1986) and another track have analysed perceptual and representational scales for relational information displays (Petersen & May, in press; Zhang, 1996). A third track can be identified around the problem of indexing images as well as indexing and sequencing video for databases (Dorai & Venkatesh, 2001, 2003), but this work has mainly dealt with low-level features, that is, features of the graphical substance of expression. Even when explicitly addressing the "semantic gap" between the physical substance of recorded film and the semantic issues of content management systems, the kind of semantics that is proposed is based on feature extraction from low-level data, although guided by film grammar. The focus on feature extraction is grounded in the practical interest in automatic segmentation and indexing of video. A fourth track is linked to the attempt to define semantic content for multimedia and hypermedia on "the semantic Web", but the standards developed for Web-based multimedia and hypermedia also have its main focus on the substance of expression and to a lesser extend on the form of expression. A good example is the advanced XML-based spatial and temporal layout facilities for media objects realised in the Synchronized Multimedia Integration Language (SMIL 2.0) (Bulterman, 2001).

From SMIL Encoding of Multimedia to Learning Object Metadata

The SMIL specification language for multimedia on the Web and for mobile devices can in many ways be seen as the realisation of the potential inherent in object-oriented multimedia with regard to the substance of expression. With SMIL 2.0, authors can specify the temporal behavior of individual elements of a presentation, their spatial layout, the hypermedia structure of the distributed objects included in the presentation, and its overall compositional structure.

The basic syntax of SMIL presentations is very similar to HTML syntax and is easy to read and edit. For a simple example, look at the following presentations that could be part of a multimedia instruction in how to play chess. Here a chess game has been visualised through a video recording of a particular game. The SMIL object is played back through the RealPlayer, but could also be embedded on a Web page. The first piece of code simply presents the recording of the board as captured in the file "board.rm".

Copyright © 2006, Idea Group Inc. Copying or distributing in print or electronic forms without written permission of Idea Group Inc. is prohibited.

A more advanced example is shown below, where the space within the player window is subdivided into four regions, one of which is used to display a textual comment synchronised with the moves of the particular chess game. The other three regions are used to display different visual perspectives on the game: the "black" player, the board, and the "white" player. In parallel with the synchronised graphical channels, an audio channel is added to present additional speech comments on the game.

The chess example above has been produced in a project on digital video in educational research and it is accordingly not meant as an instructional object for teaching or learning how to play chess. We can, however, use it here as a point of departure for discussing the relation between the encoding of multimedia documents and the metadata required for their use in instructional design of learning objects.

A learning object is a modularized self-contained unit of instructional material that has been given a metadata description of its own content and use. Ideally these instructional materials are available in digital electronic form through a learning object repository (shared over the Internet, for example). As described by David Wiley, "learning objects are elements of a new type of computer-based instruction grounded in the object-oriented paradigm of computer science. Object-orientation highly values the creation of components (called "objects") that can be reused ... in multiple contexts. This is the fundamental idea behind learning objects: instructional designers can build small (relative to the size of an entire course) instructional components that can be reused a number of times in different learning contexts." (Wiley, 2000, chap. 3, p. 1). There are however, as Wiley also points out, important theoretical problems to be addressed with regard to the granularity of these objects (their minimal "size" within the instructional material of a whole course) and with regard to semantic and cognitive constraints (what makes sense, what gives cognitive support) as well as pragmatic constrains on compositionality (what has practical value within a context-of-use). We need to go beyond a naïve "Lego"-conception of learning objects.

The chess video example could be considered as a learning object provided that we associate it with relevant metadata describing its intended use within an instructional setting. It is in fact already a learning object within its original context, where it is used as an illustration of how to use SMIL for editing and presenting digital video. The very re-use of the chess example in the present context illustrates both the potential of learning objects and an inherent weakness: By completely changing the context and reusing the object with a purpose, for which it was not designed (chess instruction), it will be difficult to determine in advance, if the object is really useful and adequate in the new context. In any case, we will need detailed metadata describing the learning object and its intended

Figure 13. SMIL code use to present a recorded movie sequence

```
<smil>
 <body>
  <video src="board.rm" clip-begin="16s" clip-end="27s"/>
 </body>
</smil>
```

Copyright © 2006, Idea Group Inc. Copying or distributing in print or electronic forms without written permission of Idea Group Inc. is prohibited.

Figure 14. The RealPlayer used to play back a SMIL presentation of a chess game using four graphical regions as well as an audio channel for speech comments; the corresponding SMIL code is shown in Figure 15; this Public Domain example is provided by the Cognitive Development Lab of the University of Illinois at Urbana-Champaign

Figure 15. SMIL code to present the RealPlayer presentation through an external URL (http://www.psych.uiuc.edu/~kmiller/smil/examples/voiceover.smil)

```
<smil>
<head>
 <layout>
  <root-layout width="1080" height="350"/>
  <region id="video_left" width="360" height="240" left="0" top="0"/>
  <region id="video_center" width="360" height="240" left="360" top="0"/>
  <region id="video_right" width="360" height="240" left="720" top="0"/>
  <region id="text_subtitle" width="560" height="100" left="260" top="250"/>
 </layout>
</head>
<body>
 <par dur="55s">
  <video src="black.rm" begin="5s" clip-begin="1.09s" region="video_left"/>
  <video src="board.rm" begin="5s" clip-begin="0s" region="video_center"/>
  <video src="white.rm" begin="5s" clip-begin="1.10s" region="video_right"/>
  <textstream src="text.rt" region="text_subtitle"/>
  <audio src="voicetrack.rm"/>
 </par>
</body>
</smil>
```

Copyright © 2006, Idea Group Inc. Copying or distributing in print or electronic forms without written permission of Idea Group Inc. is prohibited.

use, but we cannot expect instructional designers to anticipate all possible context of reuse. If learning objects are to be retrieved and reused, they will therefore have to be designed with some level of flexibility and a combination of detailed information about the intended context-of-use (to support identification of similar learning situations) and abstract compositional descriptions of their form and content independent of any specific use (to support transformations of the object and its transfer to other learning situations). Before discussing these demands on metadata, let us stay with the fictive example of the chess learning object.

As a first observation, it should be noted that video playback only provides a limited level of interactivity. A learner would very soon in the learning process need to be actively engaged in playing chess (i.e., using a chess simulator or playing chess with a human opponent), rather than watching a game passively through recorded sequences of game play. On the other hand, even this simple example illustrates the learning potential of using multiple representations within multimedia learning objects. The chess example uses the representational form of language expressed as graphical text (for comments in the subtitle region) and as acoustic speech (for synchronised "voice over") in combination with dynamic graphics (the video). Combining a graphical display of the game with voice could be an effective support for basic learning according to the cognitive theory of multimedia learning proposed by Mayer (2001). According to Mayer's so-called "modality principle", students will learn better when the language part of a multimedia document is presented in the acoustic media (i.e., as voice) rather than in the graphical media (i.e., as text), because this use of voice for comments within the multimedia presentation will be less demanding on cognitive processing (since the visual-graphical channel is already used for static graphics, animation, or video). This principle has also been referred to as "dual coding theory", that is, referring to the utilisation of the capacity of visual as well as auditory working memory.

There is another important issue, however, that is not addressed adequately by the cognitive theory of multimedia learning. The "dual coding" theory stays within a media-specific dichotomy between the visual and the auditory channel of communication, and representational forms are more or less reduced to "pictures" and "words". Although Mayer (2001) uses different diagrams in his empirical investigations, they are treated as "pictures" at the theoretical level. This makes it difficult to address didactic aspects of learning with different forms of representation. In our small example with the hypothetical learning object for multimedia support of chess learning (Figure 14), it is obvious that chess learning beyond the initial stages of getting acquainted with the context of chess playing would benefit from relying on "abstract-iconic" representations of the relational structure of the game and its strategic situations rather than on "concrete iconic" representations like images (animated or not). In other words, we need the familiar chess diagrams to support game analysis, and this requires that we abstract from the irrelevant features of the concrete situation of game play (unless our focus is the psychology of game players rather than the game itself).

In learning chess, we would in fact abstract diagrams representing the strategic situations from observing a video, but the mental work required to do so would be less demanding if these abstractions were supported by a sequence of external diagrams. This is an important educational implication of the variable cognitive support provided by

Copyright © 2006, Idea Group Inc. Copying or distributing in print or electronic forms without written permission of Idea Group Inc. is prohibited.

different external representations according to the theory of distribution between external and internal (cognitive) representations (Petersen & May, in press; Zhang, 1996)). Image type representations will be less effective and less adequate as cognitive support for chess learning and understanding compared to diagrams (including their algebraic notation) for two different reasons: Images will necessarily qua images represent features that are irrelevant to the relational structure of the game (demanding us to abstract from them) and the symbolic structure of chess will require us to see in the image "something more" than is directly visible, that is, the diagrammatic form (May, 1999). This is why the external diagram and its algebraic equivalent are more effective, that is, less cognitively demanding.

Professional chess players will require more interactive features and more functionality to support further learning than can be realised with playback of recorded games. With the full simulation provided by a game engine, the expert will not only use a program like Fritz to play against a virtual opponent, but also to perform dynamic thought experiments on the external diagrammatic representations in order to explore the game space and evaluate different strategies. This exploration corresponds to what the founding father of logical semiotics, C. S. Peirce, called "diagrammatic reasoning" (Peirce, 1906).

Figure 16. The MyChess Java applet illustrates a more powerful form of playback, where chess diagrams are used in combination with their algebraic notation to support analysis of the game (here Anand - Ivanchuk in Linares, 1993). The applet also presents a database from which the current game is selected (source: http://www.mychess.com)

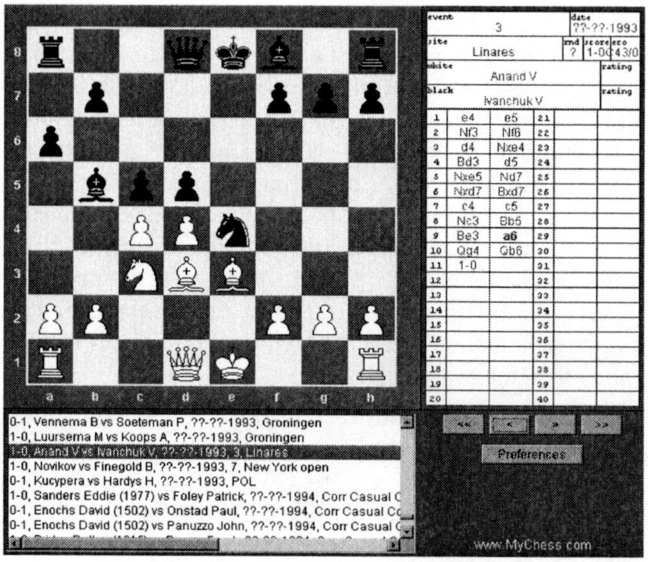

Copyright © 2006, Idea Group Inc. Copying or distributing in print or electronic forms without written permission of Idea Group Inc. is prohibited.

Extending Learning Object Metadata

As shown in the previous example, it is generally not enough to know which media are utilised by some learning objects in order to evaluate their usefulness and their cognitive support for specific activities and contexts-of-use. We also need to know details about the representational forms used by learning objects. This type of information has not yet been included in the design of metadata for learning object repositories. If we again depart from a SMIL example of a learning object, but this time one that has been give a metadata description, we can take the SMIL presentation of a lecture in "Membranes and Cell Surfaces" from a BioScience course given at University of Queensland.

The metadata given is primarily formal (author, version, department, institution, etc.), deictic (date, place, presenter) and superficial content descriptions (subject, description). An important information about the level of learning that the learning object is assumed to address is given by the statement of the audience being "First Year Biology Students" at a university, but it is not stated what the specific learning objectives for the lesson is (what are students supposed to be able to do after using the object or after taking the course?), or indeed what form of learning and instruction is supported by the lesson? On inspecting the online lesson, it turns out to be a recorded audio and video sequence of a traditional lecture, with added notes and diagrams in a separate region. This type of presentation can be presented with a SMIL 2.0 player or embedded within a website. Alternatively a similar presentation can be made with Microsoft Power Point Producer (available for Office 2003). In any case there is no metadata standard for describing the representational forms used by learning objects and what kind of cognitive support this gives for the intended activity within a context-of-use.

Learning objects have been given several competing XML-based standards or specifications in the form of the IEEE Learning Object Metadata (LOM), the Instructional Management System (IMS) Metadata, and the Sharable Content Object Reference Model (SCORM), to mention just three different models. Although these standards and specifications are supposed to be neutral to different contexts-of-use, there have been many objections towards their implementation in higher education. The SCORM standard, for example, promoted by the Advanced Distributed Learning initiative (ADL), sponsored by the U.S. Department of Defense, has a bias toward distributed learning situations with single learners engaged in self-paced and self-directed learning (Friesen, 2004). This is well suited for instructional design of individual Web-based training, but less so for the pedagogical use of learning objects within group-based and class-based teaching and learning at the university level.

The metadata specified by the standards focus on formal document information, technical requirements, superficial content descriptions (like keywords), and general educational information (like interactivity level, resource type, duration of use), but generally neglect more specific information about the content. Content specifications should include more detailed information about the scientific domain to which the object refers, the prior knowledge required by the learner to benefit from its use, the learning objectives for the use of the object, and its intended scenarios of use. This type of information is included within the Dutch proposal for a more elaborated Educational Modelling Language (EML), which goes beyond learning objects in order to model the

Copyright © 2006, Idea Group Inc. Copying or distributing in print or electronic forms without written permission of Idea Group Inc. is prohibited.

Figure 17. Example of metadata given for a SMIL learning object presenting a Bioscience lecture; the information shown here is from the Australian DSTC repository (Source: http://maenad.dstc.edu.au/demos/splash/search.html#browse)

Subject:	Structure and properties of biological membranes
Description:	This is a lecture about the properties of membranes and cell surfaces
Presenter:	Dr Susan Hamilton (University of Queensland)
Date:	08-22 on the 22-08-2000
Place:	University of Queensland Saint Lucia
Audience:	First Year Biology Students at UQ
Duration:	0:18:52
Subject:	BL114 (Semester One): Introduction to Biological Science
Course:	B.Sci: Bachelor of Science
Department:	Biological Sciences
Institution:	University of Queensland
Editor:	Suzanne Little
Date.Modified:	15-11-2001
Version:	1.0
Source Presentations:	biosciences1

whole activity of teaching and learning, the methods used, and the different roles of the social agents involved (Koper, 2001; Koper & van Es, 2004). EML is proposed as "a major building block of the educational Semantic Web" (Anderson & Petrinjak, 2004, p. 289).

This extension of the modelling perspective from the objects of instructional design, that is, teaching materials, to include the whole activity of teaching and learning, might however be counter-productive and impractical. The perspective of learning objects should not be to formalise the whole activity of teaching and learning, but to package electronically-available learning objects with metadata relevant for teaching and learning, and to embed them in object repositories for future reuse. They should, however, be tagged with some of the information provided with EML, but only as tools for reuse, not as a part of a codification of the whole activity.

There are also alternative conceptions of learning objects like the Reusable Learning Assets (RLA) proposed within the Open Course initiative (http://opencourse.org), which are defined as open source, non-proprietary learning objects.

What is lacking in all projects, however, even in the detailed modelling of "units of study" in EML, is the relevant metadata information about the media types and representational forms utilised by the learning object. This information is relevant because there is evidence that variance and integration of representational forms are important for student learning, even if there has been some debate over the specific "modality effects" of media on learning. Early experimental work on the "effects of media" on learning did not produce consistent findings (Hede, 2002), but conflicting conceptions of learning have also confused the issue. In the "effects of media" debate there was originally a

Copyright © 2006, Idea Group Inc. Copying or distributing in print or electronic forms without written permission of Idea Group Inc. is prohibited.

Figure 18. EML specification of a unit of study

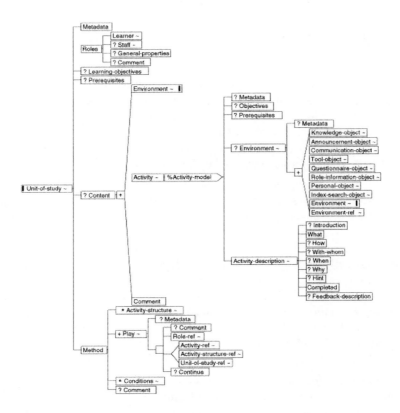

conception of learning as transmission of information, and teaching was accordingly seen as a simple matter of communication, whereas learning in modern conceptions involves constructive conceptual activity by students. Even in modern studies it is, however, not always clear whether the "modality effects" claimed for multimedia learning is due to the duality of audio and visual coding, that is, a consequence of the use of acoustic and graphic media types, or due to the multiplicity of different representational forms such as language, diagrams, graphs, and images.

The Example of Graph Comprehension

The theoretical critique of Mayer's cognitive theory of multimedia learning (CTML) has been backed up by empirical studies. Schnotz and Bannert (2003) found that adding graphical representations to text presentations did not always result in improved

Copyright © 2006, Idea Group Inc. Copying or distributing in print or electronic forms without written permission of Idea Group Inc. is prohibited.

understanding. In some cases the "pictures" could actually have negative effects by interfering with the construction of mental models. They conclude that the "dual coding theory" of multimedia learning does not take into account the effects that different forms of visualisation will have on the construction of mental models from the presentation in the process of comprehension. In another study it was found that the "affordances" of graphical representations for improving conceptual understanding should be seen in the context of specific scientific domains and also as dependant on familiarity with the specific "iconic sign system" from which the representations are derived (De Westelinck, Valcke, De Craene, & Kirschner, 2005). It seems, for instance, that students in the social sciences are less familiar with graphs and conceptual diagrams than students of natural sciences, and they will therefore not have the same benefits from these abstract-iconic forms. Learners can only benefit from representations if they understand the semantics of the representational forms to which they belong (Cox, 1999).

A striking example is the problems of graph comprehension reported in many studies of science learning in high school as well as in higher learning. An early example is the studies of students' understanding of kinematics and problems of associating graphs with their physical interpretation (Beichner, 1994; McDermott, Rosenquist, & van Zee, 1987). A recent review of the literature on graph comprehension (Shah & Hoeffner, 2002) supports these early findings and lists four implications for teaching graphical literacy: (1) graphical literacy skills should be taught explicitly (and in the context of different scientific domains), (2) activities where students have to translate between different types of representations will improve their understanding, (3) students need to pay attention to and be trained in the relation between graphs and the models they express, and (4) students should be encouraged to make graph reading a "meta-cognitive" activity, that is, to pay attention to and reflect critically on the interpretation of graphs.

One of the typical problems found in graph comprehension, even in higher education, is the interpretation of graphs as "concrete iconic" objects, that is, images rather than relational structures. An example found in students of chemical engineering at DTU were errors in reasoning about a test on Fourier's law for heat conduction, where some students, who tried to remember the graph by its shape (a graphical image feature) rather than by the conceptual model it expresses, reported the correct shape of a curved graph, but as a mirror image of its correct orientation (May, 1998).

The students who answered (c) to the test in Figure 19 could not give any physical interpretation to the graph since it was remembered as a mere image of a textbook (or lecture) presentation. This exemplifies a "typological error" within the representational forms used in reasoning by the students: treating a graph as if it was an image (or effectively reducing it to a mental image).

In order to support the didactic suggestions by Shah and Hoeffner (2002) and others regarding the need to train graph comprehension skills in higher education, learning objects need to be annotated with metadata on their level of interactivity (to distinguish learning objects with active exploration of models from more or less static illustrations, for example) and on their support for active student learning through translations between different representational forms (as in conceptual tests motivating students to transform graph representations, algebraic expressions, or natural language descriptions of models).

Copyright © 2006, Idea Group Inc. Copying or distributing in print or electronic forms without written permission of Idea Group Inc. is prohibited.

Figure 19. A conceptual test (to the left + caption) given to students in a course in chemical engineering at DTU (1995); three types of answers were given by students: the correct answer (a), the linear answer (b), which is motivated by the original context where students learned about the law (where it is in fact linear, if the area is not considered), and (c) where student have remembered the shape of the image, but without any physical understanding of the model represented by the graph (May, 1998)

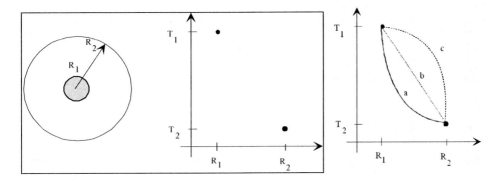

[Original caption to the test:] Consider a cylinder with radius R_2. Inside the cylinder is a kernel with radius R_1 and which has been heated to the temperature T_1. At the surface (with radius $r = R_2$) the temperature is held constant at T_2 where $T_2 < T_1$. Show in the diagram a qualitative sketch of the graph $T(r)$ for values between R_1 and R_2 (i.e., you do not have to compute any values for points on the curve).

Figure 20. Conceptual diagram to indicate the "desymbolisation" of graphs reduced to concrete-iconic images by some students and the didactic importance of training students actively in "translating" between graphs and their "diagrammatic" interpretation in terms of their underlying physical models as well as between graphs and the formal algebraic models which they express

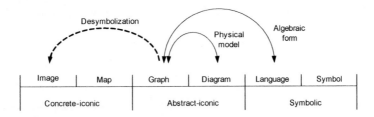

The taxonomy of representational forms introduced in this chapter can be used to extend existing standards for learning object metadata with this type of information (as well as specifications of learning objectives).

In a recent study, it was found that students working actively with integrating multiple static representations, before working with animations and simulations, have better learning outcomes than students who directly explore dynamic and interactive

Copyright © 2006, Idea Group Inc. Copying or distributing in print or electronic forms without written permission of Idea Group Inc. is prohibited.

visualisations (Bodemer, Ploetzner, Bruchmüller, & Häcker, 2005). This is again empirical evidence that indicates the importance of the integration of multiple forms of representation for conceptual understanding. In the context of instructional design of learning objects for object repositories, it will be necessary to extend current XML-based standards for metadata with information about the representational forms utilised by learning objects. This will enable teachers engaged in instructional design of learning objects to make better choices in setting up conditions for active learning, since students will benefit from being exposed to many different representations of the same content. In engineering education, for instance, it is important to set up learning environments where students can work actively with static graphs, algebraic forms, explanations in natural language, and conceptual diagrams, as well as animated graphs and interactive simulations: Conceptual understanding emerges from the discovery of invariant forms across all of these media and "modalities."

Summary

A better understanding of the invariant semantic properties of representational forms across different media will be important in the future support for flexible configurable interfaces and for supporting the design of device-independent multimedia and other forms of adaptive multimedia.

In the domain of instructional design and in the redesign of instructional material as learning objects for Web-based learning and for learning repositories, it is of primary importance to extend current standards for learning objects with (a) didactic specifications of the intended use of these objects and (b) semiotic and semantic information about the media and representational forms supported by the learning objects.

An approach to the taxonomy of media types and representational forms has been suggested, based on feature structures and Formal Concept Analysis. More research is needed in order to combine the analysis of media types and representational forms with the analysis of scale types and scale transformations (Petersen & May, in press), and to apply both to the domain of adaptive and configurable interfaces as well as to the domain of instructional design.

References

Anderson, T., & Petrinjak, A. (2004). Beyond learning objects to educational modelling languages. In R. McGreal (Ed.), *Online education using learning objects* (pp. 287-300). London; New York: RoutledgeFalmer.

Bateman, J., Delin, J., & Henschel, R. (2002). *XML and multimodal corpus design: Experiences with multi-layered stand-off annotations in the GeM corpus.* ELSNET Workshop (Towards a Roadmap for Multimodal Language Resources and Evaluation) at the LREC 2002 Conference, Las Palmas, Spain.

Copyright © 2006, Idea Group Inc. Copying or distributing in print or electronic forms without written permission of Idea Group Inc. is prohibited.

Bateman, J., Kamps, T., Kleinz, J., & Reichenberger, K. (2001). Towards constructive text, diagram, and layout generation for information presentation. *Computational Linguistics, 27*(3), 409-449.

Beichner, R. J. (1994). Testing student interpretation of kinematics graphs. *American Journal of Physics, 62*(8), 750-762

Bernsen, N. O. (1993, November). *Modality theory: Supporting multimodal interface design.* ERCIM Workshop on Multimodal Human-Computer Interaction (Report ERCIM-94-W003), Nancy, France.

Bernsen, N. O. (1994). Foundations of multimodal representations: A taxonomy of representational modalities. *Interacting with Computers, 6*(4), 347-371.

Bodemer, D., Ploetzner, R., Bruchmüller, K., & Häcker, S. (2005). Supporting learning with interactive multimedia through active integration of representations. *Instructional Science, 33*, 73-95.

Bulterman, D. (2001). SMIL 2.0. Part 1: Overview, concepts, and structure. *IEEE Multimedia, 8*(4), 82-88.

Chatman, S. (1978). *Story and discourse: Narrative structure in fiction and film.* Ithaca, New York: Cornell University Press.

Cox, R. (1999). Representation construction, externalised cognition, and individual differences. *Learning and Instruction, 9*, 343-363.

Davey, B. A., & Priestley, H. A. (1990). *Introduction to lattices and order.* Cambridge, UK: Cambridge University Press.

de Saussure, F. (1993). *Course in general linguistics.* London: Dockworth.

De Westelinck, K., Valcke, M., De Craene, B., & Kirschner, P. (2005). Multimedia learning in the social sciences: Limitations of external graphical representations. *Computers in Human Behavior, 21*, 555-573.

Dorai, C., & Venkatesh, S. (2001, September 10-12). Bridging the semantic gap in content management systems: Computational media aesthetics. *Proceedings of the First Conference on Computational Semiotics, COSIGN-2001*, Amsterdam. Retrieved from http://www.cosignconference.org/cosign2001

Dorai, C., & Venkatesh, S. (2003). Bridging the semantic gap with computational media aesthetics. *IEEE Multimedia, 10*(2), 15-17.

Friesen, N. (2004). Three objections to learning objects. In R. McGreal (Ed.), *Online education using learning objects* (pp. 59-70). RoutledgeFalmer.

Ganter, B., & Wille, R. (1999). *Formal concept analysis: Mathematical foundations.* Berlin: Springer.

Gourdol, A., Nigay, L., Salber, D., & Coutaz, J. (1992). Two case studies of software architecture for multimodal interactive systems. *Proceedings of the IFIP Working Conference on Engineering for Human Computer Interaction* (pp. 271-284). Ellivuori, Finland: Amsterdam: North-Holland.

Heller, R. S., & Martin, C. D. (1995). A media taxonomy. *IEEE Multimedia, 2*(4), 36-45.

Copyright © 2006, Idea Group Inc. Copying or distributing in print or electronic forms without written permission of Idea Group Inc. is prohibited.

Heller, R. S., Martin, C. D., Haneef, N., & Gievska-Krliu, S. (2001). Using a theoretical multimedia taxonomy framework. *ACM Journal of Educational Resources in Computing, 1*(1), 1-22.

Hende, A. (2002). An integrated model of multimedia effects on learning. *Journal of Educational Multimedia and Hypermedia, 11*(2), 177-191.

Hjelmslev, L. (1961). *Omkring Sprogteories Grundlæggelse. Travaux du Cercle Linguistique du Copenhague, XXV (Reissued 1993, Reitzel).* English translation in L. Hjelmslev, *Prolegomena to a theory of language.* Bloomington: Indiana University Press.

Johnson, M. (1987). *The body in the mind: The bodily basis of meaning, imagination, and reason.* Chicago: University of Chicago Press.

Koper, R. (2001). *Modelling units of study from a pedagogical perspective: The pedagogical meta-model behind EML.* The Educational Technology Expertise Centre of the Open University of the Netherlands. Retrieved from http://eml.ou.nl/introduction/docs/ped-metamodel.pdf

Koper, R., & van Es, R. (2004). Modelling units of study from a pedagogical perspective. In R. McGreal (Ed.), *Online education using learning objects* (pp. 43-58). London; New York: RoutledgeFalmer.

Lakoff, G. (1987). *Women, fire, and dangerous things: What categories reveal about the mind.* Chicago: University Of Chicago Press.

Lakoff, G. (1990). The invariance hypothesis: Is abstract reasoning based on image-schemas? *Cognitive Linguistics, 1*(1), 39-74.

Lohse, G. L., Biolsi, K., Walker, N., & Rueter, H. (1994). A classification of visual representations. *Communications of the ACM, 37*(12), 36-49.

Lohse, G. L., Walker, N., Biolsi, K., & Rueter, H. (1991). Classifying graphical information. *Behaviour & Information Technology, 10*(5), 419-436.

Mackinlay, J. (1986). Automating the design of graphical presentations of relational information. *ACM Transactions on Graphics, 5*(2), 110-141.

Mackinlay, J., & Genesereth, M. R. (1985). Expressiveness and language choice. *Data & Knowledge Engineering, 1*(1), 17-29.

Manning, A. D. (1989). The semantics of technical graphics. *Journal of Technical Writing and Communication, 19*(1), 31-51.

May, M. (1993). A taxonomy of representations for HCI, Parts 1-3. In N. O. Bernsen (Ed.), *Taxonomy of HCI systems: State of the art.* Esprit basic research project GRACE working papers. Deliverable 2.1. Edinburgh: Human Communication Research Centre.

May, M. (1998). Images, diagrams, and metaphors in science and science education: The case of chemistry. *Almen Semiotik, 14*, 77-102.

May, M. (1999). Diagrammatic reasoning and levels of schematization. In T. D. Johansson, M. Skov, & B. Brogaard (Eds.), *Iconicity: A fundamental problem in semiotics.* Copenhagen: NSU Press.

Copyright © 2006, Idea Group Inc. Copying or distributing in print or electronic forms without written permission of Idea Group Inc. is prohibited.

May, M. (2001, June 25-27). Semantics for instrument semiotics. In M. Lind (Ed.), *Proceedings of the 20th European Annual Conference on Human Decision Making and Manual Control (EAM-2001)*, Kongens Lyngby, Denmark (pp. 29-38).

May, M. (2001). Instrument semiotics: A semiotic approach to interface components. *Knowledge-Based Systems, 14*(2001), 431-435.

May, M. (2004). Wayfinding, ships, and augmented reality. In P. B. Andersen & L. Qvortrup (Eds.), *Virtual applications: Applications with virtual inhabited 3D worlds*. Berlin; New York: Springer Verlag.

May, M., & Andersen, P. B. (2001). Instrument semiotics. In K. Liu, R. J. Clarke, P. B. Andersen, & R. K. Stamper (Eds.), *Information, organisation, and technology: Studies in organisational semiotics*. Boston; Dordrecht; London: Kluwer Academic Publishers.

Mayer, R. E. (2001). *Multimedia learning*. Cambridge, UK: Cambridge University Press.

McDermott, L. C., Rosenquist, M. L., & van Zee, E. H. (1987). Student difficulties in connecting graphs and physics: Examples from kinematics. *American Journal of Physics, 55*(6), 503-513.

Nack, F., & Hardman, L. (2002). *Towards a syntax for multimedia semantics* (Rep. No. INS-RO204). Amsterdam: Information Systems (INS) at CWI.

Nigay, L., & Coutaz, J. (1993). A design space for multimodal systems: Concurrent processing and data fusion. *INTERCHI'93 Proceedings*. Amsterdam: ACM Press; Addison Wesley.

Peirce, C. S. (1906). Prolegomena for an apology for pragmatism. In C. Eisele (Ed.), *The new elements of mathematics, Vol. IV*. The Hague: Mouton.

Petersen, J., & May, M. (in press). Scale transformations and information presentation in supervisory control. *International Journal of Human-Machine Studies*. Retrieved from www.sciencedirect.com

Purchase, H. C., & Naumann, D. (2001). A semiotic model of multimedia: Theory and evaluation. In S. M. Rahman (Ed.), *Design and management of multimedia information systems: Opportunities and challenges* (pp. 1-21). Hershey, PA: Idea Group Publishing.

Quek, F., McNeill, D., Bryll, R., Duncan, S., Ma, X. -F., Kirbas, C., McCullough, K. E., & Ansari, R. (2002). Multimodal human discourse: Gesture and speech. *ACM Transactions on Computer-Human Interaction, 9*(3), 171-193.

Recker, M. M., Ram, A., Shikano, T., Li, G., & Stasko, J. (1995). Cognitive media types for multimedia information access. *Journal of Educational Multimedia and Hypermedia, 4*(2-3), 183-210.

Schnotz, W., & Bannert, M. (2003): Construction and interference in learning from multiple representations. *Learning and Instruction, 13*, 141-156.

Shah, P., & Hoeffner, J. (2002). Review of graph comprehension research: Implications for instruction. *Educational Psychology Review, 14*(1), 47-69

Copyright © 2006, Idea Group Inc. Copying or distributing in print or electronic forms without written permission of Idea Group Inc. is prohibited.

Stenning, K., Inder, R., & Neilson, I. (1995). Applying semantic concepts to analyzing media and modalities. In J. Glasgow, N. H. Narayanan, & B. Chandrasekaran (Eds.), *Diagrammatic reasoning: Cognitive and computational perspectives*. Menlo Park, CA: AAAI Press; MIT Press.

Stockinger, P. (1993). Multimedia and knowledge based systems. *S – European Journal of Semiotic Studies, 5*(3), 387-424.

Troncy, R., & Carrive, J. (2004). A reduced yet extensible audio-visual description language. *ACM Symposium on Document Engineering* (pp. 87-89), Milwaukee, Wisconsin.

Twyman, M. (1979). A schema for the study of graphic language. In P. A. Kolers, M. E. Wrolstad, & H. Bouma (Eds.), *Processing of visible language, Vol. 1* (pp. 117-150). New York: Plenum Press.

Ware, C. (2000). *Information visualization: Perception for design*. San Francisco: Morgan Kaufman.

Wiley, D. A. (2000). Connecting learning objects to instructional design theory: A definition, a metaphor, and a taxonomy. In D. A. Wiley (Ed.), *The instructional use of learning objects*. Retrieved from http://reusability.org/read/

Zhang, J. (1996). A representational analysis of relational information displays. *International Journal of Human-Computer Studies, 45*, 59-74.

Copyright © 2006, Idea Group Inc. Copying or distributing in print or electronic forms without written permission of Idea Group Inc. is prohibited.

Chapter IV

Incorporating and Understanding the User Perspective

Stephen R. Gulliver, Brunel University, UK

Abstract

This chapter introduces a selection of studies relating to each of the multimedia senses — olfactory (smell), tactile/haptic (touch), visual (sight), and auditory (sound) — and how such studies impact user perception and ultimately user definition of multimedia quality. A model of distributed multimedia is proposed, to allow a more structured analysis of the current literature concerning video and audio information. This model segregates studies implementing quality variation and/or assessment into three discrete information abstractions (the network, media, and content levels) and from two perspectives (the technical and user perspectives). It is the objective of the author that, by placing current research in context of a quality structure, the need for fuller incorporation of the user perspective in multimedia quality assessment will be highlighted.

Copyright © 2006, Idea Group Inc. Copying or distributing in print or electronic forms without written permission of Idea Group Inc. is prohibited.

Introduction

Multimedia quality is a multi-faceted concept that means different things to different people (Watson & Sasse, 1997). Multimedia quality definition involves the integration of quality parameters at different levels of abstraction and from different perspectives. Indeed, the perception of multimedia quality may be affected by numerous factors, for example, delay or loss of a frame, audio clarity, lip synchronisation during speech, video content, display size, resolution, brightness, contrast, sharpness, colourfulness, as well as naturalness of video and audio content, just to name a few (Ahumada & Null, 1993; Apteker, Fisher, Kisimov, & Neishlos, 1995; Klein, 1993; Martens & Kayargadde, 1996; Roufs, 1992). Moreover, as multimedia applications reflect the symbiotic *infotainment* duality of multimedia, that is, the ability to transfer information to the user while also providing the user with a level of subjective satisfaction, incorporating the user perspective in a multimedia quality definition is further complicated since a comprehensive quality definition should reflect both how a multimedia presentation is understood by the user, yet also examine the user's level of satisfaction. Interestingly, all previous studies fail to either measure the infotainment duality of distributed multimedia quality or comprehensively incorporate and understanding the user-perspective.

Inclusion of the user-perspective is of paramount importance to the continued uptake and proliferation of multimedia applications since users will not use and pay for applications if they are perceived to be of low quality. In this chapter, the author aims to introduce the reader to work relating to each of the multimedia senses and how such studies impact user perception and definition of multimedia quality. The author proposes a model in which quality is looked at from three distinct levels: the *network-*, the *media-* and the *content-levels*; and from two views: the *technical-* and the *user-perspective*. This model is used to help structure, specifically current sight and sound literature, in order to help outline the diverse approaches used when varying and assessing multimedia quality, and ultimately to emphasize the need for fuller incorporation of the user perspective in multimedia quality assessment.

Perceptual Studies and Implications

In this section we aim to introduce the reader to the studies relating to the four multimedia senses that lie at the core of the human perceptual/sensory experience.

Olfactory

Research in the field of olfaction is limited, as there is no consistent method of testing user capability of smell. The first smell-based multimedia environment (sensorama) was developed by Heilig (1962, 1992), which simulated a motorcycle ride through New York and included colour 3D visual stimuli, stereo sound, aroma, and tactile impacts (wind from fans, and a seat that vibrated).

Copyright © 2006, Idea Group Inc. Copying or distributing in print or electronic forms without written permission of Idea Group Inc. is prohibited.

A major area of olfactory research has been to explore whether scent can be recorded and therefore replayed to aid olfactory perceptual displays (Davide, Holmberg, & Lundstrom, 2001; Ryans, 2001). Cater (1992, 1994) successfully developed a wearable olfactory display system for a fire fighters training simulation with a Virtual Reality (VR) oriented olfactory interface controlled according to the users location and posture. In addition, researchers have used olfaction to investigate the effects of smell on a participant's sense of presence in a virtual environment and on their memory of landmarks. Dinh, Walker, Bong, Kobayashi, and Hodges (2001) showed that the addition of tactile, olfactory, and/or auditory cues within a virtual environment increased the user's sense of presence and memory of the environment.

Tactile/Haptics

Current research in the field of haptics focuses mainly on either sensory substitution for the disabled (tactile pin arrays to convey visual information, vibro-tactile displays for auditory information) or use of tactile displays for teleoperation (the remote control of robot manipulators) and virtual environments. Skin sensation is essential, especially when participating in any spatial manipulation and exploration tasks (Howe, Peine, Kontarinis, & Son, 1995). Accordingly, a number of *tactile display* devices have been developed that simulate sensations of contact. While "tactile display" describes any apparatus that provides haptic feedback, tactile displays can be subdivided into the follow groups:

- **Vibration** sensations can be used to relay information about phenomena, such as surface texture, slip, impact, and puncture (Howe et al. 1995). Vibration is experienced as a general, non-localised experience, and can therefore be simulated by a single vibration point for each finger or region of skin, with an oscillating frequency range between 3 and 300 Hz (Kontarinis & Howe, 1995; Minsky & Lederman, 1996).

- **Small-scale shape or pressure** distribution information is more difficult to convey than that of vibration. The most commonly-used approach is to implement an array of closely aligned pins that can be individually raised and lowered against the finger tip to approximate the desired shape. To match human finger movement, an adjustment frequency of 0 to 36 Hz is required, and to match human perceptual resolution, pin spacing should be less than a few millimetres (Cohn, Lam, & Fearing, 1992; Hasser & Weisenberger, 1993; Howe et al., 1995).

- **Thermal displays** are a relatively new addition to the field of haptic research. Human fingertips are commonly warmer than the "room temperature". Therefore, thermal perception of objects in the environment is based on a combination of thermal conductivity, thermal capacity, and temperature. Using this information allows humans to infer the material composition of surfaces as well as temperature difference. A few thermal display devices have been developed in recent years that are based on Peltier thermoelectric coolers, solid-state devices that act as a heat pump, depending on direction of current (Caldwell & Gosney, 1993; Ino, Shimizu, Odagawa, Sato, Takahashi, Izumi, & Ifukube, 2003).

Copyright © 2006, Idea Group Inc. Copying or distributing in print or electronic forms without written permission of Idea Group Inc. is prohibited.

Many other tactile display modalities have been demonstrated, including *electrorheological devices* (a liquid that changes viscosity electroactively) (Monkman, 1992), *electrocutaneous stimulators* (that covert visual information into a pattern of vibrations or electrical charges on the skin), *ultrasonic friction displays*, and *rotating disks* for creating slip sensations (Murphy, Webster, & Okamura, 2004).

Sight and Sound

Although the quality of video and audio are commonly measured separately, considerable findings shows that audio and video information is symbiotic in nature, that one medium can have an impact on the user's perception of the other (Rimmel,, Hollier, & Voelcker, 1998; Watson & Sasse, 1996). Moreover, the majority of user multimedia experience is based on both visual and auditory information. As a symbiotic relationship has been demonstrated between the perception of audio and video media, in this chapter we consider multimedia studies concerning the variation and perception of sight and sound together.

Considerable work has been done looking at different aspects of perceived audio and video quality at many different levels. Unfortunately, as a result of multiple influences on user perception of distributed multimedia quality, providing a succinct, yet extensive review of such work is extremely complex.

To this end, we propose an extended version of a model initially suggested by Wikstrand (2003), in which quality is segregated into three discrete levels: the *network-level*, the *media-level* and the *content-level*. Wikstrand showed that all factors influencing distributed multimedia quality (specifically audio and/or video) can be categorised by assessing and categorising the specific information abstraction. The network-level concerns the transfer of data and all quality issues related to the flow of data around the network. The media-level concerns quality issues relating to the transference methods used to convert network data to perceptible media information, that is, the video and audio media. The content-level concerns quality factors that influence how media information is perceived and understood by the end user.

In our work, and in addition to the model proposed by Wikstrand, we incorporated two distinct quality perspectives, which reflect the infotainment duality of multimedia: the user-perspective and the technical-perspective.

- **User-Perspective:** The user-perspective concerns quality issues that rely on user feedback or interaction. This can be varied and measured at the media- and content-levels. The network-level does not facilitate the user-perspective since user perception cannot be measured at this low level abstraction.

- **Technical-Perspective:** The technical-perspective concerns quality issues that relate to the technological factors involved in distributed multimedia. Technical parameters can be varied and measured at all quality abstractions.

Copyright © 2006, Idea Group Inc. Copying or distributing in print or electronic forms without written permission of Idea Group Inc. is prohibited.

Figure 1. Quality model incorporates network (N), media (M), and content-level (C) abstractions and technical- and user-perspectives dimensions

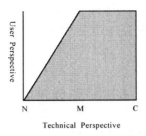

At each quality abstraction defined in our model, quality parameters can be varied, for example, jitter at the network-level, frame rate at the media-level, and the display-type at the content-level. Similarly, at each level of the model, quality can be measured, for example, percentage of loss at the network-level, user mean opinion score (MOS) at the media-level, and task performance at the content-level. By determining the abstraction and perspective at which experimental variables are both varied and measured, it is possible to place any multimedia experiment in context of our model.

Using the above model we aid to produce a succinct yet extensive summary of video/audio multimedia research. The subsequent sections describe this video/audio literature in context of the identified quality definition model (see Figure 1). Each section concerns a quality abstraction level (network, media, or content-level) and includes work relating to studies that adapt and measure quality factors at the defined perspectives (both technical- or media-perspective). Each subsection includes a summary of all relevant studies. A detailed description of new studies is given at the end of the relevant section.

The Network-Level

Network-Level Quality (Technical-Perspective)

- Claypool and Tanner (1999) manipulated jitter and packet loss to test the impact on user quality opinion scores.
- Ghinea and Thomas (2000) manipulated bit error, segment loss, segment order, delay, and jitter in order to test the impact of different media transport protocols on user perception understanding and satisfaction of a multimedia presentation.
- Procter, Hartswood, McKinlay, and Gallacher (1999) manipulated the network load to provoke degradations in media quality.
- Loss occurs due to network congestion and can therefore be used as an indication of end-to-end network performance or "quality" (Ghinea, Thomas, & Fish, 2000; Koodli & Krishna, 1998).

Copyright © 2006, Idea Group Inc. Copying or distributing in print or electronic forms without written permission of Idea Group Inc. is prohibited.

- A delay is always incurred when sending distributed video packets; however, the delay of consecutive packets is rarely constant. The variation in the period of delay is called the jitter. Wang et al. (2001) used jitter as an objective measure of video quality.

- The end-to-end bandwidth is defined as the network resource that facilitates the provision of these media-level technical parameters. Accordingly, the available end-to-end bandwidth is important since it determines the network resource available to applications at the media-level. Wang, Claypool, and Zuo (2001) measured bandwidth impact for their study of real-world media performance.

Claypool and Tanner (1999): Claypool and Tanner measured and compared the impact of jitter and packet loss on perceptual quality of distributed multimedia video.

Results showed that:

- Jitter can degrade video quality nearly as much as packet loss. Moreover, the presence of even low amounts of jitter or packet loss results in a severe degradation in perceptual quality. Interestingly, higher amounts of jitter and packet loss do not degrade perceptual quality proportionally.

- The perceived quality of low temporal aspect video, that is, video with a small difference between frames, is not impacted as much in the presence of jitter as video with a high temporal aspect.

- There is a strong correlation between the average number of quality degradation events (points on the screen where quality is affected) and the average user quality rating recorded. This suggests that the number of degradation events is a good indicator of whether a user will like a video presentation affected by jitter and packet loss.

Ghinea and Thomas (2000): Ghinea and Thomas tested the impact on user Quality of Perception (QoP) of an adaptable protocol stacks geared towards human requirements (RAP - Reading Adaptable Protocol), in comparison to different media transport protocols (TCP/IP, UDP/IP). RAP incorporates a mapping between QoS parameters (bit error rate, segment loss, segment order, delay and jitter) and Quality of Perception (the user understanding and satisfaction of a video presentation). Video material used included 12- windowed (352*288 pixel) MPEG-1 video clips, each between 26 and 45 seconds long, with a consistent objective sound quality, colour depth (8-bit) and frame rate (15 frames per second). The clips were chosen to cover a broad spectrum of subject matters, while also considering the dynamic, audio, video, and textual content of the video clip.

Results showed that:

- RAP enhanced user understanding, especially if video clips are highly dynamic. TCP/IP can be used for relatively static clips. UDP/IP performs badly, in context of information assimilation.

Copyright © 2006, Idea Group Inc. Copying or distributing in print or electronic forms without written permission of Idea Group Inc. is prohibited.

- Use of RAP successfully improves user satisfaction (10/12 videos). TCP/IP received the lowest associated satisfaction ratings.

- RAP, which incorporates the QoS to QoP mapping (Ghinea et al., 2000), is the only protocol stack used, which was not significantly different to those identified when video were shown on a standalone system. Accordingly, RAP effectively facilitates the provision of user QoP.

Koodli and Krishna (1998): Koodli and Krishna describe a metric called noticeable loss. Noticeable loss captures inter-packet loss patterns and can be used in source server and network buffers to pre-emptively discard packets based on the "distance" to the previous lost packet of the same media stream. Koodli and Krishna found that incorporation of noticeable loss greatly improves the overall QoS, especially in the case of variable bit rate video streams.

Procter et al. (1999): Procter et al. focus on the influence of different media content and network Quality of Service (QoS) variation on a subject's memory of, and comprehension of, the video material. In addition, Procter et al. focus on the impact of degraded visual information on assimilation of non-verbal information.

A simulation network was used to facilitate QoS variation. Two 4-Mbs token rings were connected by a router, so that packets had to pass through the router before being received by the client. Two background traffic generators were used to generate two network conditions: i) no load (with no traffic), and ii) load (with simulated traffic). During the load condition, packet loss was "bursty" in character, varying between zero and one hundred percent, yet had an overall average between 30% and 40%. Audio quality was subsequently dependent on network conditions. Video material used consisted of two diametrically- opposed presentations: a bank's annual report and a dramatized scene of sexual harassment. Two experiments were used:

- The first experiment was designed to investigate the effects of network QoS on a subject's assessment of quality. Two questionnaires were used: The first concerned the subjective evaluation of quality (with scores 1-5 representing very poor and very good respectively) as well as factors that had impaired quality (caption quality, audio quality, video quality, audio/video synchronisation, and gap in transmission); the second questionnaire measured a participants' comprehension of the material based upon their recall of its factual content. Results showed that subjects rated the quality higher in the non-load than in the load condition, with overall impression, ease of understanding, and technical quality being the significantly better quality with no network load. Two factors were found to significantly impair quality in the no-load network condition: video quality and audio/video synchronisation. When a network load was added, audio quality, video quality, audio/video synchronisation, as well as transmission gaps were found to significantly impair user perception of multimedia quality. No difference was measured in the level of factual information assimilated by users.

- The second experiment investigated the impact of visual degradation of the visual channel on the uptake of non-verbal signals. Again, two network load conditions

Copyright © 2006, Idea Group Inc. Copying or distributing in print or electronic forms without written permission of Idea Group Inc. is prohibited.

were used: i) no load, and ii) load, to simulate network traffic. The same test approach was used; however the second questionnaire considered: a) factual questions, b) content questions relating to what they thought was happening in a dramatised section of the video, that is, the participants' ability to judge the emotional state of people, and c) questions asking the user to specify his/her confidence with his/her answers. Results showed that subjects rated quality higher in the no-load condition, with overall impression, content, ease of understanding, and technical quality rating being significantly higher under no-load conditions. In the no-load condition, audio, video quality, and audio-video synchronisation were considered to have an effect on user perception of multimedia quality. In the load network condition, caption quality, audio quality, video quality, audio/video synchronisation, and gap in transmission were all shown to have an impairing impact on user perception of multimedia quality. No significant difference was measured between the level of factual information assimilated by users when using load and non-load conditions. In conclusion, Procter et al. (1999) observed that degradation of QoS has a greater influence on a subject's uptake of emotive/affective content than on their uptake of factual content.

Wang, Claypool, and Zuo (2001): Wang et al. presented a wide-scale empirical study of RealVideo traffic from several Internet servers to many geographically-diverse users. They found that when played over a best effort network, RealVideo has a relatively reasonable level of quality, achieving an average of 10 video frames per second and very smooth playback. Interestingly, very few videos achieve full-motion frame-rates. Wang et al. showed that: With low-bandwidth Internet connection, video performance is most influenced by the user's limited bandwidth connection speed; with high-bandwidth Internet connection, the performance reduction is due to server bottlenecks. Wang et al. (2001) used level of jitter and video frame rates as objective measures of video quality.

The Media-Level

Media-Level Quality (Technical-Perspective)

* Apteker et al. (1995), Ghinea and Thomas (1998), Kawalek (1995), Kies, Williges, and Rosson (1997), Masry, Hemami, Osberger, and Rohaly (2001), Wilson and Sasse (2000a, 2000b), and Wijesekera, Srivastava, Nerode, and Foresti (1999) manipulated video or audio frame rate.

* Ardito, Barbero, Stroppiana, and Visca (1994) developed a metric, which aimed to produce a linear mathematical model between technical and user subjective assessment.

* Gulliver and Ghinea (2003) manipulated use of captions.

* Kies et al. (1997) manipulated image resolution.

* Quaglia and De Martin (2002) used *Peak Signal-to-Noise Ratio* (PSNR) as an objective measure of quality.

Copyright © 2006, Idea Group Inc. Copying or distributing in print or electronic forms without written permission of Idea Group Inc. is prohibited.

- Steinmetz (1996) and Wijesekera et al. (1999) manipulated video skew between audio and video, that is, the synchronisation between two media. In addition, Steinmetz (1996) manipulated the synchronisation of video and pointer skews.

- Teo and Heeger (1994) developed a normalised model of human vision. Extensions to the Teo and Heegar model included: the *Colour Masked Signal to Noise Ratio* metric (CMPSNR) - Van den Branden Lambrecht and Farell (1996), used for measuring the quality of still colour pictures, and the *Normalised Video Fidelity Metric* (NVFM) - Lindh and van den Branden Lambrecht (1996), for use with multimedia video. Extensions of the CMPSNR metric include: the *Moving Picture Activity Metric* (MPAM) - Verscheure and Hubaux (1996), the *Perceptual Visibility Predictor* metric (PVP) - Verscheure and van den Branden Lambrecht (1997), and the *Moving Pictures Quality Metric* (MPQM) - van den Branden Lambrecht and Verscheure (1996).

- Wang, Claypool, and Zuo (2001) used frame rate as an objective measure of quality. Although the impact of frame rate is adapted by Apteker et al. (1995); Ghinea and Thomas (1998); Kawalek (1995); Kies et al. (1997); Masry et al. (2001); Wijesekera et al (1999), Wang et al. (2001) is, to the best of our knowledge, the only study that used output frame rate as the quality criterion.

- Wikstrand and Eriksson (2002) used various animation techniques to model football matches for use on mobile devices.

Apteker et al. (1995): Apteker et al. defined a *Video Classification Scheme* (VCS) to classify video clips (see Table 1), based on three dimensions, considered inherent in video messages: the temporal (T) nature of the data, the importance of the auditory (A) components and the importance of visual (V) components.

"High temporal data" concerns video with rapid scene changes, such as general sport highlights, "Low temporal data" concerns video that is largely static in nature, such as a talk show. A video from each of the eight categories was shown to users in a windowed multitasking environment. Each multimedia video clip was presented in a randomised order at three different frame rates (15, 10, and 5 frames per second). The users then rated the quality of the multimedia videos on a seven-point graded scale. Apteker et al. showed that:

Table 1. Video classification examples - adapted from Apteker et al (1995)

Category Number	Video Information	*Definition*
1	Logo/ Test Pattern	Tlo Alo Vlo
2	Snooker	Tlo Alo Vhi
3	Talk Show	Tlo Ahi Vlo
4	Stand-up Comedy	Tlo Ahi Vhi
5	Station Break	Thi Alo Vlo
6	Sporting Highlights	Thi Alo Vhi
7	Advertisements	Thi Ahi Vlo
8	Music Clip	Thi Ahi Vhi

Copyright © 2006, Idea Group Inc. Copying or distributing in print or electronic forms without written permission of Idea Group Inc. is prohibited.

- Video clips with a lower video dependence (Vlo) were considered as more watchable than those with a high video dependence (Vhi).

- Video clips with a high level of temporal data (Thi) were rated as being more watchable than those with a low level of temporal data (Tlo).

- Frame-rate reduction itself leads to progressively lower ratings in terms of "watchability".

- There exists a threshold, beyond which no improvement to multimedia quality can be perceived, despite an increase in available bandwidth, which is supported by Fukuda, Wakamiya, Murata, and Miyahara, 1997; Ghinea, 2000; Steinmetz, 1996; van den Branden Lambrecht, 1996).

Apteker et al. expressed human receptivity as a percentage measure, with 100% indicating complete user satisfaction with the multimedia data, and showed that the dependency between human receptivity and the required bandwidth of multimedia clips is non-linear (Apteker et al). In the context of bandwidth-constrained environments, results suggest that a limited reduction in human receptivity facilitates a relatively large reduction in bandwidth requirement (the *asymptotic property* of the VCS curves) (Ghinea, 2000).

Ardito et al. (1994): The RAI Italian Television metric attempts to form a linear objective model from data representing subjective assessments concerning the quality of compressed images (Ardito et al., 1994; Ghinea, 2000). During subjective assessment, the participants were presented with a sequence of pairs of video clips, one representing the original image and the other showing the degraded (compressed) equivalent. The user is not told which of the two is the original, yet is asked to categorise the quality of the two images using a five- point Likert double-stimulus impairment scale classification similar to the CCIR Rec. 500-3 scale (CCIR, 1974), with scores of one and five representing the "very annoying" and, respectively, "imperceptible" difference between the original and degraded images. All results are then normalised with respect to the original.

The RAI Italian Television metric initially calculates the SNR for all frames of the original and degraded video clips. To enable processing over time (the temporal variable) the SNR values are calculated across all frames, as well as in subsets of specified length *l*. Minimum and maximum values of the SNR are then determined for all groups with *l* frames, thus highlighting noisy sections of video. RAI Italian Television metric considers human visual sensitivity, by making use of a Sobel operator. A Sobel operator uses the luminance signal of surrounding pixels, in a 3x3 matrix, to calculate the gradient in a given direction. Applied both vertically and horizontally, Sobel operators can identify whether or not a specific pixel is part of an edge. The 3x3 matrix returns a level of luminance variation greater or smaller, respectively, than a defined threshold. An "edge image" for a particular frame can be obtained by assigning a logical value of 1 or 0 to each pixel, depending on its value relative to the threshold. An edge image is defined for each frame, facilitating the calculation of SNR for three different scenarios: across the whole frame, only for areas of a frame belonging to edges, and, finally, across the whole frame but only for those areas not belonging to edges. Results show that, the RAI Italian Television linear model can successfully capture 90% of the given subjective information. However, large errors occur if the subjective data is applied across multiple video clips (Ardito et al., 1993), implying high content dependency.

Copyright © 2006, Idea Group Inc. Copying or distributing in print or electronic forms without written permission of Idea Group Inc. is prohibited.

Ghinea and Thomas (1998): To measure the impact of video Quality of Service (QoS) variation on user perception and understanding of multimedia video clips, Ghinea and Thomas presented users with a series of 12 windowed (352x288 pixel) MPEG-1 video clips, each between 26 and 45 seconds long, with a consistent objective sound quality. The clips were chosen to cover a broad spectrum of subject matter, while also considering the dynamic, audio, video, and textual content of the video clip. They varied the frame per second (fps) QoS parameters, while maintaining a constant colour depth, window size, and audio stream quality. Frame rates of 25 fps, 15 fps and 5 fps were used and were varied across the experiment, yet for a specific user they remained constant throughout. Ten users were tested for each frame rate. Users were kept unaware of the frame rate being displayed. To allow dynamic (D), audio (A), video (V) and textual (T) considerations to be taken into account, in both questionnaire design and data analysis, characteristic weightings were used on a scale of 0-2, assigning importance of the inherent characteristics of each video clip. Table 2 contains the characteristic weightings, as defined by Ghinea and Thomas (1998). The clips were chosen to present the majority of individuals with no peak in personal interest, which could skew results. Clips were also chosen to limit the number of individuals watching the clip with previous knowledge and experience.

After the user had been shown a video clip, the video window was closed, and questions were asked about the video that they had just seen. The number of questions was dependent on the video clip being shown and varied between 10 and 12. Once a user had answered all questions relating to the video clip, and all responses had been noted, users were asked to rate the quality of the clip using a six-point Likert scale, with scores one and six representing the worst and, respectively, the best possible perceived level of quality. Users were instructed not to let personal bias towards the subject matter influence their quality rating of the clip. Instead they were asked to judge a clip's quality by the degree to which they, the users, felt satisfied with the service of quality. The questions, used by Ghinea and Thomas, were designed to encompass all aspects of the information being presented in the clips: *D, A, V, T.* A number of questions were used to

Table 2. Video characteristics as defined by Ghinea and Thomas (1998)

Video Category	Dynamic (D)	Audio (A)	Video (V)	Text (T)
1 – Commercial	1	2	2	1
2 – Band	1	2	1	0
3 – Chorus	0	2	1	0
4 – Animation	1	1	2	0
5 – Weather	0	2	2	2
6 – Documentary	1	2	2	0
7 – Pop Music	1	2	2	2
8 – News	0	2	2	1
9 – Cooking	0	2	2	0
10 – Rugby	2	1	2	1
11 – Snooker	0	1	1	2
12 – Action	2	1	2	0

Copyright © 2006, Idea Group Inc. Copying or distributing in print or electronic forms without written permission of Idea Group Inc. is prohibited.

Figure 2. Effect of varied QoS on user quality of perception (1998)

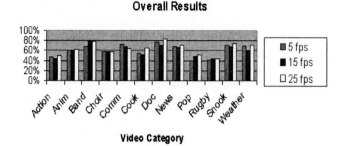

analyse a user's ability to absorb multiple media at one point in time, as correct answers could only be given if a user had assimilated information from multiple media. Lastly, a number of the questions were used that couldn't be answered by observation of the video alone, but by the users making inference and deductions from the information that had just been presented.

The main conclusions of this work include the following:

- A significant loss of frames (that is, a reduction in the frame rate) does not proportionally reduce the user's understanding and perception of the presentation (see Figure 2). In fact, in some instances the participant seemed to assimilate more information, thereby resulting in more correct answers to questions. Ghinea and Thomas proposed that this was because the user has more time to view a specific frame before the frame changes (at 25 fps, a frame is visible for only 0.04 sec, whereas at 5 fps a frame is visible for 0.2 sec), hence absorbing more information.

- Users have difficulty in absorbing audio, visual, and textual information concurrently. Users tend to focus on one of these media at any one moment, although they may switch between the different media. This implies that critical and important messages in a multimedia presentation should be delivered in only one type of medium.

- When the cause of the annoyance is visible (such as lip synchronisation), users will disregard it and focus on the audio information if considered contextually important.

- Highly dynamic scenes, although expensive in resources, have a negative impact on user understanding and information assimilation. Questions in this category obtained the least number of correct answers. However, the entertainment values of such presentations seem to be consistent, irrespective of the frame rate at which they are shown. The link between entertainment and content understanding is therefore not direct.

Copyright © 2006, Idea Group Inc. Copying or distributing in print or electronic forms without written permission of Idea Group Inc. is prohibited.

Ghinea and Thomas's method of measuring user perception of multimedia quality, later termed Quality of Perception (QoP), incorporates both a user's capability to understand the informational content of a multimedia video presentation, as well as his/her satisfaction with the quality of the visualised multimedia. QoP has been developed at all quality abstractions of our model (network-, media- and content-level) in cooperation with a number of authors: Fish (Ghinea et al. 2000), Gulliver (Gulliver and Ghinea, 2003; 2004), Magoulas (Ghinea and Magoulas, 2001) and Thomas (Ghinea and Thomas, 1998; 2000; 2001) concerning issues including the development of network protocol stacks, multimedia media assessment, attention tracking, as well user accessibility.

Gulliver and Ghinea (2003): Gulliver and Ghinea used an adapted version of QoP to investigate the impact that hearing level has on user perception of multimedia, with and without captions. They showed users a series of 10 windowed (352x288 pixel) MPEG-1 video clips, each between 26 and 45 seconds long, with a consistent sound quality and video frame rate. Additional captions were added in a separate window, as defined by the experimental design.

Results showed that deafness significantly impacts a user's ability to assimilate information (see Figure 3). Interestingly, use of captions does not increase deaf information assimilation, yet increases quality of context-dependent information assimilated from the caption/audio.

To measure satisfaction, Gulliver and Ghinea (2003) used two 11-point scales (0-10) to measure Level of Enjoyment (QoP-LoE) and user self-predicted level of Information Assimilation (QoP-PIA). A positive correlation was identified between QoP-LoE and QoP-PIA, independent of hearing level or hearing type, showing that a user's perception concerning their ability to assimilate information is linked to his/her subjective assessment of enjoyment.

Kawalek (1995): Kawalek showed that loss of audio information has a more noticeable effect on the assimilation of informational content than video frame loss. Users are therefore less likely to notice degradation of video clips if shown low quality audio media (see Figure 4).

Figure 3. A detailed breakdown of deaf /hearing information assimilation (%): (CA) caption window / audio, (D) dynamic information, (V) video information,(T) textual information and (C) captions contained in the video window

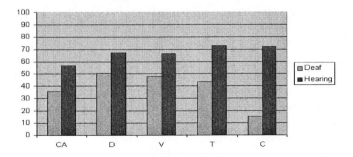

Copyright © 2006, Idea Group Inc. Copying or distributing in print or electronic forms without written permission of Idea Group Inc. is prohibited.

Figure 4. Perceived effect of transmission quality on user perception - Adapted from Kawalek (1995)

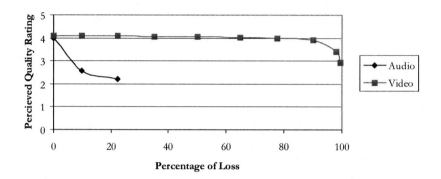

Kies et al. (1997): Kies et al. conducted a two-part study to investigate the technical parameters affecting a *Desktop Video Conferencing system* (DVC). Consequently, three frame rate conditions (1, 6 and 30 fps), two resolution conditions (160x120 and 320x240), and three-communication channel conditions were manipulated. Dependent measures included the results of a questionnaire and subjective satisfaction, specifically concerning the video quality. Like Ghinea and Thomas (1998) and Procter et al. (1999), results suggest that factual information assimilation does not suffer under reduced video QoS, but subjective satisfaction is significantly decreased. In addition, a field study was used to look at the suitability of DVC for distance learning. Interestingly, field studies employing similar dependent measures indicated that participants may be more critical of poor video quality in laboratory settings.

Quaglia and De Martin (2002): Quaglia and De Martin describe a technique for delivering "nearly constant" perceptual QoS when transmitting video sequences over IP Networks. On a frame-by-frame basis, allocation of premium packets (those with a higher QoS priority) depends upon on the perceptual importance of each MPEG *macroblock*, the desired level of QoS, and the instantaneous network state. Quaglia and De Martin report to have delivered nearly constant QoS; however, constant reliance on PSNR and use of frame-by-frame analysis raises issues when considering the user perception of multimedia quality.

Steinmetz (1990, 1992, 1996): Distributed multimedia synchronisation comprises both the definition and the establishment of temporal relationships amongst media types. In a multimedia context this definition can be extended such that synchronisation in multimedia systems comprises content, spatial, and temporal relations between media objects. Perceptually, synchronisation of video and textual information or video and image information can be considered as either: *overlay*, which is information that is used in addition to the video information; or *no overlay*, which is information displayed, possibly in another box, to support the current video information. Blakowski and Steinmetz (1995) distinguished two different types of such media objects:

Copyright © 2006, Idea Group Inc. Copying or distributing in print or electronic forms without written permission of Idea Group Inc. is prohibited.

- **Time-dependent media objects:** These are media streams that are characterised by a temporal relation between consecutive component units. If the presentation duration of all components of an object is equal, then it is called a *continuous media object*.
- **Time-independent media objects:** These consist of media such as text and images. Here the semantics of their content does not depend on time-structures.

An example of synchronisation between continuous media would be the synchronisation between the audio and video streams in a multimedia video clip. Multimedia synchronisation can, however, comprise temporal relations between both time-dependent and time-independent media objects. An example of this is a slide presentation show, where the presentation of the slides has to be synchronised with the appropriate units of the audio stream. Previous work on multimedia synchronisation was done in Blakowski and Steinmetz, (1995) and Steinmetz, (1990, 1992, 1996). as well as on topics devoted to device synchronisation requirements. Steinmetz (1990, 1992, 1996) primarily manipulated media skews to measure how lip pointer and non-synchronisation impacted user perception of what is deemed "out of synch" (Steinmetz, 1996). A presentation was considered as being "in synch" when no error was identified, that is, a natural impression. A presentation was considered as being "out of synch" when it is perceived as being artificial, strange, or even annoying. Table 3 summarises the minimal synchronisation errors, proposed by Steinmetz, that were found to be perceptually acceptable between media (see Table 3).

Further experiments incorporating variation in video content, as well as the use of other languages (Spanish, Italian, French, and Swedish) showed no impact on results. Interestingly, Steinmetz did measure variation as a result of the participant group, implying that user experience of manipulating video affects a user's aptitude when noticing multimedia synchronisation errors.

Table 3. Minimal noticeable synchronisation error Steinmetz (1996)

Media		Mode, Application	QoS
Video	Animation	Correlated	±120ms
	Audio	Lip synchronisation	±80ms
	Image	Overlay	±240ms
		Non-overlay	±500ms
	Text	Overlay	±240ms
		Non-overlay	±500ms
Audio	Animation	Event correlation (e.g. dancing)	±80ms
	Audio	Tightly coupled (stereo)	±11µs
		Loosely coupled (e.g. background music)	±500ms
	Image	Tightly coupled (e.g. music with notes)	±5ms
		Loosely coupled (e.g. slide show)	±500ms
	Text	Text annotation	±240ms
	Pointer	Audio relates to showed item	(-500ms, +750ms)

Copyright © 2006, Idea Group Inc. Copying or distributing in print or electronic forms without written permission of Idea Group Inc. is prohibited.

Teo and Heeger (1994): Teo and Heeger present a perceptual distortion measure that predicts image integrity based on a model of the human visual system that fits empirical measurements of: 1) the response properties of neurons in the primary visual cortex, and 2) the psychophysics of spatial pattern detection, that is a person's ability to detect a low contrast visual stimuli.

The Teo-Heeger model consists of four steps:

- **a front-end hexagonally sampled quadrature mirror filter transform** function (Simoncellia & Adelson, 1990) that provides an output similar to that of retina, and is similarly tuned to different spatial orientations and frequencies.

- **squaring** to maximise variation.

- **a divisive contrast normalisation mechanism**, to represent the response of a hypothetical neuron in the primary visual cortex.

- **a detection mechanism** (both linear and non-linear) to identify differences (errors) between the encoded image and the original image.

Participants rated images, coded using the Teo and Heeger perceptual distortion measure, as being of considerably better "quality" than images coded with no consideration to the user-perspective. Interestingly, both sets of test images contain similar RMS and PSNR values.

van den Branden Lambrecht and Verscheure (1996): van den Branden Lambrecht and Verscheure present the Moving Picture Quality Metric (MPQM) to address the problem of quality estimation of digital coded video sequences. The MPQM is based on a multi-channel model of the human spatio-temporal vision with parameters defined through interpretation of psychophysical experimental data.

A spatio-temporal filter bank simulates the visual mechanism, which perceptually decomposes video into phenomena such as contrast sensitivity and masking. Perceptual components are then combined, or pooled, to produce a quality rating, by applying a greater summation weighting for areas of higher distortion. The quality rating is then normalised (on a scale from one to five), using a normalised conversion. van den Branden Lambrecht and Verscheure showed MPQM (moving picture quality metric) to model subjective user feedback concerning coded video quality.

Lindh and van den Branden Lambrecht (1996): Lindh and van den Branden Lambrecht introduced the NVFM model (Normalisation Video Fidelity Metric), as an extension of the normalisation model used by Teo and Heeger. The NVFM output accounts for normalisation of the receptive field responses and inter-channel masking in the human visual system and is mapped onto the one to five quality scale on the basis of the vision model used in the MPQM metric.

Lindh and van den Branden Lambrecht compared NVFM with the Moving Picture Quality Metric (MPQM) (van den Branden Lambrecht & Verscheure, 1996). Interestingly, the results of NVFM model are significantly different from the output of the MPQM, as it has a fast increase in user perceived quality in the lower range of bit rate, that is, a slight increase of bandwidth can result in a very significant increase in quality. Interestingly, saturation occurs at roughly the same bit rate for both metrics (approximately 8 Mbit/sec).

Copyright © 2006, Idea Group Inc. Copying or distributing in print or electronic forms without written permission of Idea Group Inc. is prohibited.

Lindh and van den Branden Lambrecht proposed NVFM as a better model of the cortical cell responses, compared to the MPQM metric.

van den Branden Lambrecht and Farrell (1996): van den Branden Lambrecht and Farrell introduced a computation metric, which is termed the Colour Masked Signal to Noise Ratio (CMSNR). CMSNR incorporates opponent-colour (i.e. the stimulus of P-ganglion cells), as well as other aspects of human vision involved in spatial vision including: perceptual decomposition (Gabor Filters), masking (by adding weightings to screen areas), as well as the weighted grouping of neuron outputs. van den Branden Lambrecht and Farrell subsequently validated the CMPSNR metric, using 400 separate images, thus proving the CMPSNR metric as more able to predict user fidelity with a level of accuracy greater than the mean square error.

Wijesekera et al. (1999): A number of mathematical measures of QoS (Quality of Service) models have been proposed (Towsley, 1993; Wijesekera & Stivastava, 1996). Wijesekera & Stivastava (1996) investigated the perceptual tolerance to discontinuity caused by media losses and repetition. Moreover, Wijesekera et al. considered the perceptual impact that varying degrees of synchronisation error have across different streams. Wijesekera et al. followed the methodology of Steinmetz (1996), that is, the manipulation of media skews, to measure stream continuity and synchronisation in the presence of media losses (Wijesekera & Stivastava, 1996) and consequently, quantified human tolerance of transient continuity and synchronisation losses with respect to audio and video media. Wijesekera et al. (1999) found that:

- Viewer discontent with aggregate losses (i.e. the net loss, over a defined duration) gradually increases with the amount of loss, as long as losses are evenly distributed. For other types of loss and synchronisation error, there is a sharp initial rise in user discontent (to a certain value of the defect), after which the level of discontent plateaus.

- When video is shown at 30fps, an average aggregate loss below 17% is imperceptible, between 17% and 23% it is considered tolerated, and above 23% it is considered unacceptable, assuming losses are evenly distributed.

- Losing two or more consecutive video frames is noticed by most users, when video is shown at 30fps. Losing greater than two consecutive video frames does not proportionally impact user perception of video, as a quality rating plateau is reached. Similarly, loss of three or more consecutive audio frames was noticed by most users. Additional consecutive loss of audio frames does not proportionally impact user perception of audio, as a quality rating plateau is reached.

- Humans are not sensitive to video rate variations. Alternatively, humans have a high degree of sensitivity to audio, thus supporting the findings of Kawalek (1995). Wijesekera et al. (1999) suggest that even a 20% rate variation in a newscast-type video does not result in significant user dissatisfaction, whereas a 5% rate variation in audio is noticed by most observers.

- Momentary rate variation in the audio stream, although initially considered as being amusing, was soon considered as annoying. This resulted in participants concentrating more on the audio defect than the audio content.

Copyright © 2006, Idea Group Inc. Copying or distributing in print or electronic forms without written permission of Idea Group Inc. is prohibited.

- An aggregated audio-video synchronisation loss of more that 20% frames was identified. Interestingly, consecutive synchronisation loss of more than three frames is identified by most users, which is consistent with Steinmetz (1996).

Wilson and Sasse (2000a, 200b): Bouch, Wilson, and Sasse (2001) proposed a three-dimensional approach to assessment of audio and video quality in networked multimedia applications: measuring task performance, user satisfaction, and user cost (in terms of physiological impact). Bouch et al. used their approach to provide an integrated framework from which to conduct valid assessment of perceived QoS (Quality of Service). Wilson and Sasse (2000b) used this three-dimensional approach and measured: Blood Volume Pulse (BVP), Heart Rate (HR) and Galvanic Skin Resistance (GSR), to measure the stress caused when inadequate media quality is presented to a participant. Twenty-four participants watched two recorded interviews conducted, using IP video tools, lasting fifteen minutes each. After every five minutes, the quality of the video was changed, allowing quality variation over time. Audio quality was not varied. Participants therefore saw two interviews with video frame rates of 5-25-5 fps and 25-5-25 fps respectively. While viewing the videos, participants rated the audio/video quality using the QUASS tool (Bouch, Watson, & Sasse, 1998), a *Single Stimulus Continuous Quality* (SSCQE) system where the participant continuously rated quality on an unlabelled scale. Physiological data was taken throughout the experiment. Moreover, to measure whether users perceived any changes in video quality, a questionnaire was also included. Wilson and Sasse showed that the GSR, HR, and BVP data represented significant increases in stress when a video is shown at 5fps in comparison to 25fps. Only 16% of participants noticed a change in frame rate. No correlation was found between stress level and user feedback of perceived quality.

Wilson and Sasse (2000a) showed that subjective and physiological results do not always correlate with each other, which indicates that users cannot consciously evaluate the stress that degraded media quality has placed upon them.

Wikstrand and Eriksson (2002): Wikstrand and Eriksson used animation to identify how alternative rendering techniques impact user perception and acceptance, especially in bandwidth-constrained environments. An animation or model of essential video activity demands that only context-dependent data is transferred to the user and therefore reduces data transfer. Wikstrand and Eriksson performed an experiment to contrast different animations and video coding in terms of their cognitive and emotional effectiveness when viewing a football game on a mobile phone. Results showed that different rendering of the same video content affects the user's understanding and enjoyment of the football match. Participants who preferred video to animations did so because it gave them a better "football feeling", while those who preferred animations had a lower level of football knowledge and thought that animations were best for understanding the game. Wikstrand and Eriksson concluded that more advanced rendering, at the client end, may be used to optimise or blend between emotional and cognitive effectiveness.

Copyright © 2006, Idea Group Inc. Copying or distributing in print or electronic forms without written permission of Idea Group Inc. is prohibited.

Media-Level Quality (User-Perspective)

- Apteker et al. (1995), measured "watchability" (receptivity) as a measure of user satisfaction concerning video quality.

- Ghinea and Thomas (1998) asked respondents to rate the quality of each clip on a seven- point Likert scale.

- Procter et al. (1999) asked subjects to compare the streamed video against the non-degraded original video. Quality was measured by asking participants to consider a number of statements and, using a seven-point Likert-style scale, for example, "the video was just as good as watching a live lecture in the same room" and "the video was just as good as watching a VCR tape on a normal television".

- Steinmetz (1996) used participant annoyance of synchronisation skews as a measure of quality. In both cases only identified errors are considered as being of low quality.

- Wilson and Sasse (2000a; 2000b) used the *Single Stimulus Continuous Quality* QUASS (Bouch et al., 1998) tool to allow the user to continuously rate the audio / video quality, whilst viewing a video presentation.

- Wikstrand and Eriksson (2002) measured user preference concerning the animation rendering technique.

The media-level is concerned with how the media is coded for the transport of information over the network and/or whether the user perceives the video as being of good or bad quality. Accordingly, studies varying media quality, as a direct result of the user, are limited. The best example of quality-related user media variation concerns attentive displays, which manipulate video quality around a user's point of gaze. Current attentive display techniques were first introduced in McConkie and Rayner (1975) and Saida and Ikeda (1979) and are used in a wide range of applications, including: reading, perception of image and video scenes, virtual reality, computer game animation, art creation and analysis, as well as visual search studies (Baudisch, DeCarlo, Duchowski, & Geisler, 2003; Parkhurst & Niebur, 2002; Wooding, 2002). The perceptual work relating to the use of attentive displays is of limited benefit to the aim of this work and will therefore not be considered in this chapter.

The Content-Level

Content-Level Quality (Technical-Perspective)

- Ghinea and Thomas (1998), Gulliver and Ghinea (2003), Masry et al. (2001), as well as Steinmetz (1996) all varied experimental material to ensure diverse media content.

- Procter et al. (1999) used diametrically opposed presentations: a bank's annual report and a dramatized scene of sexual harassment.

Copyright © 2006, Idea Group Inc. Copying or distributing in print or electronic forms without written permission of Idea Group Inc. is prohibited.

- Steinmetz (1996) used three different views: head, shoulder, and body, which related to the relative proportion of the newsreader shown in the video window.

- Wilson and Sasse (2000a, 2000b) measure participants Blood Volume Pulse (BVP), Heart Rate (HR) and Galvanic Skin Resistance (GSR), to measure for stress as a result of low quality video.

Content-Level Quality (User-Perspective)

- Apteker et al. (1995) measured "watchability" (receptivity) as a measure of user satisfaction concerning video content along temporal, visual, and audio dimensions. Accordingly, "watchability" covers both media- and content-levels.

- Ghinea and Thomas (1998), and Gulliver and Ghinea (2003) used questionnaire feedback to measure a user's ability to assimilate and understand multimedia information.

- Gulliver and Ghinea (2003) asked participants to predict how much information they had assimilated during IA tasks, using scores of 0 to 10 representing "none" and, respectively, "all" of the information that was perceived as being available. Gulliver and Ghinea also measured a participant's level of enjoyment, using scores of 0 to 10 representing "none" and, respectively, "absolute" enjoyment.

- Gulliver and Ghinea (2003) varied participant demographics to measure changes in multimedia perception as a result of deafness, deafness type, and use of captions.

- Procter et al. (1999) used "ease of understanding", "recall", "level of interest", and "level of comprehension" as quality measures.

- Steinmetz (1996) tested videos using a variety of languages (Spanish, Italian, French, and Swedish) in order to check lip synchronisation errors.

- Watson and Sasse (2000) varied peripheral factors, such as volume and type of microphone, to measure in a CSCW environment, the impact on user perception of audio quality.

- Wikstrand and Eriksson (2002) adapted animation rendering techniques, while maintaining important presentation content. Wikstrand and Eriksson showed that animation rendering affects user's understanding and enjoyment of a football match.

Watson and Sasse (2000): Watson and Sasse showed that volume discrepancies, poor quality microphones, and echo have a greater impact on a user's perceived quality of network audio than packet loss.

Copyright © 2006, Idea Group Inc. Copying or distributing in print or electronic forms without written permission of Idea Group Inc. is prohibited.

Incorporating the User-Perspective

Studies have shown that at the:

- **Network-Level:** Technical-perspective network-level variation of bit error, segment loss, delay, and jitter has been used to simulate QoS deterioration. Technical-perspective network-level measurements of loss, delay, and jitter, as well as allocated bandwidth have all been used to measure network level quality performance.

- **Media-Level:** Technical-perspective media-level variation of video and audio frame rate, captions, animation method, inter-stream audio-video quality, image resolution, media stream skews, synchronisation and video compression codecs have been used to vary quality definition. Technical-perspective media-level measurement is generally based on linear and visual quality models, with the exception of who uses output frame rate as the quality criterion. User-perspective media-level variation requires user data feedback and is limited to attentive displays, which manipulate video quality around a user's point of gaze. User-perspective media-level measurement of quality has been used when measuring user "watchability" (receptivity), assessing user rating of video quality, comparing streamed video against the non-degraded original video, as well as for continuous quality assessment and gauging participant annoyance of synchronisation skews.

- **Content-Level:** Technical-perspective content-level variation has been used to vary the content of experimental material as well as the presentation language. Technical-perspective content-level measurement has, to date, only included stress analysis. User-perspective content-level variation has also been used to measure the impact of user demographics, as well as volume and type of microphone on overall perception of multimedia quality. User-perspective content-level measurement has measured "watchability" (receptivity), "ease of understanding", "recall", "level of interest", "level of comprehension", information assimilation, predicted level of information assimilation, and enjoyment.

A number of studies have been considered that measure the user-perspective at the content-level (Apteker et al (1995), Ghinea and Gulliver (2003), Ghinea and Thomas (1998), Procter et al. (1999), Wilson and Sasse (2000a; 2000b). These are summarized in Table 4 which:

1. Lists the primary studies that measure the user-perspective at the content-level, stating the number of participants used in each study.

2. Identifies the adapted quality parameters, and defines the quality abstraction at which each parameter was adapted (N = Network-level, M = Media-level, C = Content-level).

Copyright © 2006, Idea Group Inc. Copying or distributing in print or electronic forms without written permission of Idea Group Inc. is prohibited.

Table 4. Comparison of user perceptual studies

Study	Participants	Adapted	Measured
Apteker et al. (1995)	60 students	• Frame rate (M) • Video content (C)	• Watchability (M)(C)
Gulliver and Ghinea (2003)	50 participants (30 hearing / 20 deaf)	• Framerate (M) • Captions (M) • Video content (C) • Demographics (C)	• Information assimilation (C) • Satisfaction (C) • Self perceived ability (C)
Procter et al. (1999)	24 participants	• Network load (N) • Video content (C)	• Comprehension (C) • Uptake of non-verbal information (C) • Satisfaction (M)
Wilson and Sasse (2000a; 2000b)	24 participants	• Frame rate (M)	• Galvanic skin resistance (C) • Heart rate (C) • Blood volume pulse (C) • QUASS (M)
Ghinea and Thomas (1998)	30 participants	• Frame rate (M) • Video content (C)	• Information assimilation (C) • Satisfaction (M)

3.   Provides a list of the measurements taken for each study and the quality level abstraction at which each measurement was taken (N = Network-level, M = Media-level, C = Content-level).

Inclusion of the user-perspective is of paramount importance to the continued uptake and proliferation of multimedia applications since users will not use and pay for applications if they are perceived to be of low quality. Interestingly, all previous studies fail to either measure the infotainment duality of distributed multimedia quality or comprehensively incorporate and understanding the user-perspective. To extensively consider distributed multimedia quality to incorporate and understand the user-perspective, it is essential that, where possible, both technical- and user-perspective parameter variation is made at all quality abstractions of our model, that is, network-level (technical-perspective), media-level (technical- and user-perspective) and content-level (technical- and user-perspective) parameter variation. Furthermore, in order to effectively measure the infotainment duality of multimedia, that is, information transfer and level of satisfaction, the user-perspective must consider both:

•   the user's ability to assimilate/understand the informational content of the video {assessing the content-level user-perspective}.

•   the user's satisfaction, both measuring the user's satisfaction with the objective QoS settings {assessing the media-level user-perspective}, and also user enjoyment {assessing the content-level user-perspective}.

Copyright © 2006, Idea Group Inc. Copying or distributing in print or electronic forms without written permission of Idea Group Inc. is prohibited.

Summary

In this chapter we set out to consider work relating to each of the multimedia senses - Olfactory (smell), Tactile / Haptic (touch), Visual (sight) and Auditory (sound) - and how this impacts user perception and ultimately user definition of multimedia quality. We proposed an extended model of distributed multimedia, which helped the extensive analysis of current literature concerning video and audio information. We have compared a number of content-level perceptual studies and showed that all previous studies fail to either measure the infotainment duality of distributed multimedia quality or comprehensively incorporate and fully understanding the user-perspective in multimedia quality definition. In conclusion, we show that greater work is needed to fully incorporate and understand the role of the user perspective in multimedia quality definition.

The author believes that a user will not continue paying for a multimedia system or device that they perceive to be of low quality, irrespective of its intrinsic appeal. If commercial multimedia development continues to ignore the user-perspective in preference of other factors, that is, user fascination (i.e. the latest gimmick), then companies ultimately risk alienating the customer. Moreover, by ignoring the user-perspective, future distributed multimedia systems risk ignoring accessibility issues, by excluding access for users with abnormal perceptual requirements.

We have shown that to extensively consider distributed multimedia quality, and to incorporate and understand the user-perspective, it is essential that, where possible, both technical- and user-perspective parameter variation is considered at all quality abstractions of our model, that is, network-level (technical-perspective), media-level (technical- and user-perspective) and content-level (technical- and user-perspective). Furthermore, in order to effectively measure the infotainment duality of multimedia, that is, information transfer and level of satisfaction, the user-perspective must consider both:

- the user's ability to assimilate/understand the informational content of the video.
- the user's satisfaction, both of the objective QoS settings, yet also user enjoyment {assessing the content-level user-perspective}.

If commercial multimedia development effectively considered the user-perspective in combination with technical-perspective quality parameters, then multimedia provision would aspire to facilitate appropriate multimedia, in context of the perceptual, hardware and network criteria of a specific user, thus maximising the user's perception of quality.

Finally, development of quality definition models, as well as user-centric video adaptation techniques (user-perspective personalisation and adaptive media streaming) offers the promise of truly user-defined, accessible multimedia that allows users interaction with multimedia systems on their own perceptual terms. It seems strange that although multimedia applications are produced for the education and/or enjoyment of human viewers, effective development, integration, and consideration of the user-perspective in multimedia systems still has a long way to go....

Copyright © 2006, Idea Group Inc. Copying or distributing in print or electronic forms without written permission of Idea Group Inc. is prohibited.

References

Ahumada, A. J., & Null Jr., C. H. (1993). Image quality: A multidimensional problem. In A. B. Watson (Ed.), *Digital images and human vision* (pp. 141-148). Cambridge, MA: MIT Press.

Apteker, R. T., Fisher, J. A., Kisimov, V. S., & Neishlos, H. (1995). Video acceptability and frame rate. *IEEE Multimedia, 2*(3), Fall, 32-40.

Ardito, M., Barbero, M., Stroppiana, M., & Visca, M. (1994, October 26-28). Compression and quality. In L. Chiariglione (Ed.), *Proceedings of the International Workshop on HDTV '94,* Torino, Italy. Springer Verlag.

Baudisch, P., DeCarlo, D., Duchowski, A. T., & Geisler, W. S. (2003). Focusing on the essential: Considering attention in display design. *Communications of the ACM, 46*(3), 60-66.

Blakowski, G., & Steinmetz, R. (1996). A media synchronisation survey: Reference model, specification, and case studies. *IEEE Journal on Selected Areas in Communications, 14*(1), 5-35.

Bouch, A., Watson, A., & Sasse, M. A. (1998). *QUASS—A tool for measuring the subjective quality of real-time multimedia audio and video.* Poster presented at HCI '98, Sheffield, UK.

Bouch, A., Wilson, G., & Sasse, M. A. (2001). A 3-dimensional approach to assessing end-user quality of service. *Proceedings of the London Communications Symposium* (pp. 47-50).

Caldwell, G., & Gosney, C. (1993). Enhanced tactile feedback (tele-taction) using a multifunctional sensory system. *Proceedings of the IEEE International Conference on Robotics and Automation,* Atlanta, GA (pp. 955-960).

Cater, J. P. (1992). The nose have it! *Presence, 1*(4), 493-494.

Cater, J. P. (1994). Smell/taste: Odours. *Virtual Reality*, 1781.

CCIR (1974). Method for the subjective assessment of the quality of television pictures. *13th Plenary Assembly, Recommendation 50: Vol. 11* (pp. 65-68).

Claypool, M., & Tanner, J. (1999). The effects of jitter on the perceptual quality of video. *ACM Multimedia'99 (Part 2),* Orlando, FL (pp. 115-118).

Cohn, M. B., Lam, M., & Fearing, R. S. (1992). Tactile feedback for teleoperation. In H. Das (Ed.), *Proceedings of the SPIE Telemanipulator Technology,* Boston (pp. 240-254).

Davide, F., Holmberg, M., & Lundstrom, I. (2001). Virtual olfactory interfaces: Electronic noses and olfactory displays. *Communications through virtual technology: Identity, community, and technology in the Internet age, Chapter 12* (pp. 193-219). Amsterdam: IOS Press.

Dinh, H. Q., Walker, N., Bong, C., Kobayashi, A., & Hodges, L. F. (1999). Evaluating the importance of multi-sensory input on memory and the sense of presence in virtual environments. *Proceedings of IEEE Virtual Reality* (pp. 222-228).

Copyright © 2006, Idea Group Inc. Copying or distributing in print or electronic forms without written permission of Idea Group Inc. is prohibited.

Fukuda, K., Wakamiya, N., Murata, M., & Miyahara, H. (1997). QoS mopping between user's preference and bandwidth control for video transport. *Proceedings of the 5th International Workshop on QoS (IWQoS),* New York (pp. 291-301).

Ghinea, G. (2000). *Quality of perception - An essential facet of multimedia communications.* Doctoral dissertation, philosophy degree, Department of Computer Science, The University of Reading, UK.

Ghinea, G., & Magoulas, G. (2001). Quality of service for perceptual considerations: An integrated perspective. *IEEE International Conference on Multimedia and Expo, Tokyo* (pp. 752-755).

Ghinea, G., & Thomas, J. P. (1998). QoS impact on user perception and understanding of multimedia video clips. *Proceedings of ACM Multimedia '98,* Bristol UK (pp. 49-54).

Ghinea, G., & Thomas, J. P. (2000). Impact of protocol stacks on quality of perception. *Proceedings of the IEEE International Conference on Multimedia and Expo,* New York (Vol. 2, pp. 847-850).

Ghinea, G., & Thomas, J. P. (2001). Crossing the man-machine divide: A mapping based on empirical results. *Journal of VLSI Signal Processing, 29*(1/2), 139-147.

Ghinea, G., Thomas, J. P., & Fish, R. S. (2000). Mapping quality of perception to quality of service: The case for a dynamically reconfigurable communication system. *Journal of Intelligent Systems, 10*(5/6), 607-632.

Gulliver, S. R., & Ghinea, G. (2003). How level and type of deafness affects user perception of multimedia video clips. *Universal Access in the Information Society, 2*(4), 374-386.

Gulliver, S. R., & Ghinea, G. (2004). Starts in their eyes: What eye-tracking reveal about multimedia perceptual quality. *IEEE Transaction on System, Man, and Cybernetics, Part A, 34*(4), 472-482.

Hasser, C., & Weisenberger, J. M. (1993). Preliminary evaluation of a shape memory alloy tactile feedback display. In H. Kazerooni, B. D. Adelstein, & J. E. Colgate (Eds.), *Proceedings of the Symposium on Haptic Interfaces for Virtual Environments and Teleoperator Systems, ASME Winter Annual Meeting,* New Orleans, LA (pp. 73-80).

Heilig, M. L. (1962). Sensorama simulator. U.S. Patent 3,050,870.

Heilig, M. L. (1992). El cine del futuro: The cinema of the future. *Presence, 1*(3), 279-294.

Howe, R. D., Peine, W. J., Kontarinis, D. A., & Son, J. S. (1995). Remote palpation technology. *IEEE Engineering in Medicine and Biology, 14*(3), 318-323.

Ino, S., Shimizu, S., Odagawa, T., Sato, M., Takahashi, M., Izumi, T., & Ifukube, T. (1993). A tactile display for presenting quality of materials by changing the temperature of skin surface. *Proceedings of the Second IEEE International Workshop on Robot and Human Communication,* Tokyo (pp. 220-224).

Kawalek, J. A. (1995). User perspective for QoS management. *Proceedings of the QoS Workshop Aligned with the 3rd Internations Conference on Intelligence in Broadband Services and Network (IS&N 95),* Crete, Greece.

Copyright © 2006, Idea Group Inc. Copying or distributing in print or electronic forms without written permission of Idea Group Inc. is prohibited.

Kies, J. K., Williges, R. C., & Rosson, M. B. (1997). Evaluating desktop video conferencing for distance learning. *Computers and Education, 28*, 79-91.

Klein, S. A. (1993). Image quality and image compression: A psychophysicist's viewpoint. In A. B. Watson (Ed.), *Digital images and human vision* (pp. 73-88). Cambridge, MA: MIT Press.

Kontarinis, D. A., & Howe, R. D. (1995). Tactile display of vibratory information in teleoperation and virtual environments. *Presence, 4*(4), 387-402.

Koodli, R., & Krishna, C. M. (1998). A loss model for sorting QoS in multimedia applications. *Proceedings of ISCA CATA-98: Thirteenth International Conference on Computers and Their Applications, ISCA,* Cary, NC (pp. 234-237).

Lindh, P., & van den Branden Lambrecht, C. J. (1996). Efficient spatio-temporal decomposition for perceptual processing of video sequences. *Proceedings of the International Conference on Image Processing,* Lausanne, Switzerland (Vol. 3, pp. 331-334), Lausanne, Switzerland.

Martens, J. B., & Kayargadde, V. (1996). Image quality prediction in a multidimensional perceptual space. *Proceedings of the ICIP,* Lausanne, Switzerland (Vol. 1, pp. 877-880).

Masry, M., Hemami, S. S., Osberger, W. M., & Rohaly, A. M. (2001). Subjective quality evaluation of low-bit-rate video: Human vision and electronic imaging VI. In B. E. Rogowitz & T. N. Pappas (Eds.), *Proceedings of the SPIE,* Bellingham, WA (pp. 102-113).

McConkie, G. W., & Rayner, K. (1975). The span of the effective stimulus during a fixation in reading. *Perception and Phychophysics, 17*, 578-586.

Minsky, M., & Lederman, S. J. (1996). Simulated haptic textures: Roughness. *Symposium on Haptic Interfaces for Virtual Environment and Teleoperator Systems, ASME International Mechanical Engineering Congress and Exposition, Proceedings of the ASME Dynamic Systems and Control Division,* Atlanta, GA (Vol. 8, pp. 451-458).

Monkman, G. J. (1992). Electrorheological tactile display. *Presence, 1*(2).

Murphy, T. E., Webster, R. J. (3rd), & Okamura, A. M. (2004). Design and performance of a two-dimensional tactile slip display. *EuroHaptics 2004, Technische Universität München,* Munich, Germany.

Parkhurst, D. J., & Niebur, E. (2002). Variable resolution display: A theoretical, practical, and behavioural evaluation. *Human Factors, 44*(4), 611-629.

Procter, R., Hartswood, M., McKinlay, A., & Gallacher, S. (1999). An investigation of the influence of network quality of service on the effectiveness of multimedia communication. *Proceedings of the International ACM SIGGROUP Conference on Supporting Group Work,* New York (pp. 160-168).

Quaglia, D., & De Martin, J. C. (2002). Delivery of MPEG video streams with constant perceptual quality of service. *Proceedings of the IEEE International Conference on Multimedia and Expo (ICME),* Lausanne, Switzerland (Vol. 2, pp. 85-88).

Copyright © 2006, Idea Group Inc. Copying or distributing in print or electronic forms without written permission of Idea Group Inc. is prohibited.

Rimmel, A. M., Hollier, M. P., & Voelcker, R. M. (1998). *The influence of cross-modal interaction on audio-visual speech quality perception.* Presented at the 105th AES Convention, San Francisco.

Roufs, J. A. J. (1992). Perceptual image quality: Concept and measurement. *Philips Journal of Resolution, 47*(1), 35-62.

Ryans, M. A., Homer, M. L., Zhou, H., Manatt, K., & Manfreda, A. (2001). Toward a second generation electronic nose at JPL: Sensing film organisation studies. *Proceedings of the International Conference on Environmental Systems '01.*

Saida, S., & Ikeda, M. (1979). Useful visual field size for pattern perception. *Perception and Psychophysics, 25*, 119-125.

Simoncelli, E. P., & Adelson, E. H. (1990). Non-separable extensions of quadrature mirror filters to multiple dimensions. *Proceedings of the IEEE Special Issue on Multi-Dimensional Signal Processing, 78*(4), 652-664.

Steinmetz, R. (1990). Synchronisation properties in multimedia systems. *IEEE Journal Select. Areas Communication, 8*(3), 401-412.

Steinmetz, R., (1992). Multimedia synchronisation techniques experiences based on different systems structures. *Proceedings of the IEEE Multimedia Workshop '92,* Monterey, CA.

Steinmetz, R. (1996). Human perception of jitter and media synchronisation. *IEEE Journal on Selected Areas in Communications, 14*(1), 61-72.

Teo, P. C., & Heeger, D. J. (1994). Perceptual image distortion. *Human Vision, Visual Processing and Digital Display V, IS&T / SPIE's Syposium on Electronic Imaging: Science & Technology,* San Jose, CA, 2179 (pp. 127-141).

Towsley, D. (1993). Providing quality of service in packet switched networks. In L. Donatiello & R. Nelson (Eds.), *Performance of computer communications* (pp. 560-586). Berlin; Heidelberg; New York: Springer.

van den Branden Lambrecht, C. J. (1996). Colour moving pictures quality metric. *Proceedings of the ICIP,* Lausanne, Switzerland (Vol. 1, pp. 885-888).

van den Branden Lambrecht, C. J., & Farrell, J. E. (1996). Perceptual quality metric for digitally coded color images. *Proceedings of the VIII European Signal Processing Conference EUSIPCO,* Trieste, Italy (pp. 1175-1178).

van den Branden Lambrecht, C. J., & Verscheure, O. (1996). Perceptual quality measure using a spatio-temporal model of the human visual system. *Proceedings of the SPIE,* San Jose, CA, 2668 (pp. 450-461).

Verscheure, O., & Hubaux, J. P. (1996). *Perceptual video quality and activity metrics: Optimization of video services based on MPEG-2 encoding.* COST 237 Workshop on Multimedia Telecommunications and Applications, Barcelona.

Verscheure, O., & van den Branden Lambrecht, C. J. (1997). Adaptive quantization using a perceptual visibility predictor. *International Conference on Image Processing (ICIP),* Santa Barbara, CA (pp. 298-302).

Wang, Y., Claypool, M., & Zuo, Z. (2001). An empirical study of RealVideo performance across the Internet. *Proceedings of the First ACM SIGCOMM Workshop on Internet Measurement* (pp. 295-309). New York: ACM Press.

Copyright © 2006, Idea Group Inc. Copying or distributing in print or electronic forms without written permission of Idea Group Inc. is prohibited.

Watson, A., & Sasse, M, A. (1996). Evaluating audio and video quality in low cost multimedia conferencing systems. *Interacting with Computers, 8*(3), 255-275.

Watson, A., & Sasse, M. A. (1997). Multimedia conferencing via multicasting: Determining the quality of service required by the end user. *Proceedings of AVSPN '97,* Aberdeen, Scotland (pp. 189-194).

Watson, A., & Sasse, M.A., (2000). The good, the bad, and the muffled: The impact of different degradations on Internet speech. *Proceedings of the 8ᵗʰ ACM International Conference on Multimedia,* Marina Del Rey, CA (pp. 269-302).

Wijesekera, D., & Stivastava, J. (1996). Quality of service (QoS) metrics for continuous media. *Multimedia Tools Applications, 3*(1), 127-166.

Wijesekera, D., Stivastava, J., Nerode, A., & Foresti, M. (1999). Experimental evaluation of loss perception in continuous media. *Multimedia Systems, 7,* 486-499.

Wilson, G. M., & Sasse, M. A. (2000a). Listen to your heart rate: Counting the cost of media quality. In A. M. Paiva (Ed.), *Affective interactions towards a new generation of computer interfaces* (pp. 9-20). Berlin, DE: Springer.

Wilson, G. M., & Sasse, M. A. (2000b). Do users always know what's good for them? Utilising physiological responses to assess media quality. In S. McDonald, Y. Waern, & G. Cockton (Eds.), *Proceedings of HCI 2000: People and Computers XIV - Usability or Else!* Sunderland, UK (pp. 327-339). Springer.

Wikstrand, G., & Eriksson, S. (2002). Football animation for mobile phones. *Proceedings of NordiCHI* (pp. 255-258).

Wilkstrand, G. (2003). *Improving user comprehension and entertainment in wireless streaming media.* Introducing Cognitive Quality of Service, Department of Computer Science.

Wooding, D. S. (2002). Fixation maps: Quantifying eye-movement traces. *Proceedings of the Symposium on ETRA 2002: Eye Tracking Research & Applications Symposium 2002,* New Orleans, LA (pp. 31-36).

Copyright © 2006, Idea Group Inc. Copying or distributing in print or electronic forms without written permission of Idea Group Inc. is prohibited.

Section II

Multimedia and the
Five Senses

Copyright © 2006, Idea Group Inc. Copying or distributing in print or electronic forms without written permission of Idea Group Inc. is prohibited.

Chapter V

Multimodal Expressions:
E-Motional and Non-Verbal Behaviour

Lesley Axelrod, Brunel University, UK

Kate Hone, Brunel University, UK

Abstract

In a culture which places increasing emphasis on happiness and wellbeing, multimedia technologies include emotional design to improve commercial edge. This chapter explores affective computing and illustrates how innovative technologies are capable of emotional recognition and display. Research in this domain has emphasised solving the technical difficulties involved, through the design of ever more complex recognition algorithms. But fundamental questions about the use of such technology remain neglected. Can it really improve human-computer interaction? For which types of application is it suitable? How is it best implemented? What ethical considerations are there? We review this field and discuss the need for user-centred design. We describe and give evidence from a study that explores some of the user issues in affective computing.

Copyright © 2006, Idea Group Inc. Copying or distributing in print or electronic forms without written permission of Idea Group Inc. is prohibited.

Introduction

General Perspective: The Importance of Emotions

Historically, emotions were considered irrational, but are now seen as important for human development, in social relationships and higher-order thinking such as cognition, reasoning, problem solving, motivation, consciousness, memory, learning, and creativity (Salovey & Mayer, 1990). Emotional management is central to our wellbeing and working lives in a "relationship economy" (Bunting, 2004; Hochschild, 1983), giving commercial edge by improving relationships and customer retention, for example, by diffusing conflict in complaints handling. Artists and musicians have always designed to play upon our emotions. Now from toys to business processes, emotional design is seen as critical. We use computers in everyday life, not only for work, but for fun, entertainment, and communication. Usability professionals have realised that users' emotions are critical in diverse domains such as e-learning, e-commerce and e-gaming.

Objectives: Emotional Multimedia

Emerging digital multimedia interfaces give new opportunities to fit the emotional trends of our time. "Affective computing" systems are being developed to recognize and respond to emotions with the aim of improving the quality of human-computer interaction (HCI). Research has concentrated on solving the many technical difficulties involved with innovative technologies, but has largely neglected fundamental questions about the uses and usability, ethics, and implementation of affective computing systems. There is danger of costly, technology led development of applications failing because they do not meet user expectations. We describe a study at Brunel University that exemplifies inter-disciplinary, empirical, user-centred research.

Background

Historical Perspective of Emotion

Theories of emotion, tracing back to the Greek philosophers, have fallen into several schools of thought (Cornelius, 1996). Western philosophical thought viewed emotions as being opposed to reason, and this view is embedded in popular culture and use of language.

Following the Darwinian, evolutionist theories, the concept of "basic emotions" has been very influential. Paul Ekman (2003) in his studies of facial expressions showed mappings to six "basic" emotions and to certain autonomic activities, such as increased heart rate with anger. Most emotion experts think there are a few "basic emotions",

Copyright © 2006, Idea Group Inc. Copying or distributing in print or electronic forms without written permission of Idea Group Inc. is prohibited.

although they do not all agree what they are (Ortony & Turner, 1990). Other emotions are seen as related to social rules or blended from the basic ones, or to co-occur, although the mechanisms for this are not clear.

We have many words for specific emotions. Taxonomies have been constructed to help define emotions, but practical difficulties arise because words carry different meanings depending on individual contexts. Some words translate into different languages but others are culturally dependant (Wierzbicka, 1999). Dimensional models are helpful, using continuous scales to define emotions. The two dimensions commonly used are valence (if positive or negative) and the intensity, or arousal (how strong the feeling is, from mild to intense) (Plutchik & Conte, 1997).

The Social Constructivist perspective incorporates the previous schools of thought, seeking to understand the socially-constructed nature of emotions and their fit within social practices and moral order (Cornelius, 1996). Minsky (2004) suggests emotions are complex control networks that enable us to function effectively in the world. Norman (2004) describes three levels of emotional processing, visceral, behavioural, and reflective, working on biological, neurological, and psychological systems, each with different, sometimes conflicting lessons for design. Models based on such theories have influenced development of computer systems; for example, Ortony, Clore, and Collins (1988) "OCC" model is widely used in artificial intelligence (AI) systems such as in design of embodied conversational agents (ECAs).

A growing body of work examines the neural basis for emotions (LeDoux, 1998). Studies of people with localised brain damage resulting in deficits in emotional capacity have shown that too little emotion impairs decision-making (Damasio, 1994). New techniques such as functional Magnetic Resonance Imaging (fMRI) are rapidly increasing our understanding of the complexities of neural mappings, for example, cognitive regulation of emotion (Ochsner, Bunge, Gross, & Gabrieli, 2002). In time this work may lead to better conceptualisation and labelling of emotion.

Definitions Associated with Affective Computing

"Emotion" is poorly defined, and depending on the historical theoretical and physical viewpoint of researchers, is used to describe both conscious and unconscious phenomena, as well as a spectrum of events from lifelong personality traits, sustained moods, feelings or states, and fleeting momentary experiences.

The word "affect" is used to describe a spectrum of emotions with various connotations for different disciplines. Norman (2004) uses "affect" to describe the underlying unconscious element in emotional life, and "emotion" for the rational conscious element. Picard (1997) in her book "Affective Computing", uses "affect" to encompass the entire spectrum of emotions, including feelings, moods, sentiments, and so forth. She suggests that arguments about exact definitions need not prevent research from progressing, and that a broad understanding of the concept of emotion is sufficient to discuss issues related to it. One important distinction raised by Picard is between inner "emotional experiences" and the outward "emotional expressions" that people use to convey messages about their emotional states, shown via bodily systems, "what is revealed to

Copyright © 2006, Idea Group Inc. Copying or distributing in print or electronic forms without written permission of Idea Group Inc. is prohibited.

others either voluntarily, such as by a deliberate smile, or involuntarily, such as by a nervous twitch' (p. 25).

Language is a dynamic tool, and no doubt consensus about the use of terms will be reached. For the purposes of this chapter, we will adopt Picard's use of the word "affect" to encompass the spectrum of emotions and "affective computing" to cover the spectrum of emotional systems. Picard's use of "emotional expression" is appropriate for behaviours that map directly to an inner emotional experience or are primarily intended to convey emotion. However, not all interactive behaviours map to underlying emotions and not all behaviours convey emotions. Many have a practical or social function; for example, a smile of greeting may not indicate underlying happiness. We use the term "multimodal expressions" to describe the broader range of communicative behaviours that may or may not map to emotions, and "affectemes" to describe meaningful, affective interaction units in our experimental affective computing setting.

Multimodal Human Communication

Human communication consists of both verbal communication, which can be modified by paralinguistic features such as tone of voice, and of non-verbal communication (Argyle, 1988). These can be simultaneous or sequential using multi-sensory modes such as:

- audible sound: choice of words, tone of voice, voice quality, and so forth;
- visual appearance: choice of clothing, environment, and so forth;
- visual positioning such as personal space, proximity, posture, and so forth;
- visual use of body movements, such as gestures, facial expressions, and so forth;
- touch, for example, pressure exerted during a handshake, and so forth;
- taste and smell, such as sweatiness or use of perfume.

We follow complex, culturally-dependant rules to convey emotional and other information and to serve social functions, such as control of turn taking in discourse. We use multimodal expressions to augment what is said, to alter meaning, or sometimes instead of verbal language. Bavelas (1996) points out commonly-held beliefs about the meanings of some non-verbal behaviours, which have little or no scientific basis. For example that folding the arms is a symbolic emotional barrier between speakers, when this could simply convey that the speaker is feeling physically cold.

All multimodal expressions are modified by their appearance and position in time, space, and context. A blank face can give a powerful message (Givens, 2005), for example, in response to a joke. A multimodal expression might change in meaning if it is prolonged; for example, a sustained look of anger might be interpreted as more intense than a fleeting one.

Copyright © 2006, Idea Group Inc. Copying or distributing in print or electronic forms without written permission of Idea Group Inc. is prohibited.

Modal Variables

Certain modes may attract more attention than others; for example, our cerebral cortex has far greater proportional representation of the face than other areas, so attention to facial expressions may dominate. Alternatively, particular types of information might be dominant regardless of the mode used to convey it, for example, fear-provoking events. Some modes may be more prone to conscious or unconscious control, and the extent to which emotions are revealed or hidden varies between people and between situations. Ekman (2003) describes "leakage" of underlying emotions during facial actions. Certain modes may be more important to convey certain types of information or may map more exactly to underlying emotional states. Information from different modes may be additive. Depending on the selection and synchronicity of different modes, the exact meaning of a multimodal expression may be altered. We often use mixed messages in everyday interaction. For example, a smile can alter the meaning of a harsh word, or show irony. Further research is needed in these areas.

Individual Variables

People vary in their communicative skills and strategies and in the social rules they follow, dependent upon variables such as individual physiology, personality and culture; environmental and social context; sex and gender; and age.

Someone with stronger cheek muscles may appear to smile more than an individual with other physical characteristics. Some individuals may have limited perceptual or cognitive capacities so that they cannot recognise any or all aspects of a multimodal signal. Some people are more susceptible to positive or negative affect than others (Zelenski & Larsen, 1991). Different cultures have different rules about emotional displays (Wierzbicka, 1999).

It is easier to interpret the emotions of someone socially familiar, due to more sensitive response to variations from their normal usage of expressions. Social positioning dictates rules; for example, a student might greet a teacher more formally than a peer. Environmental constraints and perspective alter perception of expressions (Lyons, Campbell, Plante, Coleman, Kamachi, & Akamatsu, 2000).

Individual differences in sex and gender identity influence emotional exchanges. Females are born with an increased tendency to attend to emotional facial displays (Baron-Cohen, 2003). Men and women have different physiological responses to certain emotions and use different communication strategies (Tannen, 1992).

Age influences emotions. Tiny infants attend to faces, recognise expressions, and mimic behaviours (Nadel & Muir, 2004). With language, cognition, and social awareness comes emotional regulation. Sociocognitive theories propose that with age there is increased ability to regulate emotions due to greater life experience of emotion control strategies; for example, older adults use more passive emotion-focused strategies, such as hiding feelings (Blanchard-Fields, 1997). In Western cultures, teenagers perform poorly on emotional processing tasks (McGivern, Anderson, Byrd, Mutter, & Reilly, 2002). Neu-

Copyright © 2006, Idea Group Inc. Copying or distributing in print or electronic forms without written permission of Idea Group Inc. is prohibited.

rological research shows development and activation of brain areas associated with emotions and cognition change at puberty and again in later life (Williams, Brown, Liddell, Peduto, & Gordon, 2004).

Communicating Emotions

Emotional awareness is basic to human life. During interaction, however superficial, we are tuned to judge the slightest emotional clues and build them into complex phenomena. Even small favours such as finding a coin in the street have been found to induce a positive affect in people who in turn are likely to "pay it on" and induce positive affect in others (Isen, 1999). Emotions transfer between people and are shared in collective climates (such as the aftermath of 9/11 when the World Trade buildings were destroyed in New York). Imitation of others is important in understanding emotions and may be diminished or absent in those who, for physical or neurological reasons, are unable to imitate (Goldman & Sripada, 2005). Social support may buffer emotional experience, and mood memory can alter experience of emotion. Judgement can be altered by mood congruency (Isen, Shalker, Clark, & Karp, 1978).

Measuring Affect

Measuring affect of real individuals in everyday real-world settings is problematic, despite a number of different approaches. People feeling the same thing might use different words in description. People using the same words, might be describing different experiences. This is analogous to colour perception, where although we can accurately measure the physical characteristics of a specific colour in terms of wavelengths and components, people may perceive different shades or colours. With emotions we cannot give exact measurements or quantifications, let alone understand differences in individuals' perceptions, and so misunderstandings arise. Measures of emotion are controversial, not comprehensive, and often take a bottom-up approach, for example, identifying fine differences in facial expression, but failing to consider the full context of multimodal dialogue exchange.

Probes or self-rating questionnaires, either concurrent (during a task) or retrospective (asked after) may interfere with normal interaction and emotional states. Recall may be biased as a function of current emotions, appraisals, and coping efforts as well as personality differences (Levine & Safer, 2002). Individual interpretations of affect words might compromise scales such as Zuckerman and Lubin's (1985) "multiple affect adjective checklist" or the Positive and Negative Affect Scale (Watson & Clark, 1988).

Physiological measures such as galvanic skin response or pupil dilation are objective but have few fixed correlates to particular emotions or dimensions. Variations in rates, due to normal individual differences, occur. Intrusive wires or sensors may affect behaviour, although less intrusive devices are under development (Vyzas & Picard, 1999).

Observations of behaviour are less intrusive, but prone to subjective bias. Behaviour coding schemes are often unimodal and time-consuming to learn and use, such as Ekman and Friesen's (1978) Facial Action Coding Scheme (FACS). Only some behaviours are

Copyright © 2006, Idea Group Inc. Copying or distributing in print or electronic forms without written permission of Idea Group Inc. is prohibited.

reliably mapped to underlying emotion. Schemes that attempt to annotate and categorise multimodal affect are not yet comprehensive (Bernsen, Dybkjaer, Martin, & Wegener, 2001; Wegener, Knudsen, Martin, & Dybkjaer, 2002). Monitoring performance, related to goals can predict some emotional states, for example, records of mouse movement (Riseberg, Klein, Fernandez, & Picard, 1998).

Machine recognition of behaviours, such as gestures, is becoming more accurate and comprehensive, for example:

- facial actions (Kaiser & Wehrle, 1992);
- speech parameters (Lee & Narayanan, 2005);
- haptics - Clyne's sentograph (1989), or IBM's sentic mouse (Kirsch, 1997);
- posture (Mota & Picard, 2003);
- gaze or pupil dilation (Partala & Surakka, 2003).

Research is just beginning to combine systems and include the social interaction history of individuals, to increase accuracy.

Affective Computing Applications

There are lots of ways that emotional recognition in computing might be fun or useful..... or threatening. It is likely that different realisations of the concept will develop in different ways in different platforms for different domains and purposes. Examples include:

- marketing matched to emotional metrics (Steinbrück, Schaumburg, Duda, & Krüger, 2002);
- games adapting pace and content or avatars reflecting users' emotions (Broersma, 2003);
- robots that recognise or show emotions (Canamero & Fredslund, 2000);
- educational applications responding to levels of engagement (D'Mello, Craig, Gholson, Franklin, Picard, & Graesser, 2005);
- smart homes that adapt to emotional states (Meyer & Rakotonirainy, 2003);
- therapeutic applications for phobias or bullying (Dias, Paiva, Aylett, Woods, Hall, & Zoll, 2004);
- health and emotional management applications, for example, clothing that senses agitation, stress, or sleepiness (Zelinsky, 2004);
- social enhancement applications, for example, facial recognition and name prompting (Scheirer, Fernandez, & Picard, 1999).

Responsive systems may involve models, artificial intelligence, neural networks, and machine learning techniques so that systems develop a degree of emotional intelligence.

Copyright © 2006, Idea Group Inc. Copying or distributing in print or electronic forms without written permission of Idea Group Inc. is prohibited.

Applications may be designed to detect or show emotions or both. They may be designed to use all emotions or just selected states. They may detect or demonstrate emotions in different ways. Emotion might be displayed through the body language or artificial personality of cartoon or realistic avatars, by verbal (written or spoken) messages, or through adaptive pace or content, or by using a shared code of behaviour or meaning.

Behaviour and Attitudes with Innovative Technologies

Feelings and attitudes to technology influence human-computer interaction. For example, people may be frustrated during interaction or may anthropomorphise technologies (Johnson, Marakas, Palmer, 2002). Even a tiny amount of human attribute, supplied intuitively by artists or designers, may trigger this, and research comparing real and virtual interaction is in its infancy. In an interesting series of experiments, Reeves and Nass (1996) showed that people treat new media as if they are real people or things. They found that participants who were asked to try a computer application and report their views were kinder when responding on the same computer than when responding on a different machine.

Users develop new methods and social rules to communicate affect when using new technologies. An example of this is the use of new conventions for spellings, formatting, layout, and emoticons in e-mailing and messaging (Wallace, 1999). Novel evolution of rule systems for emerging affective technologies is likely.

Emotion Recognition in Computing (ERIC) Project

User-Centred Research

Emphasis on solving technical difficulties has led to relative neglect of user preferences and behaviours with emerging innovative technologies. The ERIC project at Brunel has overcome this barrier by simulating the capabilities of future affective technologies using "Wizard of Oz" (WOZ) approach. In a detailed study, we simulated an affective system to test out user preferences and behaviour during a simple word game (Axelrod & Hone, 2004b). The system appeared to adapt to the users' emotional expressions, when in fact the researcher was controlling the pace and content of the game in response to actual emotional expressions observed via a two-way mirror.

Methodology: Factorial Design

The experiment had a 2 × 2 between-subjects factorial design. The factors were:

Copyright © 2006, Idea Group Inc. Copying or distributing in print or electronic forms without written permission of Idea Group Inc. is prohibited.

Figure 1. 2 x 2 factorial design

	Told affective Expect affective system	Not told affective Expect standard system
Acted affective System appears affective	Group 1	Group 2
Did not act affectively System appears standard	Group 3	Group 4

1. **Acted affective** (with two levels; "standard" vs. "system appears affective"). This refers to whether the system appeared to act affectively. In the "standard" condition, clues and messages appeared only in response to the user clicking the "help" button. In the "system appears affective" condition, if the participant was observed via the two-way mirror to use emotional expressions, the game was controlled so that appropriate clues or messages appeared to them.

2. **Told affective** (with two levels; "expect standard system" vs. "expect affective system"). This refers to whether the participant expected the system to act affectively. In the "expect standard system" condition, participants were told they would be testing the usability of a word game. In the "expect affective system" condition, they were told that they would be testing the usability of a word game on a prototype system that might recognize or respond to some of their emotions.

There were, therefore, four experimental conditions in total, representing the factorial combination, as shown in Figure 1.

The Game

A "word-ladder" problem solving game was used. Initial and target words are supplied and intervening "rungs" have to be completed, by changing only one letter each time, in response to clues, as seen in Figure 2.

The game was designed to trigger episodes of user frustration and satisfaction. The rate and type of clues and messages were manipulated. Figure 3 shows some screenshots.

Figure 2. Structure of word-ladder game

Given words	Clues	Solution
HEAD		HEAD
	To listen	HEAR
	Animal that growls	BEAR
BEER		BEER

Copyright © 2006, Idea Group Inc. Copying or distributing in print or electronic forms without written permission of Idea Group Inc. is prohibited.

Figure 3. Screen shots of Word-ladder application

 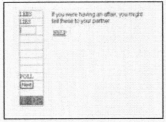

Participants

Sixty participants took part (42 male and 18 female); fifteen were allocated to each experimental condition and paid for their participation. All were currently living in the UK and reasonably fluent in English. Most were students aged 18 to 25.

Procedure

Prior to the task, informed consent and demographic data were collected. The degree of detail given about the purposes of the study was varied according to the experimental condition and later debriefing was carried out as required. Participants interacted with the game for 10 minutes, in either standard or affective mode depending on their allocation to one of the four experimental conditions.

The experiment took place in a usability laboratory where two adjoining rooms, linked by a two- way mirror and networked computers, so the researcher could observe the participant without being seen, and adjust the application responses accordingly. All interactions were video-taped for later analysis and screen activity and logs of user keystrokes and mouse movements were collected.

After undertaking the task, participants used the stylised figures of the Self Assessment Manikin (Bradley & Lang, 1994), to rate their emotional state. They completed question- naires about their interaction, including whether they believed that they had shown emotion during the experiment.

Analysis

Task performance was measured by counting the number of completed "rungs" on the ladder. We borrowed from established methods in the social sciences, such as sequential analysis and speech analysis to develop a novel, inductive, top-down coding method to assess and quantify the multimodal expressions of our users (Axelrod & Hone, 2004a).

Copyright © 2006, Idea Group Inc. Copying or distributing in print or electronic forms without written permission of Idea Group Inc. is prohibited.

From the video footage, we identified and marked the series of discrete multimodal expressions or "affectemes" used. Subjective satisfaction was measured from the ratings of emotional valence (from happy to sad) given on the Self Assessment Manikin. Behavioural responses were coded from the recorded observational data. Common behaviours were identified, counted, and rated for valence and intensity. We imported data into Statistical Package for Social Sciences (SPSS) for analysis. The experiment's main research hypotheses were tested using analysis of variance (ANOVA).

Results

Task Performance

ANOVA showed a main effect of system affective response on task performance $(F(1,56)=7.82, p<0.01)$. Participants were able, on average, to complete significantly more rungs of the puzzle with the affective intervention.

Subjective Satisfaction

There was a main effect of affective response on user subjective ratings of their affective state after the interaction $(F(1,56)=10.25, p<0.005)$. Participants reported themselves as significantly happier, on average, after interaction with the system that responded to their emotional expression.

Valence

We found that when using an apparently affective system, users' affectemes are rated as having significantly more positive valence. ANOVA showed a main effect of system affective response on ratings of valence of user's emotional expressions $F(1,56)=12.63$, $p<0.01$). Participants showed, on average, use of more positively-rated emotional expressions with the affective intervention. Participants were more likely to show positively-rated emotional expressions when they had also been told that the system was affective. There was a significant interaction effect with whether the system had provided an affective response $(F(1,56)=12.63, p<0.01)$. The most positively-rated emotional expressions were from those participants who had been told that the system might respond to their emotional expression, and where the system did in fact act affectively. This interaction effect is shown in the interaction plot in Figure 4.

Copyright © 2006, Idea Group Inc. Copying or distributing in print or electronic forms without written permission of Idea Group Inc. is prohibited.

Figure 4. Interaction plot for ratings of positive valence

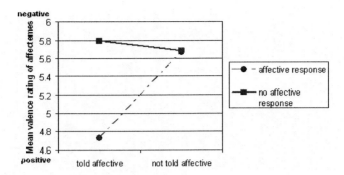

Arousal

We found that when told the system they will be using is an affective system, users are rated as having significantly more intense arousal. ANOVA showed a main effect of system affective response on ratings of valence of user's emotional expressions $F(1,56)=4.74$, $p<0.05$). Participants showed, on average, use of more intensely-rated emotional expressions with the told affective intervention. Participants were more likely to show emotional expressions rated as having higher arousal when they had been told that the system was affective.

Behaviour

Individual participants show an immensely wide array of emotional expressions during affective HCI, ranging from subtle to intense and using many modalities, in all experimental conditions during the course of a game. Interestingly, since it was a problem-solving game, the classic pose of a thinker, resting his chin on his hand, was frequently used.

Facial expressions, such as smiles and frowns were used for various durations. Postural shifts, both large and small, were noticeable, with attempts to peer closely at the screen. Some users spoke aloud or whispered to the screen. They used affect bursts such as sighs, laughs, and whistles. Mouth tension was often evident, with jaw grinding or lip pursing carried out. Grooming activities, such as tucking hair behind ears or shunting spectacles, was frequent.

There was a main effect of being told that the system was affective on participants' self ratings of the extent to which they had displayed their emotions ($F(1,56)=4.34$, $p<0.05$). They were more likely to say they showed their emotions when they had been told that the system was affective. There was also a significant interaction effect with whether the system had provided an affective response ($F(1,56)=4.34$, $p<0.05$). The highest self report of emotion was from those who were told that the system might respond to their emotional expression, but where the system had not in fact acted affectively. This is shown in the interaction plot in Figure 6.

Copyright © 2006, Idea Group Inc. Copying or distributing in print or electronic forms without written permission of Idea Group Inc. is prohibited.

Figure 5. Array of emotional expressions of one participant

Figure 6. Interaction plot for self reported emotion display

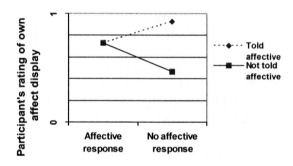

Of the macro-level affective behaviours that have been coded, only blink rate was found to vary with experimental condition. There was thus a main effect of being told the system was affective on blink rate ($F(1,56)=4.57$, $p<0.05$). Blink rates were significantly higher when participants were told that the system may respond to emotional expressions.

Limitations

The WOZ technique simulated affective responses, depending on the researcher's subjective judgments about when the system should make its affective interventions. Post hoc validation of the intervention protocols used is needed (and planned) on the basis of the detailed coding of emotional expressions. However, subjective data from experimental participants provides encouraging preliminary support for the validity of

Copyright © 2006, Idea Group Inc. Copying or distributing in print or electronic forms without written permission of Idea Group Inc. is prohibited.

the experimental manipulation, with participants' post trial belief in the affective nature of the system varying by condition in the ways that one would predict.

Our participants were mainly university students and may not be very representative of the "real world". For example, our observed variables and data collection included assessment of personality types, and many of our participants had a high extraversion score.

The lab setting was perfect for us to control the experimental conditions required for our study, but may not be the same as real-world conditions, and this might be reflected in participants' behaviour. Our results are specific to this particular setting, although some of our findings may generalise to other contexts.

Summary

We found:

- WOZ methods were successful to simulate an affective system;
- people performed significantly better when they used the responsive system;
- after the game, people reported themselves as significantly happier when they used a responsive system;
- people reported that they think they show emotions significantly more when they are told that the system will respond to them, and most of all when they are told that it will respond but it does not do so;
- people blink significantly more if they are told that the system will respond to them;
- user behaviour was rated as more positive when the system responded to them, and most positive of all for the group who were also told that the system would respond;
- behaviour was rated as more intense for participants who were told that the system was affective.

Discussion

Adaptive Behaviour

It has been suggested that, in the short term at least, emotion recognition technology will need users to exaggerate or deliberately pose their emotional expressions (Hayes-Roth, Ball, Lisetti, Picard, & Stern, 1998). Collectively, the results from this subjective data are encouraging for the success of emotional recognition in the short term, as they suggest that users will adopt behaviours that have the potential to aid the recognition process. Our users naturally adapted to the responsive system by displaying more expression, even when not specifically told that the system might respond. We found

Copyright © 2006, Idea Group Inc. Copying or distributing in print or electronic forms without written permission of Idea Group Inc. is prohibited.

that telling a user that the system might respond to their emotions lead to self reports of a greater use of emotional expression at the interface, with highest self ratings from participants who were expecting an affective response, but where the system did not adapt.

In human-to-human interaction, if one participant does not respond as expected, the other person will try to use conversational repair strategies, such as repetition, or exaggeration. It is possible that when the system failed to respond as expected, people used more behaviours in order to facilitate better interaction.

It is clear from video data that participants displayed a large number of affective signals at the interface in all conditions, not just those in which the participants were expecting, or got, an affective response. Blink rate was significantly higher in conditions where participants were expecting an affective response. This may indicate increased arousal and stress (Givens, 2005). Participants may, to some extent, misjudge the degree to which they display emotions when interacting with computer systems. A minority of participants stated categorically that, once told the system would be "looking at them", they had decided to "show no emotion". In fact, they did still show emotional expressions during interaction.

Behaviours may be strongly linked to individual differences, or sequential behaviours not yet analysed in depth. Multimodal expressions often showed a number of modes used concurrently or sequentially. Our coders noticed any sudden changes in modes and often interpreted these as signs of a change in the message. Using several modes synchronously made messages more noticeable, and increased their arousal rating. Sometimes one mode appeared dominant although several were in use.

Agreement on Definitions and Terminology

This new domain benefits from cross-disciplinary research, but suffers from different use of language, related to different disciplines; for example, terms such as "emotion" and "affect" are not used consistently. As the domain develops, clear usage should emerge. It is very difficult to discuss something meaningfully that is not adequately named or defined. If users are to be involved in decision-making, they must have words that help to conceptualise and facilitate discussion. It proved difficult for users to grasp concepts about "affective computing" and "emotional expressions"; "feelings" and "body language" were terms more readily understood. We suggest that affective computing should not be (and often is not) limited to emotional mapping, so the use of the term "emotional expressions" and even "affective computing" may be inappropriate. We should refer to and reflect the totality of human-to- human communication and involve all levels, emotional and non-emotional, superficial and deep. Terminology is needed that will be intuitive and meaningful to users and technologists alike.

We suggest that "multimodal expressions" describe the broad spectrum of communicative behaviours. In the context of affective computing, we should look for meaningful communicative units that could be called "affectemes" to reflect the domain.

Copyright © 2006, Idea Group Inc. Copying or distributing in print or electronic forms without written permission of Idea Group Inc. is prohibited.

Tools and Methods

We need to develop tools and methods to reliably and simply automate interaction that capitalises on the multimodal nature of human communication. The sequential significance of multimodal expressions and their components needs further research. It may not be necessary to recognise all multimodal expressions in every domain or application. Different behaviours may be significant for different applications. Or some aspects of a multimodal expression might be more significant than others.

The User Perspective

The user perspective is frequently neglected. We must use accessible terminology and methods such as WOZ with affective multimedia applications to establish what users want to be recognised, what providers want to recognise and what new communication rules both sides are willing to develop, either intuitively or following set guidelines. These may vary for different domains.

Ethics and Guidelines

New ethical dilemmas arise about the potential uses of affect in computing and fears that while usability and satisfaction with computer applications might be enhanced, user control might be lost. Emotional communication is so fundamental to humans that affective interfaces could be used to influence them in subtle ways that may greatly advantage them, but in some circumstances may also be subliminal, unwanted, or undesirable. Information gathered about emotional states could be linked to purchasing data and other personal details to offer businesses, at best, a powerful marketing tool and, at worst, opportunities for psychological manipulation. Disaster scenarios of a "big brother" culture or of computer systems escaping human control could arise.

Users may find very realistic systems irritating, as in the "uncanny valley" syndrome that arises with avatars that are very realistic and yet leave the user feeling uncomfortable (Mori, 1982).

There is a danger that some individuals will be unduly susceptible to use of empathic displays that may not be in their best interests, resulting in depression, revisiting past affective experiences, or confusion about their own views. Empathy can lead to unplanned disclosure later viewed as embarrassing. Discriminatory behaviour could arise from increased knowledge about users, described as "weblining" (Stepanek, 2000). There are particular concerns about child interaction, for example, learning to form only shallow emotional bonds with robotic pets (Kahn, Batya, Friedman, Perez-Granados, & Freier, 2004).

Organisations already demand emotional labour from employees, and excessive demands may be deleterious to workers, for example, danger of emotional "burnout". This raises the question of training in affective skills so as not to disadvantage sections of the population. There is currently little preparation for emotional labour in our educa-

Copyright © 2006, Idea Group Inc. Copying or distributing in print or electronic forms without written permission of Idea Group Inc. is prohibited.

tional system, although elements of the British National Curriculum now include empathy.

Affective interfaces may disadvantage individuals with difficulties in interpreting or displaying emotional expressions, for example, due to learning disabilities, autism spectrum disorders, or traumatic head injury.

People should have choices about whether to use affective systems, and in what conditions. We can design for different levels of affective response, used appropriately, and with user awareness of them, such as:

- reflect user's affect
- acknowledge/show empathy to user
- design to alter/enhance user affect
- design to develop application's own affective state

Multimodal multimedia design and affective computing must:

- meet commercial potential;
- meet potential for increased accessibility/usability for all users including disabled or elderly;
- develop basis in old interaction rules, and added new ones;
- consider user preferences and enjoyment, or systems will not be adopted and no benefits made;
- use evidence based on user-led and user-centred approaches to develop relevant applications and design tools.

Recommendations for Future Work

There is more analysis to be carried out on our data. Data mining and sequential analysis may establish links between particular variables and particular behaviours and correlate components of multimodal expressions. The results may help to design systems so resources are directed to the most relevant sensory modes or behavioural aspects.

We have used a structured rating scale to categorise perceived emotional behaviours, but we made no attempt to link the observed behaviours to specific inner emotional experiences of participants. More research is needed to establish components of multimodal expressions and map them to specific meanings. We are experimenting with adaptation of an Appraisal Analysis Framework (White, 2005) to examine the use of behaviours in context and assess their function. It seems from assessment of the video data that various behaviours are used when a participant is having difficulty in progressing the game. We observed a number of behaviours that are commonly used in human-to-human communication to manage discourse. These include sudden shifts in posture, shaking of the head, and negative facial expressions to promote turn taking.

Copyright © 2006, Idea Group Inc. Copying or distributing in print or electronic forms without written permission of Idea Group Inc. is prohibited.

We are now carrying out a further experiment to see if these behaviours are increased when the affective system responds specifically to them, or when the participants are trained to use them. Half of our participants will receive a training session to learn to use these indicators. We will compare satisfaction, performance, and usability of the game in these conditions.

Further research is needed to establish if the benefits found here extend to other applications, platforms, and types of adaptation, and if so, to specify which ones.

There is much to be done in this emerging field, particularly in considering the user perspective. As affective systems are implemented, users will need education about emotional computing, and their attitudes, preferences, and willingness to share emotions using multimodal communication with or via machines should be assessed.

Summary

We demonstrated empirically that affective systems improve user performance and satisfaction, and we have reported behavioural responses of users. These results have relevance to the designers of affective systems. Detailed analysis of the observational data is ongoing and has the potential to further inform future design work. The Wizard of Oz technique used here appeared to work successfully to simulate the capabilities of future affective systems and could be used in the future to explore further questions within this domain. Further research is needed to establish if the benefits found here extend to other applications, platforms, and types of adaptation, and if so, to specify which ones, and the best ways to allocate multi-sensory resources.

References

Argyle, M. (1988). *Bodily communication*. London: Routledge.

Axelrod, L., & Hone, K. (2004a). Affecteme, affectic, affecticon: Measuring affective interaction with standard and affective systems. *Proceedings of BCS HCI 2004*, Leeds, UK (pp. 5-8).

Axelrod, L., & Hone, K. (2004b). Smoke and mirrors: Gathering user requirements for emerging affective systems. *Proceedings of 26th International Conference on Information Technology Interfaces, ITI 2004,* Dubrovnik, Croatia (pp. 323-328).

Baron-Cohen, S. (2003). *The essential difference: The truth about the male and female brain.* London: Allen Lane

Bavelas, J. B. (1996). *Debunking body language: New research on non-verbal communications.* Lecture given at University of Victoria. Retrieved from http://novaonline.nv.cc.va.us/eli/spd110td/interper/message/linksnonverbal.html

Copyright © 2006, Idea Group Inc. Copying or distributing in print or electronic forms without written permission of Idea Group Inc. is prohibited.

Bernsen, N. O., Dybkjær, L., Martin, J. -C., & Wegener, M. (2001). Multimodal and natural interactivity corpora and coding schemes. *ELRA Newsletter, 6*(2).

Blanchard-Fields, F. (1997). The role of emotion in social cognition across the adult life span. In K. W. Schaie & M. P. Lawton (Eds.), *Annual review of gerontology and geriatrics: Vol. 17* (pp. 238-265). New York: Springer.

Bradley, M. M., & Lang, P. J. (1994). Measuring emotion: The self-assessment manikin and the semantic differential. *Journal of Behavioral Therapy and Experimental Psychiatry, 25*(1), 49-59.

Broersma, M. (December 1, 2003). Future PlayStations to read hand gestures. *CNET News.com.* Retrieved from http://news.com.com/2100-1043-5112295.html?tag=cd_top

Bunting, M. (2004). *Willing slaves: How the overwork culture is ruling our lives.* London: Harper Collins.

Canamero, D., & Fredslund, J. (2000). *I show you how I like you: Human-robot interaction through emotional expression and tactile stimulation.* Retrieved from http://www.daimi.au.dk/PB/544/PB-544.pdf

Clyne, M. (1989). *Sentics.* Bridport, UK: Prism Press.

Cornelius, R. R. (1996). *The science of emotion: Research and tradition in the psychology of emotion.* Upper Saddle River, NJ: Prentice-Hall.

Damasio, A. R. (1994). *Descartes' error: Emotion, reason, and the human brain.* New York: Grosset; Putnam.

Dias, J., Paiva, A., Aylett, R., Woods, S., Hall, L., & Zoll, C. (2004). *Learning by feeling: Evoking empathy with synthetic characters.* Retrieved from http://www.victec.org

D'Mello, S. K., Craig, S. D., Gholson, B., Franklin, S., Picard, R. W., & Graesser, A. C. (2005). Integrating affect sensors in an intelligent tutoring system. In *Affective Interactions: The Computer in the Affective Loop Workshop at 2005 International Conference on Intelligent User Interfaces* (pp. 7-13). New York: AMC Press.

Ekman, P. (2003). *Emotions revealed: Understanding faces and feelings.* New York: Times Books.

Ekman, P., & Friesen, W. V. (1978). *Facial affect coding system: A technique for the measurement of facial movement.* Palo Alto, CA: Consulting Psychologists Press.

Givens, D. B., (2005). *Non-verbal dictionary of gestures, signs, & body language cues.* Spokane, WA: Center for Nonverbal Studies Press http://members.aol.com/non-verbal2/blank.htm

Goldman, A. I., & Sripada, C. S. (2005). Simulationist models of face-based emotion recognition. *Cognition, 94*(3), 193-213. Retrieved from http://ruccs.rutgers.edu/~karin/GoldmanSri.pdf

Hayes-Roth, B., Ball, G., Lisetti, C., Picard, R., & Stern, A. (1998). Panel on affect and emotion in the user interface. *Proceedings of the International Conference on Intelligent User Interfaces* (pp. 91-94).

Copyright © 2006, Idea Group Inc. Copying or distributing in print or electronic forms without written permission of Idea Group Inc. is prohibited.

Hochschild, A. R. (1983). *The managed heart: Commercialization of human feeling.* Berkeley, CA: University of California Press.

Isen, A. M. (1999). Positive affect. In T. Dalgleish & M. Power (Eds.), *The handbook of cognition and emotion* (pp. 521-539). London: Wiley.

Isen, A. M., Shalker, T. E., Clark, M., & Karp, L. (1978). Resources required in the construction and reconstruction of conversation. *Journal of Personality and Social Psychology, 36*, 1-12.

Johnson, R. D., Marakas, G. M., & Palmer, J. W. (2002). *Tool or social actor? Factors contributing to differential social attributions toward computing technology* (Tech. Rep. No. TR129-1). Kelly School of Business, Indiana University.

Kahn, P. H., Jr., Friedman, B. Perez-Granados, D. R., & Freier, N. G. (2004). Robotic pets in the lives of preschool children *Proceedings of CHI 04,* Vienna.

Kaiser, S., & Wehrle, T. (1992). Automated coding of facial behavior in human-computer interactions with FACS. *Journal of Nonverbal Behavior, 16*(2), 67-83.

Kirsch, D. (May, 1997). *The sentic mouse: Developing a tool for measuring emotional valence.* S.M. thesis, MIT, Media Arts and Sciences. Retrieved from http://vismod.media.mit.edu/tech-reports/TR-495/

Ledoux, J. E. (1998). *The emotional brain: The mysterious underpinnings of emotional life.* New York: Simon & Schuster.

Lee, M. C., & Narayanan, S. S. (2005). Toward detecting emotions in spoken dialogs. *IEEE Transactions on Speech and Audio Processing, 13*(2), 293.

Levine, L. J., & Safer, M. A. Sources of bias in memory for emotions. *Current Directions in Psychological Science, 11*(5), 169-173.

Lyons, M., Campbell, R., Plante, A., Coleman, M., Kamachi, M., & Akamatsu, S. (2000). The Noh mask effect: Vertical viewpoint dependence of facial expression perception. *Proceedings of the Royal Society of London* B 267 (pp. 2239-2245). Retrieved from http://www.mis.atr.co.jp/~mlyons/Noh/noh_mask.html

McGivern, R. F., Andersen, J., Byrd, D., Mutter, K. L., & Reilly, J., (2002). Cognitive efficiency on a match to sample task decreases at the onset of puberty in children. *Brain Cogn., 50*(1), 73-89.

Meyer, S., & Rakotonirainy, A. (2003). A survey of research on context-aware homes. *Proceedings of the Australasian Information Security Workshop Conference on ACSW Frontiers 2003: Vol. 21.*

Minsky, M. (2004). The emotion machine (draft). Retrieved from http://web.media.mit.edu/~minsky/

Mori, M. (1982). *The Buddha in the robot.* Boston: Charles E. Tuttle Co.

Mota, S., & Picard, R. (2003). Automated posture analysis for detecting learner's interest level. *Proceedings of the IEEE Workshop on Computer Vision and Pattern Recognition for Human Computer Interaction (CVPRHCI, Madison, WI, June, 2003).*

Nadel, J., & Muir, D. (Eds.) (2004). *Emotional development: Recent research advances.* Oxford, UK: Oxford University Press.

Copyright © 2006, Idea Group Inc. Copying or distributing in print or electronic forms without written permission of Idea Group Inc. is prohibited.

Norman, D. A. (2004). *Emotional design: Why we love (or hate) everyday things*. New York: Basic Books.

Ochsner, K. N., Bunge, S. A., Gross, J. J., & Gabrieli, J. D. E. (2002). Rethinking feelings: An fMRI study of the cognitive regulation of emotion. *Journal of Cognitive Neuroscience, 14*(8), 1215-1229.

Ortony, A., Clore, G., & Collins, A. (1988). *The cognitive structure of emotions*. Cambridge, UK: Cambridge University Press.

Ortony, A., & Turner, T. J. (1990). What's basic about basic emotions? *Psychological Review, 97,* 315-331.

Partala, T., & Surakka, V. (2003). Pupil size variation as an indication of affective processing. *International Journal of Human-Computer Studies, 59*(1-2), 185-198.

Picard, R. (1997). *Affective computing*. Cambridge, MA: MIT Press.

Plutchik, R., & Conte, H. R. (Eds.). (1997). *Circumplex models of personality and emotions*. Washington, DC: American Psychological Association.

Reeves, B., & Nass, C. (1996). *The media equation: How people treat computers, television, and new media like real people and places*. Stanford, CA; Cambridge, UK: Center for the Study of Language and Information; Cambridge University Press.

Riseberg, J., Klein, J., Fernandez, R., & Picard, R. (1998). Frustrating the user on purpose: Using biosignals in a pilot study to detect the user's emotional state. *Proceedings of ACM 1998 CHI: Conference on Human Factors in Computing Systems* (pp. 227-228).

Salovey, P., & Mayer, J. D. (1990). Emotional intelligence. *Imagination, Cognition, and Personality, 9,* 185-211.

Scheirer, J., Fernandez, R., & Picard, R. W. (1999). Expression glasses: A wearable device for facial expression recognition. *Proceedings of the CHI (late-breaking paper)*, Pittsburgh, PA.

Steinbrück, U., Schaumburg, H., Duda, S., & Krüger, T. (2002, April 20-25). A picture says more than a thousand words - Photographs as trust builders in e-commerce Websites. *CHI2002 Conference Proceedings*, Minneapolis, MN (pp. 748-749).

Stepanek, M. (2000, April 3). Weblining: Companies are using your personal data to limit your choices-and force you to pay more for products. *Businessweek Online*. Retrieved from http://www.businessweek.com/@4WJy4cASJ*2SwAA/2000/00_14/b3675027.htm

Tannen, D. (1992). *You just don't understand: Women and men in conversation*. London: Virago Press.

Vyzas, E., & Picard, R. W. Offline and online recognition of emotion expression from physiological data. *Proceedings of the Workshop on Emotion-Based Agent Architectures, Third International Conference on Autonomous Agents*, May, 1999 (pp. 135-142).

Copyright © 2006, Idea Group Inc. Copying or distributing in print or electronic forms without written permission of Idea Group Inc. is prohibited.

Wallace, P. (1999). *The psychology of the Internet*. Cambridge, UK: Cambridge University Press.

Watson, D., & Clark, L. A. (1988). Development and validation of brief measures of positive and negative affect: The PANAS scales. *Journal of Personality and Social Psychology, 54,* 1063-1070.

Wegener Knudsen, M., Martin, J. C., & Dybkjaer, L. (Eds.). (2002). Survey of multimodal annotation schemes and best practice. *ILE Natural Interactivity and Multimodality Working Group*. Retrieved from http://www.ilc.cnr.it/EAGLES96/isle/ISLE_Home_Page.htm

White, P. R. R. (2005). *The Appraisal Analysis Website*. Retrieved from http://www.grammatics.com/appraisal/

Wierzbicka, A. (1999). *Emotions across languages and cultures*. Cambridge, UK: Cambridge University Press.

Williams, L. M., Brown, K. J., Liddell, B. J., Peduto, A., & Gordon, E. (2004). Developmental changes in limbic-prefrontal responses to emotion across seven decades. *Neuroimage, 22,* 28.

Zelenski, J. M., & Larsen, R. J. (1999). Susceptibility to affect: A comparison of three personality taxonomies. *Journal of Personality, 67*(5), 761-791.

Zelinsky, A. (2004). *Computer vision technologies for driver safety applications*. Retrieved from http://www.volpe.dot.gov/opsad/saveit/docs/may04/zelinsky.pdf

Zuckerman, M., & Lubin, B. (1985). *The multiple affect adjective check list revised*. San Diego, CA: Educational and Industrial Testing Service.

Copyright © 2006, Idea Group Inc. Copying or distributing in print or electronic forms without written permission of Idea Group Inc. is prohibited.

Chapter VI

Learning Science Concepts with Haptic Feedback

Linda D. Bussell, Kinder Magic Software, USA

Abstract

This chapter examines the use of haptic feedback in a multimedia simulation as a means of conveying information about physical science concepts. The case study presented herein investigates the effects of force feedback on children's conceptions of gravity, mass, and related concepts following experimentation with a force-feedback-enabled simulation. Two groups of 17 children conducted experiments with the simulation; the experimental group used both visual and force feedback, and the control group used visual feedback only. Evidence of positive gains by the experimental group who used the simulation with force feedback is presented. Guidelines for applying these technologies effectively for educational purposes are discussed. This chapter adds to the limited research on the application of haptic feedback for conceptual learning and provides a basis for further research into the effects of computer-based haptic feedback on children's cognition.

Copyright © 2006, Idea Group Inc. Copying or distributing in print or electronic forms without written permission of Idea Group Inc. is prohibited.

Introduction

Force feedback can add realism and enjoyment to our computer gaming experiences, and haptic (tactile and force) feedback is used for applications that have a critical psycho-motor aspect, such as flight and surgical training. Haptic feedback has been used to adapt educational software for students with visual impairment, and there is evidence that it can enhance the usability of software for those with normal vision.

This chapter will investigate the use of multimedia simulations with haptic feedback for conceptual learning. Theories of cognition that may illuminate the process of learning through multimedia simulations and haptic interfaces are considered, and a case study of children's experimentation with a simulation incorporating force feedback is presented. Implications of the research and principles that may serve to guide the effective use of multimedia and haptic feedback are discussed.

Background

The following literature review presents evidence that force feedback within a simulation may hold potential to promote conceptual development in children. Relevant theories of cognition will provide a framework for considering the process of learning through multimedia simulations and haptic interfaces. Next, related research on the use of multimedia, simulation, and haptic technology for educational applications is reviewed.

Physical Interaction, Embodied Knowledge, and Cognition

Touch is essential to normal human development (Blackwell, 2000). As we strive to make sense of our experiences, our understanding of forces plays an essential role (Johnson, 1987; Lakoff, 1987; Piaget, 1930). We learn about our world through this interaction. Our sense of touch helps us to form mental models of the shape and structure of objects; we learn to recognize many objects by touch alone. Through these direct experiences, our bodies become more aware of natural forces than our conscious minds do. We learn to hit a moving target such as a ball by trying repeatedly until we are successful. Our muscles learn how much force to apply and in what direction to place the ball where we want it. This embodied knowledge is implicit, automated expertise gained through years of elaboration and practice under countless conditions, and so imposes little cognitive load (Anderson, 2000).

Copyright © 2006, Idea Group Inc. Copying or distributing in print or electronic forms without written permission of Idea Group Inc. is prohibited.

Embodied Knowledge and Science Education

Embodied knowledge is essential to our successful day-to-day living, but can also impede learning. This is a notorious problem in physical science education, as many concepts run counter to our common-sense understanding of the world. Alternative conceptions or misconceptions of science often continue to persist even after considerable time and effort have been spent to dislodge them.

These alternative conceptions or misconceptions can perhaps be viewed most charitably in the context of ecological physics. Gibson (1982) defined ecological physics as what there is in the environment to be directly perceived. An individual senses the environment and perceives *affordances*, invariant properties of objects that afford opportunities for action. We attend to affordances or not, as our goals change. An individual's goal-directed exploratory actions drive selective perception, and attention and perception are honed and tuned through multiple interactions with environmental stimuli. In this way, learning often takes place directly through perceptual means.

Learning to perceive specific affordances becomes automated through repeated interactions in similar situations. In other words, this automation is situated; it is unlikely that improving your tennis shots will improve your billiards shots. When an affordance disappears from awareness, it becomes "transparent." The tennis racket disappears, becoming an extension of the player's intention.

Children's embodied knowledge of physical forces is well-developed by the time they begin school. A number of researchers have investigated children's conceptions about force and gravity. Piaget (1930) found that young children often equated force with being alive, being large, or having a useful purpose. The concept of gravity has been shown to be difficult for children to comprehend adequately. Palmer (2001) interviewed 112 students about gravity. He found that only 11% of the sixth graders and 29% of the tenth graders in the study held scientifically-acceptable ideas. Some children expressed the belief that gravity is a force that acts only in the presence of air. Some believed that gravity acts only on a falling object or that heavier objects fall faster. All of these are misconceptions.

Media Attributes and Cognitive Load

Different kinds of objects and media representations provide different benefits, constraints, and affordances. Both real-time multimedia simulations and video offer the benefit of motion, which is difficult to ignore. A video can be paused at a crucial point or played in slow motion in order to study a process more closely. Physical objects provide opportunities to learn through manipulation and tactile sense, although virtual reality technologies such as haptic feedback are beginning to afford users some of these opportunities as well.

Working memory capacity is limited (Miller, 1956) and can easily become overloaded, particularly if the material being learned is inherently difficult (Sweller, 1988). Research has shown that it is best to present only essential information, as extraneous information

Copyright © 2006, Idea Group Inc. Copying or distributing in print or electronic forms without written permission of Idea Group Inc. is prohibited.

increases cognitive load and may limit learning retention and transfer (Cooper, 1990, 1998; Mayer, 2001; Sweller & Chandler, 1991). People who use an unfamiliar interface to accomplish a goal or task must also learn how to use the software to achieve the desired end (Gee, 2003). If the cognitive demands of this secondary task are too great, they may adversely impact the attainment of the primary goal (Moreno & Mayer, 2004).

Promoting Conceptual Development through Simulation

Computer-based simulations can help learners visualize and comprehend difficult concepts. Simulations can serve as virtual laboratories and provide learners with real-time feedback as they interact, modify variables, and repeat experiments that are difficult and slow to perform under normal classroom conditions. For these reasons, simulations can serve as useful tools in science education.

Simulations provide a way of experiencing relationships among variables more directly than can be done in a textbook. Real-time feedback can help learners link cause and effect, action and perception of results. Feedback informs us when we are performing adequately and guides us to adjust our performance when necessary. Both real and simulated experimentation may afford this immediate feedback that is conducive to maintaining an optimal experience for constructing knowledge.

User-controlled simulations afford the opportunity for reflection. The learner can stop, modify, and restart the simulation, actions that often are not possible in actual laboratory experiments. Effective simulations focus on the essential information and may mitigate the problem of cognitive overload.

However, computer-based simulations are neither a cure-all nor a substitute for instruction. Gredler (1996) noted that a systematic strategy for conducting investigations is important for the effective use of simulation. Unstructured investigation using simulation is as ineffective as unstructured experimentation in a laboratory. Instructors should assess the suitability of a simulation and monitor student learning as they would with any classroom activity.

Haptic Interfaces: An Overview

High-end haptic interfaces are used for applications such as flight simulators, surgical simulators and robot-assisted surgery, manufacturing, and three-dimensional computer modeling. Inexpensive haptic devices have been available in the mainstream consumer market for several years, mostly to enhance feelings of presence or immersion in computer and video games. These devices take various forms such as joysticks, steering wheels, mice, and game controllers, and the haptic feedback they provide varies in quality and sophistication. Different kinds of haptic devices have different sets of features that may make them more appropriate for one application than another. Some merely vibrate or "rumble." Others simulate finer textures, and still others deliver realistic force feedback sensations.

Copyright © 2006, Idea Group Inc. Copying or distributing in print or electronic forms without written permission of Idea Group Inc. is prohibited.

Force feedback devices must be attached to a base to give the user something to push against. Tactile feedback mice are free-moving but there is nothing to push against, so these mice can provide only tactile sensations.

Uses and Benefits of Haptic Interfaces

Additional modes of sensory feedback can help to improve performance, particularly in situations where the primary mode (usually vision) is otherwise occupied (Norman, 1988). Hasser and Massie (1996) identified a number of benefits associated with haptic feedback. These include reduced training time, reduced task completion time, reduced dependence on vision for some tasks, reduced errors, and an increased sense of immersion in virtual environments.

A number of studies support the conclusion that haptic technology can improve performance by reducing task completion time and errors dramatically, sometimes by up to half (Burdea, 1996; Richard, Birebent, Coiffet, Burdea, Gomes, & Langrana, 1996; Sallnäs, 2000).

Educational Applications of Haptic Feedback

Various studies with visually-impaired mathematics and science students have shown that haptic technology holds potential for making computer- and Web-based instructional materials more accessible (Van Scoy, Kawai, Darrah, & Rash, 2000; Yu, Ramloll, & Brewster, 2000). Wies, Gardner, O'Modhrain, Hasser, and Bulatov (2000) developed a Web-based instructional module with a haptic interface to help blind physics students learn about electric fields. The application allowed students to test the electric charge on a sphere by feeling the force attract or repel their hand.

Some educational content readily lends itself to haptic treatment. The force sensations themselves comprise part of the content. Reiner (1999) studied the application of embodied knowledge to physics instruction for students with a limited physics background. They used a haptic interface to sense fields of forces that were not represented visually on the computer screen. The subjects then drew diagrams depicting what they felt. These diagrams were remarkably similar and accurate when compared with those that would result from complex physics calculations.

A potentially powerful application of haptics may be to help students overcome misconceptions, such as the notion that constant force results in constant motion rather than acceleration. A computer simulation with haptic feedback could, for instance, simulate a frictionless environment to provide a basic sensory experience that may help to counteract ingrained misconceptions.

In summary, the theories of cognition presented above may serve to illuminate the process of learning through multimedia and force feedback. The theories of embodied mind and ecological psychology in particular point to the importance of tactile and kinesthetic sensory perception to the most foundational, concrete experiences that underlie, and sometimes compete with, abstract thought.

Copyright © 2006, Idea Group Inc. Copying or distributing in print or electronic forms without written permission of Idea Group Inc. is prohibited.

Well-designed computer simulations employing visual and haptic feedback may offer features and benefits that can make them effective tools for helping students to construct more powerful understandings of physical science concepts. Simulations can provide a safe, multi-sensory environment in which to experiment, perceive immediate results, pause to reflect, modify variables, and refine the experiment.

The aims of the preceding sections were to provide a basis and rationale for the use of force feedback simulations in elementary science education. The following sections will lay the groundwork for the research questions that were investigated in the case study.

Can Force Feedback Promote Conceptual Development in Children?

This question is central to the purpose of the case study to follow, and research in this area is limited. In their review of literature on the broader topic of virtual reality and children's education, Strangman and Hall (2003) found three studies and limited evidence of its effectiveness for the classroom. Indeed, Moreno and Mayer (2004) found that a virtual environment could interfere with learning retention, perhaps by distracting learners from the educational content.

In studies of educational applications of haptic interfaces with visually-impaired subjects, haptic feedback is compensating for a lack of visual feedback. From a theoretical perspective, it is unclear whether and under what conditions haptic feedback may interfere with visual feedback in learners with normal vision.

People process information through verbal and non-verbal channels, and memory can be enhanced if both are employed in instruction (Paivio, 1986). However, an overload of information in one channel may interfere with cognitive processing (Mayer, 2001). Can memory be enhanced if the haptic sensory mode is employed along with the visual mode, or is memory hindered due to interference with visual stimuli in the non-verbal channel? For instance, if a simulation presents a visual object moving on screen together with the feel of the object's movement, will the two sensations utilizing the non-verbal channel create cognitive overload? Or, since the two sensations are coming through separate senses, may the effects be complementary? More research is needed to answer these questions.

In summary, there is little extant research regarding the application of haptic technology for educational purposes that do not have an essential psychomotor or adaptive aspect. Mainstream educational or edutainment titles that employ haptic feedback have done so primarily to boost the entertainment value. And yet, research has shown that extraneous visual and auditory stimuli can negatively impact learning (Mayer, 2001; Sweller & Chandler, 1991). The same may be true for non-essential haptic stimuli. Knowing the answer to this question would enable the technology to be applied appropriately to support learning. The following study will examine whether, in the balance, the effects of haptic feedback served as an aid or obstacle to students' learning.

Copyright © 2006, Idea Group Inc. Copying or distributing in print or electronic forms without written permission of Idea Group Inc. is prohibited.

Learning Styles and Modal Preferences

Another question to be examined in this study is whether force feedback is more effective for some learners than others. Although there has been interest in individual learning styles since the early Greeks, in recent decades there has been increased attention paid to meeting the needs of individual learners with varying styles or modes of learning (Lemire, 1998). There is evidence that learning styles are developmental (Lemire, 1996), and that even in adults they are not fixed and may be culturally based (Reid, 1987).

One aspect of a learning experience or environment is the sensory mode or modes (visual, auditory, haptic, or multi-modal) in which learners process information. Lecture or discussion may be more effective for those with an auditory preference, while reading or viewing graphics may benefit those with a visual preference, and hands-on work may benefit those with a haptic preference. Best practices dictate that a classroom should offer learning opportunities in a variety of modes to meet the needs of most learners. Perhaps a simulation that provides both visual and haptic feedback will be more effective with haptic learners than a simulation that provides only visual feedback. Conversely, it is possible that the added stimulation of force feedback could detract from learning for those with a predominantly visual preference. A limitation of this construct is a lack of consensus both on the dimensions of the construct and on a valid instrument to measure modal preferences. This study will investigate possible interactions of modal preference and treatment.

Methods of Investigating Children's Conceptions of Science

Interview is often used as a method of investigating children's understandings of science concepts. Developmental psychologist Jean Piaget (1930) used interview methods to investigate young children's ideas about physical causality, including force. Interviewing or questioning students about their reasoning can uncover some of the more subtle conceptual relationships that may not be apparent initially.

Graphic representations of a concept or event can be used in conjunction with interview or other methods to elicit children's ideas (McCloskey, Caramazza, & Green, 1980; Palmer, 2001). Palmer used diagrams to probe children's beliefs about gravity and then interviewed them to uncover their reasoning.

Case study research can reveal potentially useful ideas and hypotheses for further investigation. In a descriptive case study, Monaghan and Clement (1999) used interviews to investigate the impact of a computer simulation on students' understanding of relative motion and whether they were able to use what they had learned to generate mental imagery. They used a think-aloud protocol with a predict-observe-explain format, wherein subjects predicted what would happen in a particular scenario, articulating the reasoning they used to reach that conclusion. Students then watched the simulation run and explained whether the results were as predicted or not, and why. This approach is similar to the method that is used in the following case study.

Copyright © 2006, Idea Group Inc. Copying or distributing in print or electronic forms without written permission of Idea Group Inc. is prohibited.

Methods

As little research exists related to force feedback and conceptual reasoning among children, this study is exploratory. Both qualitative and quantitative methods of data collection and analysis were employed, and statistical tests were two-tailed.

Research Questions

Research questions that were addressed include:

- Does work with the simulation (regardless of feedback modes) have an effect on learners' conceptions of key concepts?
- Do differences exist between the reasoning of learners who practice with a simulation with visual and force feedback and those who practice with visual feedback only?
- Does the effectiveness of force feedback vary depending on the subject's modal learning preference, gender, or prior knowledge or experience?

This chapter will focus primarily on the second question, while summarizing the result of the first and third.

The Simulation Experiments

Students' main interaction with the simulation was through conducting a series of four structured yet open-ended experiments intended to organize their experience with the simulation. These experiments were not designed to teach the concepts, but to investigate how children learned and reasoned while working through the experiments with the software. Students followed a worksheet to conduct the experiments and record their observations.

The Logitech® WingMan® Force Feedback Mouse (WFFM) was used in this study. It is an inexpensive haptic device designed for the consumer market. It is capable of providing both force and tactile feedback, though force feedback alone was used in this study. The paddleball simulation used in the study, *FEELitPaddle.exe,* was created by Immersion Corporation to demonstrate the capabilities of the WFFM. The interface consists of two interactive windows. The main window contains the paddle, which users can control and manipulate with the WFFM, and the ball, which users can manipulate indirectly by striking it with the paddle. The smaller window contains slider controls, two of which were used in this study to control the mass of the ball and the gravitational force. When force feedback is enabled, the simulation conveys forces to the user through the mouse. The slider controls range on a scale of 5 to 100, signifying the strength of the force feedback. Five is a barely noticeable effect and 100 is a very strong effect.

Copyright © 2006, Idea Group Inc. Copying or distributing in print or electronic forms without written permission of Idea Group Inc. is prohibited.

Figure 1. Key features of experiments and resting positions

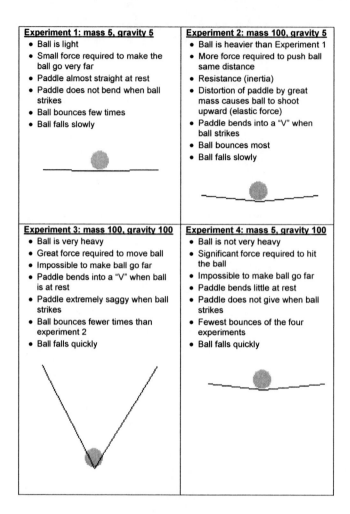

Experiment 1: mass 5, gravity 5
- Ball is light
- Small force required to make the ball go very far
- Paddle almost straight at rest
- Paddle does not bend when ball strikes
- Ball bounces few times
- Ball falls slowly

Experiment 2: mass 100, gravity 5
- Ball is heavier than Experiment 1
- More force required to push ball same distance
- Resistance (inertia)
- Distortion of paddle by great mass causes ball to shoot upward (elastic force)
- Paddle bends into a "V" when ball strikes
- Ball bounces most
- Ball falls slowly

Experiment 3: mass 100, gravity 100
- Ball is very heavy
- Great force required to move ball
- Impossible to make ball go far
- Paddle bends into a "V" when ball is at rest
- Paddle extremely saggy when ball strikes
- Ball bounces fewer times than experiment 2
- Ball falls quickly

Experiment 4: mass 5, gravity 100
- Ball is not very heavy
- Significant force required to hit the ball
- Impossible to make ball go far
- Paddle bends little at rest
- Paddle does not give when ball strikes
- Fewest bounces of the four experiments
- Ball falls quickly

Figure 1 presents some key features and the positions of the simulation at rest under the four experimental conditions. Some of the effects of mass and gravity can be seen by the position and appearance of these elements at rest, while others are only apparent when the simulation is in motion.

From the evidence provided by the changing positions and behavior of the simulation elements, participants drew conclusions about the effects of mass and gravity. With force feedback, effects such as changes in weight could be felt as well as seen. For each experiment, participants were asked to describe the following:

Copyright © 2006, Idea Group Inc. Copying or distributing in print or electronic forms without written permission of Idea Group Inc. is prohibited.

- Speed of ball?
- How hard did you hit the ball?
- Describe what you see and feel when the ball hits the paddle.
- Number of bounces? (hit ball to top of screen)

For experiments 2 through 4, participants were asked what had changed from the previous experiment.

Participants

Participants were 34 fifth-grade students (18 boys and 16 girls) from three public elementary schools in San Diego County, California. Their ages ranged from 9 to 11 years with a mean age of 10 years. All students who returned a signed parental consent form elected to participate in the study.

Data Collection

Each participant completed the Learning Channel Preference Checklist (LCPC), a modal (visual, auditory, and haptic) learning preference assessment prior to group placement. The reliability of the LCPC, calculated with Cronbach's alpha (for consistency) corrected by Spearman-Brown Prophecy Formula is .98. Individual scale reliabilities are visual: .62, auditory: .62, and haptic: .69.

Participants completed a pre-test prior to working with the simulation. This asked them to provide definitions for gravity, mass, and force. Additional items asked whether they had a computer at home, how often they played video games, and how often they played games with force feedback (this was described).

At each of the test sites, subjects were assigned to one of two treatment groups based on gender and modal learning preference as determined by their LCPC scores. Boys and girls were distributed equally between groups. An effort was made to distribute the various modal learning preferences equally between the groups, though this was not always possible due to the other selection constraints. Due to these constraints, no attempt was made to assign by prior knowledge or game experience. However, group assignment was done on a per school basis, and students' prior instruction in key concepts correlated with the school attended.

Working in individual sessions, the experimental group used the simulation with both visual and force feedback, and the control group used the simulation with visual feedback only. Participants each conducted four experiments, observing and recording the results on their worksheets. Then they were asked to develop rules to describe the behavior of the simulation.

During follow-up interviews, students were asked to define gravity, mass, and weight. They were asked to predict the outcomes of two additional experiments and then to explain the results after completing the experiments.

Copyright © 2006, Idea Group Inc. Copying or distributing in print or electronic forms without written permission of Idea Group Inc. is prohibited.

The sessions concluded with a multiple-choice written post-test of concepts that were encountered during the subjects' experimentation with the software. No prior knowledge was assumed. The individual sessions took nearly an hour in most cases.

Treatment of the Data

The data sources were the pre- and post-tests, student-recorded data collection worksheets, audio-recordings of the sessions, and observation notes.

The audio recordings were transcribed. The transcripts were reviewed against the original audio files and then analyzed using QSR's N6 qualitative analysis software.

The transcripts were coded and analyzed to understand the nature of the subjects' reasoning about key concepts within the context of the simulation. Codes were developed from the data using inductive analysis, allowing categories to emerge. Emergent categories were compared to existing categories, so categories and themes evolved through an iterative process of refinement by breaking apart and combining into new categories and themes. Examples of some codes describing gravity are:

- /Gravity/surrounds us, in the air
- /Gravity/holds us, keeps us from falling over
- /Gravity/makes heavier

The pre-test and post-test quantitative and demographic data were analyzed with SPSS statistical analysis software. These data were imported into N6 so that qualitative responses could be compared with demographic data and post-test responses.

Results

Did work with the simulation (regardless of feedback modes) have an effect on learners' conceptions of key concepts?

To investigate this question, participants' prior knowledge as evidenced by their responses on the pre-test and their developing understanding as evidenced by their answers during the interview and post-test were compared. The participants expressed understandings of the key concepts that were varied, often fascinating, and sometimes contradictory. To illustrate, a brief overview of participants' ideas about gravity follows.

All but one participant indicated some prior understanding of the concept of gravity, though the degree of accuracy varied. Gravity's role in keeping us on the ground or keeping us from floating off into space was most frequently mentioned in the pre-test, and was often mentioned again in the interview. One stated that gravity makes us float in space, and another stated that she has a hard time remembering whether gravity keeps us down or makes us float. Several students said that gravity is a force in the air or atmosphere that pushes down on us, or surrounds us "like a curtain" and holds us down.

Copyright © 2006, Idea Group Inc. Copying or distributing in print or electronic forms without written permission of Idea Group Inc. is prohibited.

When asked to define gravity during the pre-test, one student mentioned weight. During the interview, few students explicitly mentioned a connection between gravity and weight when defining gravity. However, 21 participants connected the two at some point during the experiments or interview, and seven equated weight with the pull of gravity. During the interview, when asked how to make the ball as heavy as possible, the most common response was some variation of "set mass and gravity to 100" (the maximum setting for both).

For a detailed discussion of participants' ideas about gravity, mass, weight, speed/acceleration, and force, see Bussell (2004). The most relevant findings are summarized.

Regardless of treatment group, work with the simulation was found to increase students' understanding of key concepts. Even those without prior knowledge were able to draw conclusions and make predictions based on their experiments with the simulation.

During work with the simulation, participants with prior knowledge noted aspects of concepts that they had not conveyed in the pre-test, such as changes in the weight or speed of the ball. During the post-experiment interview, these participants tended to articulate definitions of concepts that were similar to those that they had given in the pre-test. However, when prompted, they could discuss the key concepts in terms of the experiments. They had gained knowledge related to the experiments, but in most cases there was little evidence that it had been integrated with their prior knowledge. Students also had a difficult time generalizing rules from their work with the simulation without guidance; in a classroom, the teacher would normally provide this.

Do differences exist between the reasoning of learners who practice with a simulation with visual and force feedback and those who practice with visual feedback only?

Force feedback provides additional information about the changing effects of mass and gravity on the state of the simulation, but it can also distract from the visual information display. An important question is whether the positive effects of force feedback outweigh possible negative effects. Much of the essential feedback within the simulation can be discerned from the visual information alone. For instance, most participants in both groups indicated, either verbally or by post-test responses, that the ball changed weight under different simulation conditions.

To examine the effects of force feedback, post-test scores were analyzed for differences between the control and experimental groups. Figure 2 shows the post-test items and frequency of responses by group. True statements are shaded.

The experimental group achieved a higher mean total score (16.9412 versus 15.5294), though the difference is not significant ($t=1.595$, $df=32$, $p=.121$, two-tailed).

The post-test scores were divided into two groups, those above and those below the mean of 16.2353. There were 19 in the Low score group and 15 in the High score group. Ten experimental group participants were in the High scoring group compared with five control group participants in the High scoring group. A Pearson Chi-square test was performed ($\times_2=2.982$, $df=1$, $p=.084$).

After the total scores were analyzed, the four items for each question were grouped and a total score based on a four-point scale was computed for each question. Of the six questions, the only scores that showed significant differences between groups were those for Question 3. The experimental group scored significantly higher on this question ($t=2.689$, $df=32$, $p=.011$, two-tailed).

Copyright © 2006, Idea Group Inc. Copying or distributing in print or electronic forms without written permission of Idea Group Inc. is prohibited.

Figure 2. Frequency of post-test responses by group and total

Items	Exper.	Control	Total
1. How can you increase the weight of the ball?			
a. Increase the mass of the ball	15	12	27
b. Increase the gravity	11	13	24
c. Decrease the gravity	2	1	3
d. Decrease the mass of the ball	2	3	5
2. What affects the speed that the ball falls?			
a. The mass of the ball	6	10	16
b. The force of gravity	14	12	26
c. The distance the ball falls	1	3	4
d. None of the above	0	0	0
3. What happens when you increase the gravity?			
a. The mass of the ball increases	1	3	4
b. The ball becomes heavier	13	7	20
c. The ball falls faster	11	9	20
d. The ball falls more slowly	0	3	3
4. What happens when you increase the mass?			
a. The ball becomes heavier	7	5	12
b. The ball falls faster	8	8	16
c. The ball becomes harder to move	4	8	12
d. The ball falls more slowly	4	4	8
5. What happens when you decrease the gravity?			
a. The ball becomes heavier	3	1	4
b. The ball becomes lighter	11	13	24
c. The ball falls faster	3	5	8
d. The ball falls more slowly	11	7	18
6. What happens when you decrease the mass?			
a. The gravity increases	3	2	5
b. The ball falls more slowly	7	11	18
c. The ball becomes lighter	11	8	19
d. The ball falls faster	3	3	6

There were differences between groups on two of the four individual items under Question 3. On the first item, significantly more participants in the experimental group agreed that increasing gravity makes the ball heavier ($t=2.173$, $df=31.318$, $p=.037$, two-tailed). The second item was the false statement that increasing gravity makes the ball fall more slowly. There was a weak difference between groups ($t=1.852$, $df=16.000$, $p=.083$, two-tailed), though fewer than 10% of participants agreed with this statement.

Qualitative evidence supported the idea that force feedback may have influenced student reasoning in certain situations. The following excerpt from an interview illustrates how one student reasoned that the weight of the ball remained the same, but the stiffness of the paddle decreased when the mass variable was increased.

Copyright © 2006, Idea Group Inc. Copying or distributing in print or electronic forms without written permission of Idea Group Inc. is prohibited.

Interviewer:	When the paddle's bending down, do you think that's a heavy ball?
Participant:	Yeah.
Interviewer:	Is that heavier than when the paddle is straight?
Participant:	I think they're kind of the same it's just that yeah the mass gets stiffer a lot. And it holds the heavy ball.
Interviewer:	Okay.
Participant:	Like the ball weighs the same thing.

This participant was in the control group. If she had been able to feel weight changes that accompanied changes in mass it is unlikely she would have concluded that "the ball weighs the same thing."

In summary, though the visual display usually provided adequate information with which to draw a correct conclusion, evidence suggests that the experimental group sometimes benefited from the supplementary information that force feedback provided.

Does the effectiveness of force feedback vary depending on the subject's modal learning preference, gender, or prior knowledge or experience?

No conclusive evidence was found to support the proposition that force feedback is either more or less effective for different kinds of learners.

Modal preference was distributed unevenly among participants (Table 1). The overwhelming majority had a visual preference, and only one participant indicated a clear haptic preference. Because of this, it was not possible to address the question of whether the haptic learners benefited more from force feedback.

The post-test scores of those with a visual preference were analyzed by treatment group. There is no evidence that the visual-preferring students in the experimental group were disadvantaged. In fact they performed better on the post-test than those in the control group, though the difference was not significant ($t=1.586$, $df=19$, $p=.129$, two-tailed). Tot

No significant differences among treatment groups based on other learner characteristics were found, though students from the school where mass was being taught concurrently tended to do better than those from another school that had no prior knowledge of the concept ($t=-1.906$, $df=17$, $p=.074$, two-tailed).

Table 1. Modal learning preferences by group

	Modal learning preference						Total
	Visual	Vis./Aud.	Vis./ Aud./Hap.	Auditory	Haptic	Hap./Vis.	
Exper.	11	1	1	2	1	1	17
Control	10	4		3			17
Total	21	5	1	5	1	1	34

Copyright © 2006, Idea Group Inc. Copying or distributing in print or electronic forms without written permission of Idea Group Inc. is prohibited.

On the other hand, prior knowledge sometimes caused students to interpret the results of experiments through a lens of partial understanding. For instance, one participant determined that the more massive ball had increased in size, even though there was no visual evidence to support this conclusion. She had learned that mass is related to the volume an object occupies, and so therefore the ball must have increased in volume even though it did not appear to do so. She decided this was "an illusion."

There was no difference between treatment groups in post survey responses of satisfaction. During the interview, all 34 participants stated that they liked the activity and gave various reasons for this. Five participants identified the haptic effects as the reason they liked the experience.

Summary of Key Findings

Students could reason based on their experimentation with the simulation, but this new knowledge tended to be specific to the simulation. In most cases, there was no evidence that it had been integrated with their prior knowledge of the concepts.

Participants who used the simulation with force feedback tended to score higher on the post-test, largely due to a significantly-higher score on a question about the effects of increased gravity. The force feedback effect of weight under maximum gravity was arguably the most conspicuous effect among all of the force feedback effects experienced by the experimental group.

The effectiveness of force feedback on concept development was not shown to vary significantly with regard to learner attributes.

All participants found the experience engaging. The participants who used force feedback expressed a high degree of interest and enthusiasm for the novel technology and effects.

Solutions and Recommendations

The results of this exploratory study provide evidence that haptic feedback may promote conceptual learning if the feedback is directly related to the content. Any negative effects of increased cognitive demands imposed by the force feedback interface were offset by the positive effects of additional relevant information, and the result was a slightly positive impact on learning.

Participants in the experimental group did not appear to struggle unduly with the interface. Drawbacks of force feedback that were noted by some participants included fatigue and the necessity of keeping a hand on the mouse while the ball was in motion. With force feedback, the paddle was pushed to the bottom of the screen when the ball struck if it was not actively controlled. Although this meant that more attention was required to monitor the simulation, it also meant that participants could not allow their attention to waver from the task at hand.

Copyright © 2006, Idea Group Inc. Copying or distributing in print or electronic forms without written permission of Idea Group Inc. is prohibited.

Certainly an immersive interface should be considered if there is a compelling reason. If the learning task depends being able to perform within a particular environment, or if essential information to be conveyed is haptic in nature, an immersive interface may be warranted. If the environment is an important condition of successful performance, practice within a virtual environment can be justified, since dealing with accompanying distractions and overload are part of what must be learned. If the primary goal is to learn concepts, however, even within a simulation or game format, it is best to curtail extraneous information and sensory overload.

Participants who rated themselves high on visual or visual-auditory preference performed significantly better on the post-test than those with other preferences ($t= 2.253$, $df=32$, $p=.031$, two-tailed). Spatial reasoning is important for both scientific reasoning and learning from multimedia (Mathewson, 1999; Mayer, 2001), and those with greater spatial ability learn more from well-designed multimedia (Mayer, 2001). In the current study, modal preference may have served as a proxy for spatial ability. Visual preference does not ensure strong visual reasoning skills, but it is likely that there is overlap between preference and skill because people often prefer to use their more-developed skills. Evidence from the current study indicates that students without strong visual reasoning skills may be at a disadvantage when working with simulation software. For this reason, it may be advisable to turn off force feedback to focus learner attention on crucial visual information, particularly for those who find it difficult to focus on visual information.

Limitations of the Study

One limitation of the study was the brevity of the post-test. As a result, it may not have been powerful enough to pick up other differences that existed. Also, it did not include all of the concepts that the participants described during their experimentation. It would have been useful to have more concepts included in the post-test to compare with participants' verbally-expressed understandings.

Another limitation was that the distribution of the visual and visual-auditory participants between the control and experimental group may have inadvertently statistically advantaged the control group. In an effort to distribute at least one of each learning style in the experimental group and other group assignment constraints, more visual-auditory subjects were placed in the control group. The combined number of visual and visual-auditory students in the control group was 14, compared with 12 in the experimental group.

Future Trends

In most virtual learning environments, learners are learning things other than the content. They are working at the same time to master the environment and the interface. As immersive technologies and conventions evolve, and if and when people become expert users of these interface conventions (as with point-and-click interfaces), the extraneous cognitive load imposed by these interfaces will diminish. We tend not to worry much

Copyright © 2006, Idea Group Inc. Copying or distributing in print or electronic forms without written permission of Idea Group Inc. is prohibited.

about the negative impact of point-and-click interfaces on conceptual learning these days.

Could some of the adverse effects of extraneous cognitive load be offset by increased time spent within the activity due to increased motivation? If virtual forces were implemented in a game format, might learners unconsciously acquire new embodied knowledge that could serve to undermine the tenacious hold of misconceptions? For instance, a well-designed space game that required the manipulation of objects under different gravitational forces could give learners a more intuitive understanding of the effects of gravity. It may be that the added motivation of a novel interface coupled with a compelling game would motivate students to spend more time with the content in an informal way. Even if time spent learning the salient content was greater than with a more traditional approach, the combination of the two approaches may serve as powerful means to assist teachers in promoting conceptual change.

Further Research

Clearly, much research remains to be done in this area. Following are some ideas that ensued from the results of this study.

The study was conducted with few participants in a small geographic region, and it should be repeated.

The intervention was concise and intense, considering the number and complexity of the concepts. Ideally, participants would have been able to use the simulation over a number of sessions to build on their understanding. Participants may have been better able to connect their experiences to their prior knowledge if their participation was spread out over a number of sessions. A similar study done over a longer period of time could investigate this.

As discussed previously, participants who expressed a visual or visual-auditory modal preference also scored significantly higher on the post-test. Since visual reasoning skills were essential to reasoning from the experiments, subjects who have strong visual reasoning skills would be expected to perform better than those who do not. A similar study that grouped participants by visual/spatial reasoning skills rather than modal preference could investigate this directly.

This study examined one content domain; there are many more to be investigated. It would be useful to know when and in what ways haptic technology can assist in conveying other difficult concepts, and when haptic technology does not benefit learning.

Summary

Multimedia simulations with relevant force feedback show potential as an aid to promote the development of physical science concepts in children. To promote learning, force

Copyright © 2006, Idea Group Inc. Copying or distributing in print or electronic forms without written permission of Idea Group Inc. is prohibited.

feedback is used to best advantage where it can impart essential information about the concepts being learned. Researchers need to build upon the limited existing knowledge base to determine the best ways to employ this and other new technologies to support learning.

Few studies have been undertaken to study the effects of computer-based haptic feedback on children's cognition. The current study adds to the limited research in this area and provides a starting point for further research into the effects of haptic feedback on children's reasoning. It suggests several potentially fruitful directions for research that could further knowledge in this virtually unexplored area.

References

Anderson, J. R. (2000). *Cognitive psychology and its implications* (5th ed.). New York: Worth Publishers.

Blackwell, P. L. (2000). The influence of touch on child development: Implications for intervention. *Infants and Young Children, 13*(1), 25-39.

Burdea, G. (1996). *Force and touch feedback for virtual reality.* New York: John Wiley & Sons.

Bussell, L. (2004). *The effect of force feedback on student reasoning about gravity, mass, force, and motion.* Doctoral dissertation, University of San Diego and San Diego State University, San Diego, CA.

Cooper, G. (1990). Cognitive load theory as an aid for instructional design. *Australian Journal of Educational Technology, 6*(2), 108-113.

Cooper, G. (1998, December). *Research into cognitive load theory and instructional design at UNSW.* Retrieved March 23, 2003, from www.arts.unsw.edu.au/ CLT_NET_Aug_97.HTML

Gee, J. P. (2003). *What video games have to teach us about learning and literacy.* New York: Palgrave Macmillan

Gibson, J. J. (1982). *Reasons for realism: Selected essays of James J. Gibson.* Hillsdale, NJ: Lawrence Erlbaum Associates.

Gredler, M. E. (1996). Educational games and simulations: A technology in search of a (research) paradigm. In D. H. Jonassen (Ed.), *Handbook of research for educational communications and technology* (pp. 521-540). New York: Macmillan Library Reference USA.

Hasser, C. J., & Massie, T. H. (1996). The haptic illusion. In C. Dodsworth (Ed.), *Digital illusion: Entertaining the future with high technology* (pp. 287-310). Addison-Wesley Publishing Co.

Johnson, M. (1987). *The body in the mind: The bodily basis of meaning, imagination, and reason.* Chicago: The University of Chicago Press.

Copyright © 2006, Idea Group Inc. Copying or distributing in print or electronic forms without written permission of Idea Group Inc. is prohibited.

Lakoff, G. (1987). *Women, fire, and dangerous things*. Chicago: The University of Chicago Press.

Lemire, D. (1996). Using learning styles in education: Research and problems. *Journal of Accelerated Learning and Teaching, 21*(1 & 2), 45-59.

Lemire, D. (1998). Three learning styles models: Research and recommendations for developmental education. *The Learning Assistance Review: The Journal of the Midwest College Learning Center Association, 3*(2), 26-40.

Mathewson, J. H. (1999). Visual-spatial thinking: An aspect of science overlooked by educators. *Science Education, 83*(1), 33-54.

Mayer, R. E. (2001). *Multimedia learning*. Cambridge, UK: Cambridge University Press.

McCloskey, M., Caramazza, A., & Green, B. (1980). Curvilinear motion in the absence of external forces: Naive beliefs about the motion of objects. *Science, 210*, 1139-1141.

Miller, G. A. (1956). The magical number seven, plus or minus two: Some limits on our capacity for processing information. *The Psychological Review, 63*, 81-97.

Monaghan, J. M., & Clement, J. (1999). Use of a computer simulation to develop mental simulations for understanding relative motion concepts. *International Journal of Science Education, 21*(9), 921-944.

Moreno, R., & Mayer, R. E. (2004). Personalized messages that promote science learning in virtual environments. *Journal of Educational Psychology, 96*(1), 165-173.

Norman, D. A. (1988). *The psychology of everyday things*. New York: Basic Books.

Paivio, A. (1986). *Mental representations: A dual coding approach*. New York: Oxford University Press.

Palmer, D. H. (2001). Students' alternative conceptions and scientifically acceptable conceptions about gravity. *International Journal of Science Education, 23*(7), 691-706.

Piaget, J. (1930). *The child's conception of physical causality*. New York: Harcourt, Brace, and Company.

Reid, J. M. (1987). The learning style preference of ESL students. *TESOL Quarterly, 21*(1), 87-111.

Reiner, M. (1999). Conceptual construction of fields through tactile interface. *Interactive Learning Environments, 7*(1), 31-55.

Richard, P., Birebent, G., Coiffet, P., Burdea, G., Gomes, D., & Langrana, N. (1996). Effect of frame rate and force feedback on virtual object manipulation. *Presence, 5*(1), 95-108.

Sallnäs, E. L. (2000). *Supporting collaboration in distributed environments by haptic force feedback*. Retrieved August 18, 2005, from http://www.dcs.gla.ac.uk/%7Estephen/workshops/haptic/papers/sallnas.pdf

Strangman, N., & Hall, T. (2003, December 22, 2003). *Virtual reality and computer simulations*. Retrieved March 3, 2005, from http://www.cast.org/ncac/index.cfm?i=4832.

Copyright © 2006, Idea Group Inc. Copying or distributing in print or electronic forms without written permission of Idea Group Inc. is prohibited.

Sweller, J. (1988). Cognitive load during problem solving: Effects on learning. *Cognitive Science, 12*(2), 257-285.

Sweller, J., & Chandler, P. (1991). Evidence for cognitive load theory. *Cognition & Instruction, 8*(4), 351-362.

Wies, E. F., Gardner, J. A., O'Modhrain, M. S., Hasser, C. J., & Bulatov, V. L. (2000). *Web-based touch display for accessible science education.* Retrieved September 18, 2000.

Yu, W., Ramloll, R., & Brewster, S. (2000). *Haptic graphs for blind computer users.* Retrieved August 18, 2005, from http://www.dcs.gla.ac.uk/%7Estephen/work-shops/haptic/papers/yu.pdf

Copyright © 2006, Idea Group Inc. Copying or distributing in print or electronic forms without written permission of Idea Group Inc. is prohibited.

Chapter VII

Perceptual-Based Visualization of Auditory Information Using Visual Texture

Kostas Giannakis, Society for Excellence and Innovation in Interactive
Experience Design, Greece

Abstract

This chapter investigates the use of visual texture for the visualization of multi-dimensional auditory information. Twenty subjects with a strong musical background performed a series of association tasks between high-level perceptual dimensions of visual texture and steady-state features of auditory timbre. The results indicated strong and intuitive mappings between (a) texture contrast and sharpness, (b) texture coarseness-granularity and compactness, and (c) texture periodicity and sensory dissonance. The findings contribute in setting the necessary groundwork for the application of empirically-derived auditory-visual mappings in multimedia environments.

Copyright © 2006, Idea Group Inc. Copying or distributing in print or electronic forms without written permission of Idea Group Inc. is prohibited.

Introduction

Auditory information is an integral part of multimedia environments. This is an area where information visualization can be employed to make auditory information more useful and accessible in a variety of multimedia contexts, ranging from entertainment and creativity tools to assistive tools for the disabled. However, the effective visualization of auditory information depends on the type of auditory-visual *mappings*, that is, the associations between auditory and visual elements.

It can be argued that the graphical representation of auditory information poses a different challenge than ordinary data (e.g., a series of stock market prices) in the sense that it could be directly related to a different modality. As such, there might be perceptually-based correspondences between different modalities, in this case hearing and vision, which are hard to break and need to be identified. Partial evidence for such perceptual-based mappings comes from the investigations of *synesthesia* and cross-modal associations in the field of cognitive psychology (Cytowic, 1993; Harrison & Baron-Cohen, 1997).

Common perceptual-based mappings include the use of height and color brightness to represent changes in frequency and amplitude respectively, as in *sonogram* representations (Roads, 1996). However, prominent dimensions of sound, such as *timbre*, have been largely neglected or oversimplified in existing approaches to describe sound in visual terms (Barrass, 1997; Caivano, 1994). Moreover, the proposed auditory-visual associations are primarily based on subjective judgments rather than on empirical evidence. A comprehensive review of auditory-visual associations in the area of computer music research can be found in Giannakis and Smith (2000).

As discussed in more detail later, timbre, or the quality of sound, is a multi-dimensional attribute of auditory perception. Recently, visual texture has been used effectively in studies such as the visualization of multi-dimensional data sets (Healey & Enns, 1998; Ware, 2000; Ware & Knight, 1995) and visual information retrieval (Bimbo, 1999; Gupta & Jain, 1997; Niblack, Barber, Equitz, Flickner, Glasman, Petkovic, Yanker, Faloutsos, & Taubin, 1993). The motivation behind these studies is to exploit the sensitivity of the human visual system to texture in order to overcome the limitations inherent in the display of multi-dimensional data and provide more intuitive ways for searching and retrieving information from image and video databases (Rao, 1990). The postulate here is that visual texture can form an adequate sensory channel for the graphical representation of multi-dimensional auditory information, such as timbre, in an intuitive and comprehensible manner.

In the sections that follow, a critical review of research efforts that attempt to identify the high-level perceptual dimensions of auditory timbre and visual texture is presented first, followed by the design of an experiment for the empirical investigation of the cognitive associations between these two sensory percepts. The results are presented and discussed.

Copyright © 2006, Idea Group Inc. Copying or distributing in print or electronic forms without written permission of Idea Group Inc. is prohibited.

The Perception of Timbre

The American National Standards Institute defines timbre as "that attribute of auditory sensation in terms of which a listener can judge that two sounds similarly presented and having the same loudness and pitch are dissimilar" (ANSI, 1960). This definition has been strongly criticized for being too general and ill-defined (Bregman, 1990; Slawson, 1985). In fact, the term "timbre" is used in a variety of contexts, and it is extremely difficult to agree on a single definition. For example, timbre may refer to a class of musical instruments (e.g., string instruments as opposed to brass instruments), a particular instrument in this class (e.g., violin), a particular type of this instrument (e.g., Stradivarius), the quality of playing this instrument (e.g., bad or good quality), and so on. Therefore, any attempts to investigate the dimension of timbre should clarify which aspects of timbre are addressed.

For the purposes of this study, timbre is defined as that perceptual attribute which pertains to the steady-state characteristics of sound. Although temporal characteristics are equally important and necessary for a complete description of timbre (Grey, 1975), they are more related with the identification of sound sources and their intrinsic behavior rather than the qualitative characteristics of timbre that are hidden in the steady-state spectrum of sounds.

Many studies attempted to identify the prominent dimensions of timbre (Bismarck, 1974a, 1974b; Ehresman & Wessel, 1978; Grey, 1975; McAdams, 1999; McAdams, Winsberg, Donnadiey, De Soete, & Krimphoff, 1995; Plomp, 1976). These studies suggest that there are a limited number of dimensions on which every sound can be given a value, and that if two sounds have similar values on some dimension, they are alike on that dimension even though they might be dissimilar on others. However, with very few exceptions, there is no agreement on the dimensions of timbre that these studies have proposed. This is mainly due to the different sets of sounds that were used as stimuli in the experiments (e.g., instrument tones as opposed to synthetic tones) and the different time portions of the sounds that were investigated (e.g., attack transients as opposed to steady states). As a result, the findings of these studies hold very well for the limited range of sounds that they refer to, but they lack generality of application (Bregman, 1990).

Prominent dimensions of timbre that have been proposed in the related literature are as follows:

•   **Sharpness.** Sharpness (other terms include auditory brightness and spectral centroid) is the most prominent dimension of timbre suggested in studies of auditory perception. For pure tones, sharpness is determined by the fundamental frequency, that is, the higher the fundamental frequency, the greater the sharpness. In the case of complex tones, the determining factors for sharpness are the upper limiting frequency and the way energy is distributed over the frequency spectrum, that is, the higher the frequency location of the spectral envelope centroid, the greater the sharpness (Bismarck, 1974a, 1974b).

•   **Compactness.** Compactness (or tonalness) is a measure of a sound on a scale between complex tone and noise, that is, the difference between discrete and

Copyright © 2006, Idea Group Inc. Copying or distributing in print or electronic forms without written permission of Idea Group Inc. is prohibited.

continuous spectra (Bismarck, 1974a). However, the formulation of such a scale has been proven difficult. Compactness is also related to the concept of periodicity, in the sense that periodic sounds are tone-like as opposed to aperiodic or noise-like sounds.

- **Spectral Smoothness.** Spectral smoothness is a dimension of timbre discussed in McAdams (1999). It describes the shape of the spectral envelope and it is a function of the degree of amplitude difference between adjacent partials in the spectrum of a complex tone. Therefore, large amplitude differences produce jagged envelopes, whereas smaller differences produce smoother envelopes.

- **Sensory Dissonance.** Sensory dissonance (or roughness) is related to the phenomenon of beats. When two pure tones with very small difference in frequency are sounded together, a distinct beating occurs that gives rise to a sensation of sensory dissonance (Sethares, 1999). In a series of experiments with pairs of pure tones, Plomp (1976) found that sensory dissonance reaches its maximal point at approximately 1/4 of the relative critical bandwidth. For complex tones, sensory dissonance can be estimated as the sum of all the dissonances between all pairs of partials (Hutchinson & Knopoff, 1978; Sethares, 1999).

The Perception of Visual Texture

Texture is a property that can be analyzed either visually or through touch. Even though texture is an intuitive concept, an exact definition of texture, either as a surface property or as an image property, has never been adequately formulated (Heaps & Handel, 1999; Rao, 1990). In the context of this research, texture is considered as a visual percept and is defined as a visual pattern that exhibits high homogeneity.

In vision research to date, there are two main computational approaches to the analysis of texture: (a) the *statistical* approach and (b) the *structural* approach. The former relies primarily on pre-attentive viewing (i.e., when textures are viewed in a quick glance) and is based on statistics and probability. In the structural approach, a texture is defined as being composed of a primitive pattern that is repeated periodically or quasi-periodically over some area. The relative positioning of the primitives in the pattern is determined by placement rules. Francos, Meiri, and Porat (1991) describe a texture model, which unifies the statistical and structural approaches. Their model allows the texture to be decomposed into three orthogonal components: a harmonic component, a global directionality component, and a purely non-deterministic component. An analogy might be drawn between this approach to visual texture and a sound synthesis technique proposed by Serra (1997) based on a deterministic plus stochastic model.

Although texture has been studied extensively in various research areas, there are a small number of studies that attempted to identify relevant perceptual dimensions of visual texture. Early studies (Amadasun & King, 1989; Tamura, Mori, & Yamawaki, 1978) focused on how to improve computer systems in order to match the human visual system in texture identification and classification experiments. To this end, these studies constructed various sets of texture dimensions based on the statistical analysis of

Copyright © 2006, Idea Group Inc. Copying or distributing in print or electronic forms without written permission of Idea Group Inc. is prohibited.

textures. The participants in these experiments were asked to rate a small number of textures along those dimensions, and their ratings were compared with the ones obtained by computer-based texture identification and classification. Tamura et al. (1978) suggest coarseness, contrast, and directionality as the most prominent perceptual dimensions of texture. Amadasun and King (1989) propose coarseness, contrast, busyness, complexity, and texture strength as appropriate texture dimensions that correspond to texture perception by humans. In a different approach, Ware and Knight (1995) propose a visualization method based on a structural model of visual texture where each texture primitive is assumed to be a Gabor function. According to Ware and Knight, various mathematical parameters of Gabor functions can be used to describe dimensions of visual texture such as orientation, size, and contrast. However, this latter approach has not been empirically validated.

A limitation of the above studies is that the proposed dimensions were not empirically derived but based on the authors' subjective views. Thus, the question of whether humans use these dimensions in texture judgments was not adequately answered. In order to address this limitation, Rao and Lohse (1996) performed a series of experiments that tried to identify the high-level dimensions of texture perception by humans. Their studies were different to earlier studies of visual texture perception in that they used a variety of experimental designs and statistical methods that have been shown to provide an objective and appropriate way of investigating multi-dimensional phenomena (e.g., multidimensional scaling techniques). Rao and Lohse confirmed some of the dimensions proposed by earlier studies as being prominent in visual texture perception and suggested a strong correlation between certain sets of dimensions.

The perceptual space proposed by Rao and Lohse comprises the following three dimensions:

- **Repetitiveness.** This dimension refers to the way primitive elements are placed and repeated over a texture image. The degree of repetition (e.g., periodic, quasi-periodic, random) can be specified and controlled by placement rules.

- **Contrast and Directionality.** Contrast is related with the degree of local brightness variations between adjacent pixels in an image (sharp vs. diffuse edges). The directionality of a texture is a function of the dominant local orientation within each region of texture.

- **Coarseness, Granularity, and Complexity.** Coarseness refers to the size (large vs. small) and density (coarse vs. fine) of the texture elements. Texture granularity corresponds to the degree of randomness with which the texture elements are distributed across the texture image. According to Rao and Lohse, "complexity is a harder feature to capture computationally" (p. 1667), and in their study they suggested the time taken for a subject to describe a texture and the suitability of the description as indicators of complexity. These dimensions are very similar to each other, a fact that is supported in Rao and Lohse's study by high correlation coefficients between these dimensions.

Copyright © 2006, Idea Group Inc. Copying or distributing in print or electronic forms without written permission of Idea Group Inc. is prohibited.

Method

This section describes the design of an experiment to investigate associations (if any) between auditory timbre and visual texture. Table 1 presents the two sets of perceptual dimensions, proposed as suitable for further investigation. Note should be made that the dimension of spectral smoothness has been excluded as it appears to be related to visual characteristics of the sound's amplitude spectrum, and it is not clear how it can be described in auditory terms. Furthermore, although Rao and Lohse (1996) suggested a strong positive correlation between contrast and directionality, it appears that the proposed correlation was a function of the texture images used in the above-described study (Giannakis, 2001). However, this is not the same for the composite dimension of coarseness, granularity, and complexity since correlation between these dimensions has been suggested in other studies as well (Tamura et al., 1978). For the purposes of this study, texture complexity is assumed to be a function of coarseness and granularity.

A within-subjects experimental design was used, where each participant performed a series of texture-timbre association tasks for all the perceptual dimensions under investigation.

Subjects

One group of 20 subjects with strong musical backgrounds participated in the experiment, comprised by 16 undergraduate students, 3 university lecturers, and 1 professional composer. The level of musical experience was determined by a screening questionnaire given to subjects prior to participation. Four subjects had previously taken part in a similar empirical investigation of auditory-visual mappings (Giannakis & Smith, 2001); however, no carry-over effects were anticipated on those subjects' responses since timbre and visual texture were not part of that study.

Apparatus and Stimuli

A prototype computer application was designed in Macromedia Director for use in this experiment comprising a texture palette and three series of sound sequences (see Figure 1).

Table 1. Perceptual models of timbre and visual texture

Timbre	Visual Texture
Sharpness	Contrast
Sensory Dissonance	Repetitiveness
Compactness	Coarseness-Granularity

Copyright © 2006, Idea Group Inc. Copying or distributing in print or electronic forms without written permission of Idea Group Inc. is prohibited.

Figure 1. The prototype application used in this experiment; texture sequences were arranged horizontally; subjects could drag images onto the empty display area (see top of image)

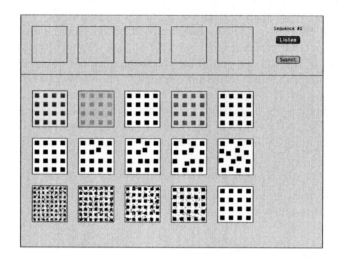

The texture palette consisted of three series of texture sequences, one for each perceptual dimension of visual texture. There were three variation modes (ascending, descending, and non-monotonic) for each series, thus giving nine sequences in total. Each texture sequence consisted of five stimuli. The stimuli were based on a simple texture pattern that was then modified to represent the variation in each perceptual dimension (see Figure 2).

Contrast manipulation was performed using the built-in functionality of standard image processing software (see Figure 2 - top). In addition, the degree of texture repetitiveness was specified by increasing or decreasing the number of displaced texture elements from an ideal symmetrical grid, as shown in Figure 2 - center. Finally, the combination of gradually smaller texture elements with increasing levels of density and visual noise provided the control for the composite dimension of coarseness and granularity (Figure 2 - bottom). Note that the texture sequences used in this experiment represented variation in a single dimension, that is, one dimension was varying, while the remaining two were kept constant.

Three series of sound sequences were designed, one for each perceptual dimension of timbre. The same variation modes were used as for the design of texture sequences, and each sequence consisted of five sound stimuli. All sound stimuli had a fundamental frequency of 220 Hz and the same amplitude level for pitch and loudness equalization respectively. The duration of each stimulus was one second, and stimuli within a sequence were separated by 0.5 seconds of silence. Sharpness stimuli were designed by gradually adding harmonic frequency components up to the sixth in the harmonic series (i.e., the highest frequency component was 1320 Hz). Dissonant stimuli consisted of six

Copyright © 2006, Idea Group Inc. Copying or distributing in print or electronic forms without written permission of Idea Group Inc. is prohibited.

Figure 2. Stimuli for visual texture; top = contrast, center = repetitiveness, bottom = coarseness-granularity

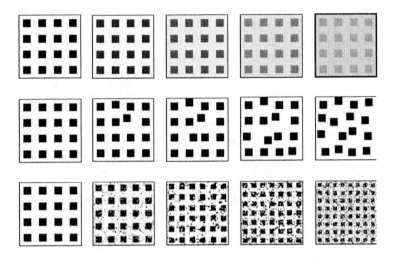

frequency components deviating from the harmonic series in order to produce beating among adjacent partials. The degree of beating was measured with the formula described in Hutchinson and Knophoff (1978). The tone-noise effect for the dimension of compactness was produced by using noise bands centered on the above six harmonic frequency components and by gradually increasing the noise bandwidth.

Procedure

The experimenter demonstrated how to use the prototype application. This was followed by a practice period of three sound sequences. The experimental task was: for the current sequence of five sounds to create a sequence of five corresponding textures. Subjects could listen to the current sequence as many times as they wished, at any point during the task. During the experiment, both sound and texture sequences were introduced in a different order for each subject (a random number generator was used to create random sequence orders) in order to control possible ordering effects. Each subject completed the task for nine sequences. Subjects performed the experiment at their own pace and times ranged from 20 to 30 minutes. The experimenter was present throughout the experiment recording observations that formed the basis for post experiment interviews with subjects. Finally, a data collection program logged texture selections in the form of screenshots, as well as completion times per sequence.

Copyright © 2006, Idea Group Inc. Copying or distributing in print or electronic forms without written permission of Idea Group Inc. is prohibited.

Experimental Environment

The experiment was conducted in a room with normal "office" lighting, and sounds were presented binaurally through headphones. The experiment was designed and run on an Apple Power Macintosh G3 personal computer. Subjects sat approximately 80 cm away from the computer screen, and the components of the interface were sized for comfortable viewing and manipulation at that distance.

Results and Discussion

The presentation of the results for the above-described experiment is based on a qualitative method supported by quantitative data. The auditory-visual mappings obtained from subjects' selections were ranked on a five-point comprehensiveness scale (see Table 2). In general, an auditory-visual mapping was considered *comprehensive* if the subject had selected at least three stimuli from a single visual dimension as a response to the corresponding sound sequence.

Table 3 shows average results of comprehensive auditory-visual mappings by variation mode. It can be noticed that for all dimensions of timbre and each variation mode, there is a clear preponderance of single auditory-visual mappings, that is, subjects preferred selecting stimuli from a single visual dimension rather than mixing stimuli from multiple dimensions. In more detail, there is a clear preponderance of sharpness-contrast associations (88.33% average) as opposed to associations between sharpness and other texture dimensions. These results suggest that for the majority of subjects, textures with low levels of contrast were associated with dull sounds, while higher levels of contrast were associated with sharper sounds. In addition, a very strong positive correlation between the variations in sharpness and contrast can be noticed in all variation modes (see Figure 3).

Table 2. The 5-point scale used to grade the comprehensiveness of subjects' auditory-visual mappings

Score	Description
Very High	Five stimuli selected from a single visual dimension
High	Four stimuli selected from a single visual dimension
Medium	Three stimuli selected from a single visual dimension
Low	No more than two stimuli from a single visual dimension
Very Low	No more than one stimulus from a single visual dimension

Copyright © 2006, Idea Group Inc. Copying or distributing in print or electronic forms without written permission of Idea Group Inc. is prohibited.

Table 3. Auditory-visual mappings by variation mode

Visual mapping	Variation mode		
	Ascending	Descending	Non-monotonic
	Sharpness		
Contrast	90	90	85
Repetitiveness	5	10	—
Coarseness-Granularity	—	—	5
Mixed	5	—	10
	S. Dissonance		
Contrast	10	20	5
Repetitiveness	85	70	85
Coarseness-Granularity	5	10	5
Mixed	—	—	5
	Compactness		
Contrast	—	—	10
Repetitiveness	—	5	—
Coarseness-Granularity	95	95	90
Mixed	5	—	—

Similarly, in the case of sound sequences with varying sensory dissonance, the results suggest a strong sensory dissonance-texture repetitiveness correspondence (80% average of all variation modes). A positive correlation between sensory dissonance and texture repetitiveness can be noticed in all variation modes (see Figure 4). Regular textures were associated with sounds composed of harmonically-related partials, while irregular textures corresponded to inharmonic sounds with increasing beating among adjacent partials.

In the case of sound sequences with varying levels of compactness, the dominant association strategy was to vary texture coarseness-granularity (93.33% average of all variation modes) as opposed to associations between compactness and other texture dimensions. Figure 5 shows a very strong positive correlation between compactness and texture coarseness-granularity in all varying modes. Overall, coarse textures with large elements were associated with tone-like sounds, while finer textures with small elements were associated with noise-like sounds.

As shown in Table 4, the sum of *very high* comprehensiveness levels for the above auditory-visual mappings is very high for monotonic variation modes, providing further evidence for the intuitiveness of subjects' responses. However, results for the non-monotonic variation mode are less satisfactory but overall within acceptable levels.

Figure 6 shows maximum, minimum, and mean response times for each dimension of timbre and each variation mode. Average times were fast (less than 60 seconds) for all three dimensions of timbre and variation modes although subjects were slower when presented with non-monotonic sound sequences. These results lead to the conclusion that subjects were at ease with the experimental task and responded instantly without having to listen many times to the sound stimuli.

Copyright © 2006, Idea Group Inc. Copying or distributing in print or electronic forms without written permission of Idea Group Inc. is prohibited.

Figure 3. Average levels of variation in texture contrast for sharpness stimuli by variation mode; the displayed levels are based on subjects' auditory-visual mappings with maximum comprehensiveness

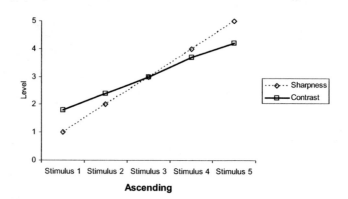

Ascending

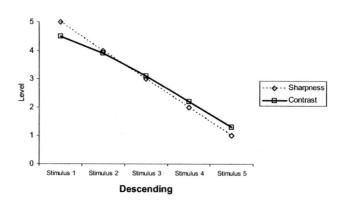

Descending

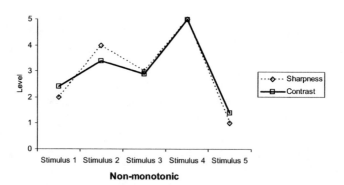

Non-monotonic

Copyright © 2006, Idea Group Inc. Copying or distributing in print or electronic forms without written permission of Idea Group Inc. is prohibited.

Figure 4. Average levels of variation in texture repetitiveness for sensory dissonance stimuli by variation mode; the displayed levels are based on subjects' auditory-visual mappings with maximum comprehensiveness

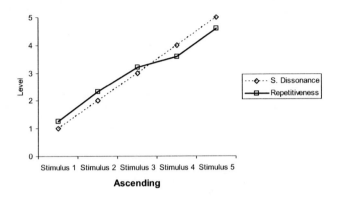

Ascending

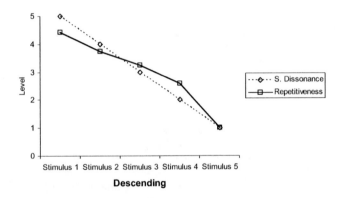

Descending

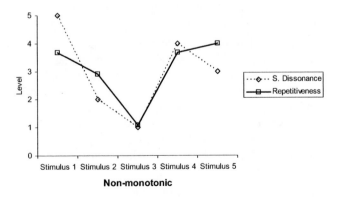

Non-monotonic

Copyright © 2006, Idea Group Inc. Copying or distributing in print or electronic forms without written permission of Idea Group Inc. is prohibited.

Table 4. Comprehensiveness levels of auditory-visual mappings

Comprehensiveness	Variation Mode		
level	Ascending	Descending	Non-monoton
	Sharpness-Contrast		
Very high	16	17	14
High	1	1	1
Medium	1	—	2
Low	—	—	2
Very low	2	2	1
	Sensory Dissonance-Repetitiveness		
Very high	15	12	12
High	2	1	3
Medium	—	1	2
Low	—	1	—
Very low	3	5	3
	Compactness-Coarseness-Granularity		
Very high	18	16	9
High	—	3	6
Medium	1	—	3
Low	1	—	—
Very low	—	1	2

Summary

Based on the above analysis and discussion, a three-dimensional space for the associations between timbre and visual texture can be constructed, as depicted in Figure 7. In this space, the dimension of sharpness is associated with variations in texture contrast, sensory dissonance corresponds to texture repetitiveness, and compactness is associated with the composite dimension of texture coarseness and granularity.

A limitation of the empirical investigation described in this chapter arises from the small dimension sets used in the investigation. Ideally, a high number of perceptual dimensions are required for the complete description of both timbre and visual texture. Therefore, the obtained results are limited to the perceptual models used in this experiment. In addition, this study did not attempt to incorporate perceptually-uniform scales for the associations between the proposed dimensions of timbre and visual texture. To develop such scales would require rigorous psychophysical experiments; such experiments would be a natural extension leading from the research described in this chapter.

In addition, note should be made that the obtained results can be only suggestive due to the small number of participants. One of the short-term goals of this research is to incorporate larger subject groups and use statistical methods that will strengthen the

Copyright © 2006, Idea Group Inc. Copying or distributing in print or electronic forms without written permission of Idea Group Inc. is prohibited.

Figure 5. Average levels of variation in texture coarseness-granularity for compactness stimuli by variation mode; the displayed levels are based on subjects' auditory-visual mappings with maximum comprehensiveness

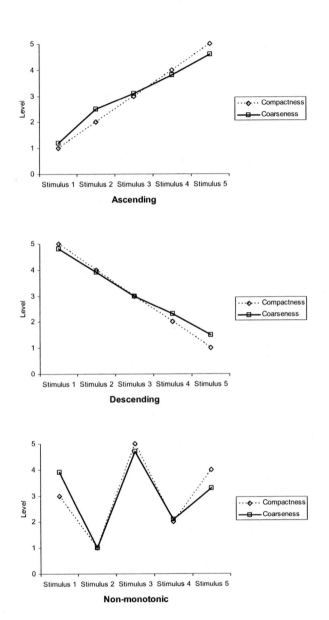

Copyright © 2006, Idea Group Inc. Copying or distributing in print or electronic forms without written permission of Idea Group Inc. is prohibited.

Figure 6. Maximum, minimum and mean completion times for each dimension of timbre by variation mode

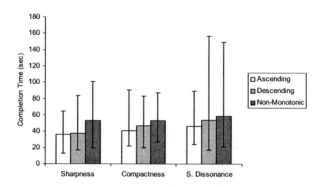

Figure 7. The 3-D space proposed for the associations between perceptual dimensions of timbre and visual texture

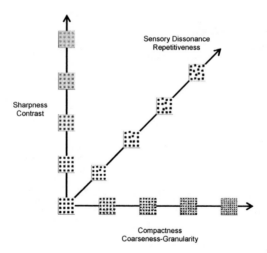

Copyright © 2006, Idea Group Inc. Copying or distributing in print or electronic forms without written permission of Idea Group Inc. is prohibited.

reliability and validity of our conclusions. It is also of further interest to investigate the validity of the above texture-timbre associations with non-music subjects.

The empirical investigation described in this chapter formed part of a series of experiments for the design of *Sound Mosaics*, a novel graphical user interface for sound synthesis based on the direct manipulation of visual representations of sound (Giannakis, 2001). However, it is anticipated that the proposed auditory-visual mappings could be of benefit in many research areas, other than computer-based sound synthesis. For example, auditory-visual mappings are also important in the development of interfaces for visually-impaired users, where the goal is to communicate information that is graphical in nature and thus non-accessible to them (Alty & Rigas, 1998; Edwards, 1989). Furthermore, ongoing research in the development of visual aids for visually-impaired users has also focused on the direct translation of real-world images to sound (Cronly-Dillon, Persaud, & Blore, 2000; Meijer, 1992).

It is hoped that the work presented in this chapter will contribute to future investigations of empirically-derived auditory-visual mappings as these apply to the development of effective multimedia information systems and related applications.

Acknowledgments

The research described in this chapter was funded by a Middlesex University (London, UK) research studentship. The author would like to thank those who kindly volunteered to participate in the described experiment. Particular thanks go to John Dack, Andrew Deakin, Martin Robinson, and Tony Gibbs at Middlesex University's Sonic Arts Department for facilitating the conduct of this study. The experiment stimuli can be downloaded from http://www.mixedupsenses.com/pub.htm

References

Alty, J. L., & Rigas, D. I. (1998). Communicating graphical information to blind users using music: The role of context. *Proceedings of the 1998 CHI - ACM Conference on Human Factors in Computing Systems*, Los Angeles (pp. 574-581).

Amadasun, M., & King, R. (1989). Textural features corresponding to textural properties. *IEEE Transactions on Systems, Man, and Cybernetics, 19*(5), 1264-1274.

ANSI (1960). *Acoustical terminology (S1.1-1960)*. New York: American National Standards Institute.

Barrass, S. (1997). *Auditory information design*. Unpublished doctoral dissertation, The Australian National University, Australia.

Bimbo, A. del. (1999). *Visual information retrieval*. San Francisco: Morgan Kaufmann Publishers.

Copyright © 2006, Idea Group Inc. Copying or distributing in print or electronic forms without written permission of Idea Group Inc. is prohibited.

Bismarck, G. von. (1974a). Timbre of steady sounds: A factorial investigation of its verbal attributes. *Acustica, 30*, 146-159.

Bismarck, G. von. (1974b). Sharpness as an attribute of the timbre of steady sounds. *Acustica, 30,* 159-172.

Bregman, A. S. (1990). *Auditory scene analysis: The perceptual organization of sound.* Cambridge, MA: MIT Press.

Caivano, J. L. (1994). Color and sound: Physical and psychophysical relations. *Color Research and Application, 1*(2), 126-132.

Cronly-Dillon, J., Persaud, K., & Blore, R. (2000). Blind subjects construct conscious mental images of visual scenes encoded in musical form. *Proceedings of the Royal Society of London, Series B—Biological Sciences: Vol. 267* (pp. 2231-2238).

Cytowic, R. (1993). *The man who tasted shapes.* London: Abacus.

Edwards, A. (1989). Soundtrack: An auditory interface for blind users. *Human-Computer Interaction, 4*(1), 45-66.

Ehresman, D., & Wessel, D. (1978). *Perception of timbral analogies* (IRCAM Rep. No. 78-13). Paris: IRCAM.

Francos, J., Meiri, A., & Porat, B. (1991). Modeling of the texture structural components using 2-D deterministic random fields. *Visual Communications and Image Processing, SPIE 1666,* 554-565.

Giannakis, K. (2001). *Sound mosaics: A graphical user interface for sound synthesis based on auditory-visual associations.* Unpublished doctoral dissertation, Middlesex University, UK.

Giannakis, K., & Smith, M. (2000). Auditory-visual associations for music compositional processes: A survey. *Proceedings of the International Computer Music Conference 2000* (pp. 12-15).

Giannakis, K., & Smith, M. (2001). Imaging soundscapes. In R. I. Godøy & H. Jørgensen (Eds.), *Musical imagery.* Lisse: Swets & Zeitlinger.

Grey, J. M. (1975). *Exploration of musical timbre* (CCRMA Rep. No. STAN-M-2). Stanford, CA: Stanford University.

Gupta, A., & Jain, R. (1997). Visual information retrieval. *Communications of the ACM, 40*(5), 71-79.

Harrison, J. E., & Baron-Cohen, S. (Eds.). (1997). *Synaesthesia.* Oxford: Blackwell Publishers.

Healey, C., & Enns, J. (1998). Building perceptual textures to visualize multidimensional datasets. *Proceedings of the IEEE Visualization 1998* (pp. 111-118).

Heaps, C., & Handel, S. (1999). Similarity and features of natural textures. *Journal of Experimental Psychology: Human Perception and Performance, 25*(2), 299-320.

Hutchinson, W., & Knopoff, L. (1978). The acoustic component of western consonance. *Interface, 7*(1), 1-29.

McAdams, S. (1999). Perspectives on the contribution of timbre to musical structure. *Computer Music Journal, 23*(3), 85-102.

Copyright © 2006, Idea Group Inc. Copying or distributing in print or electronic forms without written permission of Idea Group Inc. is prohibited.

McAdams, S., Winsberg, S., Donnadiey, S., De Soete, G., & Krimphoff, J. (1995). Perceptual scaling of synthesized musical timbres: Common dimensions, specificities, and latent subject classes. *Psychological Research, 58*(3), 177-192.

Meijer, P. B. L. (1992). An experimental system for auditory image representations. *IEEE Transactions on Biomedical Engineering, 39*(2), 112-121.

Niblack, W., Barber, R., Equitz, W., Flickner, M., Glasman, E. H., Petkovic, D., Yanker, P., Faloutsos, C., & Taubin, G. (1993). The QBIC project: Querying images by content using color, texture, and shape. *Proceedings of the SPIE Conference on Storage and Retrieval for Image and Video Databases: Vol. 1908* (pp. 173-181).

Plomp, R. (1976). *Aspects of tone sensation.* London: Academic Press.

Rao, A. R. (1990). *A taxonomy for texture description and identification.* New York: Springer-Verlag.

Rao, A. R., & Lohse, G. L. (1996). Towards a texture naming system: Identifying relevant dimensions of texture. *Vision Research, 36*(11), 1649-1669.

Roads, C. (1996). *The computer music tutorial.* Cambridge, MA: MIT Press.

Serra, X. (1997). Musical sound modelling with sinusoids plus noise. In C. Roads, S. Pope, A. Piccialli, & G. de Poli (Eds.), *Music signal processing.* Lisse: Swets & Zeitlinger.

Sethares, W. A. (1999). *Tuning, timbre, spectrum, scale.* New York: Springer-Verlag.

Slawson, W. (1985). *Sound color.* Berkeley, CA: University of California Press.

Tamura, H., Mori, S., & Yamawaki, T. (1978). Textural features corresponding to visual perception. *IEEE Transactions on Systems, Man, and Cybernetics, 8*, 460-473.

Ware, C. (2000). *Information visualization: Perception for design.* San Diego, CA: Academic Press.

Ware, C., & Knight, W. (1995). Using visual texture for information display. *ACM Transactions on Graphics, 14*(1), 3-20.

Copyright © 2006, Idea Group Inc. Copying or distributing in print or electronic forms without written permission of Idea Group Inc. is prohibited.

Chapter VIII

Using Real-Time Physiological Monitoring for Assessing Cognitive States

Martha E. Crosby, University of Hawaii, USA

Curtis S. Ikehara, University of Hawaii, USA

Abstract

This chapter describes our research focused on deriving changing cognitive state information from the patterns of data acquired from the user, with the goal of using this information to improve the presentation of multimedia computer information. Detecting individual differences via performance and psychometric tools can be supplemented by using real-time physiological sensors. Described is an example computer task that demonstrates how cognitive load is manipulated. The different types of physiological and cognitive state measures are discussed along with their advantages and disadvantages. Experimental results from eye tracking and the pressures applied to a

Copyright © 2006, Idea Group Inc. Copying or distributing in print or electronic forms without written permission of Idea Group Inc. is prohibited.

computer mouse are described in greater detail. Finally, adaptive information filtering is discussed as a model for using the physiological information to improve computer performance. Study results provide support that we can create effective ways to adapt to a person's cognition in real time and thus facilitate real-world tasks.

Introduction

The way that humans interact and absorb information delivered through technology is of interest to researchers in many fields. For the last several years, we have designed and executed experiments about individual differences in the way people perceive, search, and understand information presented in multimedia environments. Our research focuses on deriving changing cognitive state information from the patterns of data acquired from the user, with the goal of making more effective multimedia computer systems. A cognitive state, or more specifically for our research, cognitive load, is a subjective experience which cannot always be explicitly expressed or accurately evaluated based on performance. Cognitive load is limited by working memory, increases in relation to mental attention, and is impacted by the affective relationship that a user has with the task or tasks.

The ultimate goal of this research is to unobtrusively monitor humans performing tasks using multimedia systems with several levels of complexity to determine their current cognitive state. With the ability to identify a person's cognitive state, we can dynamically adapt models of their interactions with multimedia systems to produce desired learning improvements. Ideally, such a system could increase the rate of learning, comprehension, and retention, while maintaining user satisfaction.

There are three methods to assess the cognitive load of a user doing a multimedia activity (Charleton, 2002). The first method is by using performance measures. Measures of task performance can be used to assess the cognitive load (e.g., percent correct or time taken), but because tasks often differ, every performance measure needs to be customized to the task. The second method is by using rating scales. These are subjective measures usually taken after the activity is completed, although if interruptions do not degrade task performance, ratings can be taken between or during the tasks of an activity. Subjective measures are essential to assessing the user's perceptions of the task and can provide information of the cognitive state of the user, but since most rating scales are taken after the activity, real-time cognitive state assessment is not possible. The third method is by using physiological measures. Unobtrusive physiological measures are measured in the same way regardless of task, are objectively analyzed, and can be taken in real-time without impacting user performance or user ratings. Sensors can also be used to measure perceptual-motor changes during a task and, by inference, perceptual-motor changes can be related to cognitive performance (Card, Moran, & Newell, 1986).

Each of the three methods of cognitive assessment adds in overlapping and different ways to the understanding of the user's cognitive state. For example, task difficulty can be measured by the user's performance on a task, the user's rating of the difficulty or by the level of stress detected by physiological sensors. Although all three measures may

Copyright © 2006, Idea Group Inc. Copying or distributing in print or electronic forms without written permission of Idea Group Inc. is prohibited.

converge to a similar level of task difficulty, there are many cases where these measures will not coincide, such as when a user has a minimum of errors. The user's task performance implies the task was not difficult, yet the user may rate the task as difficult, and physiological measures may indicate a higher stress level.

Detecting Individual Differences via Performance and Psychometric Tools

Individual differences in people usually account for more variability in performance than variations in system design or training procedures. Using news stories, Dumais and Wright, (1986) studied how people searched for information with various combinations of names and screen locations as the search variables. They found a ratio of 7:1 to 10:1 between individual's slowest to fastest search performance. People also vary greatly in their performance using multimedia systems. Research shows that visual thinking is particularly relevant to how humans interact with multimedia systems. Two findings from the literature on visual thinking research are persistent. First, the ability to think visually (visual ability) differs from person to person. Second, the degree to which people rely on visual thinking also varies (Friedhoff & Benzon, 1989). Spatial memory may be an important component of multimedia searching tasks. In addition, several studies in human-computer interaction suggest a strong connection between visual ability and the use of computer interfaces (Egan, 1988). For example, Egan and Gomez (1985) found that variation in computer interaction performance due to age and spatial memory of subjects with similar experience was 20 times greater than variation in performance due to design of an editor. Questionnaires and psychometric tests can assess individual characteristics. For example, Egan and Gomez (1985) found that the Building Memory Test (Ekstrom, French, & Harman, 1976) that measures the ability to remember the spatial arrangement of objects, correlated significantly with the subject's ability to learn text editing. Vincente, Hayes, and Williges (1987) found two measures of spatial visualization were the best predictive measures for the time subjects took to locate target texts in a hierarchical retrieval system. Studies of text editors (Gomez, Egan, Wheeler, Sharma, & Gruchaz, 1983) indicate novice users with high visual ability perform significantly better than those with low ability. Mayer and Sims (1994) found that high spatial ability students were better able than low spatial ability students to transfer what they learned about a scientific system when verbal and visual explanations were presented contiguously, that is, coordinated animation and narration are most useful for high spatial ability students. Sein and Bostrom (1989) studied the differences in performance of undergraduate students in learning an electronic mail system, and found that both learning styles (KLSI) and visual ability (VZ-2) predicted their performance (Ekstrom, French, & Harman, 1976).

We performed a series of experiments that examined the relationships between characteristics of the user and multimedia interfaces. Along with various presentation materials as independent variables, we used individual differences as co-variates. These individual difference measures included demographics and results obtained from psychological tests such as a visual ability test (VZ-2) and the Myers-Briggs Type Indicator

Copyright © 2006, Idea Group Inc. Copying or distributing in print or electronic forms without written permission of Idea Group Inc. is prohibited.

(MBTI). The four-scale MBTI rating instrument contained preferences scales for measuring features such as introvert vs. extrovert and sensing vs. intuitive that we found to be related to student performance using multimedia courseware (Crosby & Iding, 1997a, 1997b). For dependent variables we used combinations of performance measures such as comprehension task, time on task, keystroke or mouse-click protocols, and eye-fixations (Crosby & Iding, 1997a, 1997b; Crosby, Iding, & Chin, 2001). For several experiments, we used a variety of subjects that included such diverse populations as eighth-grade Hawaiian students, tenth-grade pre-engineering science students, undergraduate and graduate university computer science students, and faculty. Most of our early experiments examined different aspects of multimedia systems (Crosby & Stelovsky, 1995; Crosby, Stelovsky, & Ashworth, 1994, 1996). As a general method, we presented data on a video screen and asked the participants to solve simple problems. Specific examples of the task domains we used included programs (Crosby & Stelovsky, 1989, 1990), maps (Crosby & Chin, 1997, 1999) and graphical representations of data models (Nordbotten & Crosby, 1997, 1999). Results from those studies helped us determine which visual cues and cognitive strategies people used to extract information, how individual differences influenced the way people get information from a video display, and how people used the information they found. In particular, we recorded eye movements to gather detailed information on viewing patterns and strategies for a range of typical computer screen types during the performance of different types of tasks. Eye-scanning patterns provided a particularly rich collection of data about how an individual searches or views specific multimedia interfaces (Howard & Crosby, 1993). Data from the eye-movement monitor helped us determine that, in addition to task variables, individual differences are related to the way students read and understand computer programs (Crosby & Stelovsky, 1989, 1990). Crosby and Peterson (1989) also used eye-movements to study how students searched lists. We found positive correlations between the participants' personality types, experience, and performance. Although previous research on individual differences informed multimedia designers that what is good for one individual at one time is not necessarily good for other individuals or even the same individual in all situations, better multimedia system and interface designs did not necessarily follow. A potential reason for this is that evaluating results from psychological tests is not feasible if the goal is to build real-time adaptable software. As a result, our research focused on creating a methodology to improve learning and task performance by optimizing the human-computer interface based on the user's cognitive state or cognitive load inferred from a suite of passive physiological measures.

The accurate assessment of the users' cognitive states is essential to identifying and testing models of how cognitive processes operate and interact. The individual and the situation can affect the measurement of a cognitive state from a single type of sensor. Measurements from multiple sensors, when combined, are expected to produce more robust measurements of the users' cognitive states. Different sets of sensors can cross-validate similar types of cognitive measures, such as stress, which is measured by both eye tracking and relative blood flow. Also, different sets of sensors can fill in expected missing data. For example, the pressures on a computer mouse are primarily available during mouse interaction, eye tracking is primarily available when there are images to observe, and some sensor may have a ceiling or floor effect (e.g., maximum or minimum). Where single sensor system could be plagued with gaps of missing data, multiple sensors

Copyright © 2006, Idea Group Inc. Copying or distributing in print or electronic forms without written permission of Idea Group Inc. is prohibited.

will be less prone to having this problem.

Our research employed a real-time suite of physiological sensors to assess these cognitive states. Specific questions we examined included: What factors contribute to the users' cognitive states? What are effective ways to assess cognitive states? How can we predict the users' cognitive states in real-time tasks? Using real-time assessment of the users' cognitive states, we plan to adaptively filter information and adapt to the users' learning experience. The issues regarding the positive and negative factors of the different physiological measures are discussed, but in general combining the different measures can fill in temporal gaps in measurement and reaffirm newer measurement methods.

Detecting Individual Differences Using Real-Time Physiological Sensors

The purpose of our current research is to create a methodology to improve learning and task performance by optimizing the human-computer interface based on a person's cognitive state obtained from passive biosensors. Biosensors are an interface media that detects, records, and transmits information regarding a physiological change or process. Picard (1999) used biosensors to measure the affect of participants using multimedia systems. Ark and her colleagues at IBM (Ark, Dryer, & Lu, 1999) built an emotion mouse that showed how using a suite of biosensors could successfully predict certain emotions. We built a functionally equivalent prototype of IBM's emotion mouse that included measures such as skin conductivity, peripheral temperature, relative blood flow, blood oxygen, and the pressures applied to a computer mouse. However, before we were able to conduct experiments using physiological monitoring, we first needed to create an environment to facilitate this research. We picked fractions as one of our task domains because previous experiments showed that fractions could be difficult for people of all ages (Crosby & Sophian, 2003). We determined that we could create an environment where we could control the difficulty, both perceptually and cognitively, by presenting participants with different formats of ratios and proportions (Aschwanden, 2001).

Moving Targets Fraction Task Testbed

The testbed software we created was called the Moving Targets Fractions (MTF) task. It presents a controlled cognitive load task to the user and adapts the presentation by adjusting the degree of information filtering based on what the biosensors indicate is the instantaneous cognitive load of the user. The MTF task presents a fixed number of oval targets containing fractions on a computer screen. These fractions float across the screen from left to right (see Figure 1). The Adaptive Multimodal Interaction Laboratory has information on the MTF task at: http://ami.ics.hawaii.edu/AMIWebsite/software.html.

Copyright © 2006, Idea Group Inc. Copying or distributing in print or electronic forms without written permission of Idea Group Inc. is prohibited.

Figure 1. Screen capture of the moving targets fraction (MTF) task

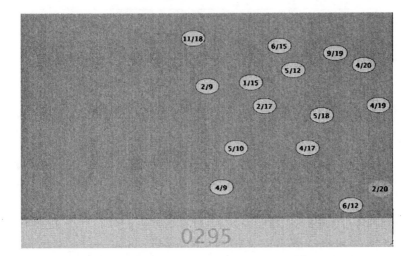

The primary goal of the user is to maximize the score by selecting the correct fractions before they reach the right edge of the screen. Cognitive load is controlled by adjusting fraction values, speed of the fractions across the screen and the number of fractions presented. Adaptive information filtering provides incomplete but helpful information to the user, and the degree of filtering is modified based on the user's cognitive state. For the participants to obtain the highest score, they must select all fractions greater than some critical value (such as 1/3) before the oval touches the right edge of the screen. The participant can maximize the score by achieving four sub-goals before taking action. These sub-goals include evaluating difficulty, fraction relationships, score, and timing. They also can take on different priorities depending on task variables such as the difficulty of evaluating the fraction, the value of the fraction, how close the fraction is to the right side of the screen and the number of fractions presented. The priorities of the sub-goals can be affected by factors such as the participants' motivation and stress. In the MTF task, cognitive load is affected by fraction values, movement speed of the fractions, and the number of fractions presented.

Biosensor Measurements

A measure of the user's cognitive load can be inferred from the cognitive measures and mental states detected with the biosensors (see Table 1). The biosensors may be able to provide both computational and visual/spatial load assessments in real-time. Computational load could be extracted from blink rate, blink duration, stress levels, and arousal state. Visual/spatial load could be extracted from gaze locations, search patterns, and eye

Copyright © 2006, Idea Group Inc. Copying or distributing in print or electronic forms without written permission of Idea Group Inc. is prohibited.

Table 1. Biosensor and potential cognitive state measures

Physiological Measures	Secondary Measures	Potential Cognitive Measures
Eye Position Tracking	Gaze Position, Fixation Number, Fixation Duration, Repeat Fixations, Search Pattern	Difficulty, Attention, Stress, Relaxation, Problem Solving, Successful Learner, Higher Reading Skill (Andreassi, 1995), (Sheldon, 2001)
Pupil Size	Blink Rate, Blink Duration	Fatigue, Difficulty, Interest, Novelty, Mental Activity, Information Processing Speed (Andreassi, 1995)
Skin Conductivity	Tonic and Phasic Changes	Arousal (Andreassi, 1995)
Finger, Wrist and Ambient Temperature		Negative Affect (Decrease), Relaxation (Increase) (Andreassi, 1995)
Relative Blood Flow	Heart Rate and Beat to Beat Heart Flow Change	Stress, Emotion Intensity (Andreassi,1995)
Mouse Pressure Sensors (Left/ Right Buttons and Case)		Stress (Lange-Küüttner, 1998), Certainty of Response
Mouse Position	Speed of Mouse Motion	Arousal, Stress, Difficulty

fixations. Numeric computation and visual/spatial cognitive abilities are of specific interest.

The first column of Table 1 lists all the physiological sensors currently being used (i.e., biosensors). Secondary measures (column 2) can be extracted from the primary physiological measures and are needed to derive some of the potential cognitive and affective states. Column 3 lists several potential cognitive measures.

Each physiological measure has its positive and negative factors as shown in Table 2. Combining the different measures can fill in temporal gaps in other measurements and reaffirm newer measurement methods.

A Defense Advanced Research Projects Agency grant in which our laboratory participated involved several teams from throughout the country using a variety of physiological sensors. Besides the sensors shown in Table 1, physiological sensors evaluated in that project from other teams included: EEG, ERP, functional near-infrared, head posture, and body posture. A description of the different teams and physiological sensors used is found in a report by St. John, Kobus, and Morrison (2003).

Copyright © 2006, Idea Group Inc. Copying or distributing in print or electronic forms without written permission of Idea Group Inc. is prohibited.

Table 2. Positive and negative factors of physiological sensors

	Positives	Negatives
Eye Tracking	Able to determine several cognitive states during task	Does not provide information when there are no visual targets. Rest periods go unmonitored.
Pupil Size	Able to determine several cognitive states during task	Affected by sudden changes of image intensity
Mouse	Provides information primarily during movement and clicking	Does not provide information during intervening periods
Heart Rate	Provides continuous information	Slow rise and decline in heart rate relative to the trigger event
Skin Conductivity	Provides continuous information	Subject to user movements and slow rise and decline in rate relative to the trigger event
Temperature	Provides continuous information	Very slow rise and decline in temperature change relative to the trigger event

A recent experiment employed user actions, physiological results, and data from eye tracking, synchronized in real time, to measure the cognitive task of calculating the values of fractions moving at various speeds across a visual display. Our test platform controlled for easy or difficult problems and allowed us to examine the interactions between perceptual, motor, and cognitive components of the moving fractions task. (Ikehara & Crosby, 2003) We have developed a gauge to show degrees of mental effort for this task (Ikehara, Chin, & Crosby, 2004).

Our previous experiments had shown eye fixations to be a good measure of task difficulty for fixed images, since the number of fixations is greatest during problem solving (Sophian & Crosby, 1999); however, extracting measures from eye-tracking data when the targets are moving and the target locations are unknown is problematic. Without knowing the location of the targets, the cluster in Figure 2 of the three targets B, C, & D could result in a measure indicating a relatively long fixation on a single target. Clustering of targets would reduce the fixation count, increase fixation duration, and reduce the amount of eye movement needed to view the targets.

With moving targets, not knowing the location of the targets further complicates the determination of what is a fixation, especially when targets cross paths (see Figure 3). In Figure 3, since the observer is free to change fixation to the other target, it is not clear whether the number of fixations is one or two. With a large number of moving targets, the formation of transient clusters due to motion and path crossing would make obtaining an accurate fixation count difficult. The measures from the eye movement monitor

Copyright © 2006, Idea Group Inc. Copying or distributing in print or electronic forms without written permission of Idea Group Inc. is prohibited.

Figure 2. Target clusters

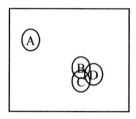

Figure 3. Targets crossing paths

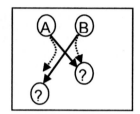

significantly distinguish between the different numbers of moving targets of unknown location. These results supported our results testing the eye-tracking measures with more dynamic visual scenes where task performance information is not available in real time (Ikehara & Crosby, 2003).

We examined the dynamics of how people extract information from moving images, especially as they progressed from novice to expert performers. We investigated several methods to instantaneously assess the users' cognitive states and their cognitive capability (Ikehara, Chin, & Crosby, 2004). We are investigating effective ways to augment cognition, taking into account the users' current cognitive state. We found that measures from the eye movement monitor significantly distinguished between the different numbers of moving targets of unknown location. These results supported our results testing the eye-tracking measures with more dynamic visual scenes where task performance information is not available in real-time.

Andreassi, in his summary of muscle activity measured by electromyography (EMG), noted that muscle tension increased with task difficulty (Andreassi, 2000, pp. 181). We asked if this would also be true for the finger pressure applied when clicking a computer mouse. A pilot experiment with participants doing a task of increasing difficulty showed that when task difficulty increased, individual click pattern pressures became more variable.

Shown are the composite views for two subjects of mouse click pressure. Each composite view contains over 80 clicks. The rear of the valley is where the first click occurs and the

Copyright © 2006, Idea Group Inc. Copying or distributing in print or electronic forms without written permission of Idea Group Inc. is prohibited.

Figure 4. The composite view of over 80 mouse clicks of a subject who has practiced the task dozens of times and finds the task easy

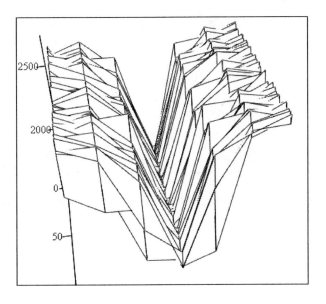

Figure 5. The composite view of over 80 mouse clicks of a subject who is performing the task for the first time

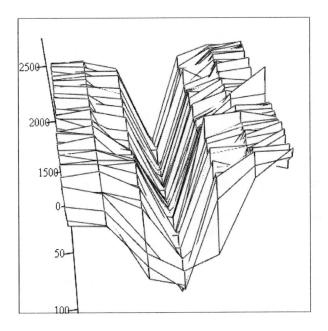

Copyright © 2006, Idea Group Inc. Copying or distributing in print or electronic forms without written permission of Idea Group Inc. is prohibited.

front is the most recent click. The task increases the difficulty of clicking a moving square on a computer screen.

Figure 4 represents a person who has practiced the task dozens of times and has no difficulty with the task. Note the deep consistent click pattern.

Figure 5 shows the clicks of a participant who is performing the task for the first time. Note that the initial clicks, when the task is easiest, in the rear of the valley are relatively consistent. The clicks in the front of the valley show the greatest inconsistency when the task is most difficult.

Adaptive Information Filter Model

Information filtering is an effective method of enhancing performance on cognitive tasks, but can be suboptimal when it does not respond to changes in the user's cognitive ability (Quiroga & Crosby, 2004). The adaptive augmented cognition research at our laboratory is targeted at collecting biosensor information from the user so that the information filtering program can optimize the presentation of information in real-time and achieve learning and performance goals. The MTF task requires the user to elicit several important abilities. These abilities include: hand-eye coordination, visual searching, mathematical problem computation, fraction estimation, strategy selection, learning and motivation. All these requirements of the task affect the user's cognitive ability and constrain cognitive load to different degrees. Manipulation of the presentation to the user is designed to control the user's cognitive load and optimize the user's cognitive ability to achieve the short-term goal of maximum performance and the long-term goal of maintaining a high level of cognitive ability.

Adaptive information filtering can be presented to the user by a combination of three methods: emphasis, de-emphasis, and deletion. Information filtering using de-emphasis is the preferred method since it will allow an incremental change in the task difficulty changing the number of filtered targets based on the user's computational cognitive load. Cognitive load is shifted from computation to visual/spatial load. The following sections discuss a potential solution using our suite of real-time passive physiological sensors to assess the cognitive ability of the user and control an adaptive information filter.

Figure 6 shows the system model used in our research. The user, Moving Target Fractions (MTF) task and adaptive information filter, are subsystems of the testbed system. What makes this system model unique is that it includes biosensors (i.e., passive physiological sensors) to monitor the user and provide input to the adaptive information filter. We examined the dynamics of how people extract information from moving images, especially as they progressed from novice to expert performers. We investigated several methods to instantaneously assess cognitive load and people's cognitive capability in order to create effective ways to augment cognition as the cognitive load increased (Ikehara, Chin, & Crosby, 2004). We are presently performing experiments to determine which combination of sensor data can assess the user's cognitive states listed in Table

Copyright © 2006, Idea Group Inc. Copying or distributing in print or electronic forms without written permission of Idea Group Inc. is prohibited.

Figure 6. Model of how the adaptive information filter changes the display of the moving target fractions (MTF) task when biosensor information indicates excessive cognitive load

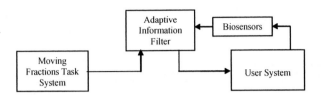

1. To calibrate the measurement of cognitive load, a test with tasks of varying cognitive load is presented to the user. Upper and lower limits are extracted, and the range of cognitive load is established for the user. Cognitive load is not a single measure, but can be partitioned as shown in Figure 7. Once the cognitive state of the user is assessed, presentation changes are made by an adaptive information filter to improve task performance.

We collect biosensor information from the user so that the adaptive information filtering program can, in real-time, optimize the presentation of information. The MTF task requires the user to elicit several important abilities including: hand-eye coordination, visual search, mathematical computation, fraction estimation, strategy selection, learning, memory, and motivation. Manipulation of the presentation to the user is designed

Figure 7. The adaptive information filter changes the MTF display depending on the computational cognitive load derived from the biosensors; in this figure, the adaptive filter shifts cognitive load S#1 and S#3 from computation to visual/spatial

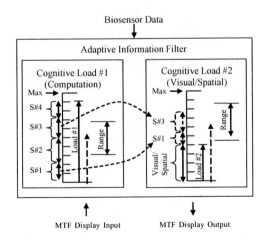

Copyright © 2006, Idea Group Inc. Copying or distributing in print or electronic forms without written permission of Idea Group Inc. is prohibited.

to optimize the user's cognitive ability to achieve the short-term goal of maximum performance and the long-term goal of maintaining a high level of cognitive ability.

In Figure 7, Cognitive Load #1 (Computation) has a graded vertical scale. The top of that scale represents the maximum cognitive ability. When the user is performing the MTF task, each sub-goal (i.e., "S#1, S#2, S#3, & S#4) increases the total computational cognitive load (i.e., "Load #1"). The solid arrow next to Load #1 shows the current computational cognitive load. The dashed arrow indicates a desired cognitive load value. The two-headed arrow denotes the desired computational cognitive load range. The figure shows the computational cognitive load is currently above the desired range. In the same figure, Cognitive Load #2 (Visual/Spatial) shows the visual/spatial cognitive load to be less than the desired range. The adaptive information filter shifts cognitive load from computational to visual to maintain the computational cognitive load within the desired range.

Summary

Our research investigated ways to instantaneously assess cognitive state and cognitive capability. The suite of physiological sensors to assess cognitive states is shown to be a viable alternative to task performance measures when performance measures are not available. Also, our research has shown that, in some cases, these physiological measures can be a more sensitive measure of cognitive state than task performance measures. These cognitive state factors can then be used to improve task performance through adaptive filtering and other techniques. Using real-time assessment of the users' cognitive states, we were able to design adaptive multimedia systems; then, depending on the situation and task, we could then determine if the system helped the user meet the desired performance improvements of making more accurate, faster, or appropriate decisions. Results from these studies provide support that we can create effective ways to adapt to a person's cognition in real-time and thus facilitate real-world tasks.

Acknowledgments

This research was supported in part by the Office of Naval Research grants no. N000149710578 and N000140310135 and DARPA grant no. NBCH1020004.

Copyright © 2006, Idea Group Inc. Copying or distributing in print or electronic forms without written permission of Idea Group Inc. is prohibited.

References

Andreassi, J. L. (1995). *Psychophysiology: Human behavior and physiological response* (3rd ed.). Hillsdale, NJ: Lawrence Erlbaum.

Andreassi, J. L. (2000). *Psychophysiology: Human behavior and physiological response* (4th ed.). Hillsdale, NJ: Lawrence Erlbaum.

Ark, W., Dryer, D., & Lu, D. (1999). The emotion mouse. In H. J. Bullinger & J. Ziegler (Eds.), *Human-computer interaction: Ergonomics and user interfaces* (pp. 818-823). London: Lawrence Erlbaum Associates.

Aschwanden, C. (2001). *Investigating user comprehension of Web-based educational applications.* Masters thesis, Dept. EE, ETH Zurich, Switzerland.

Card, S., Moran, T., & Newell, A. (1986). The model human processor. In K. Boff, L. Kaufman, & J. Thomas (Eds.), *Handbook of perception and human performance: Vol. 2* (pp. 45-1 - 45-35). New York: Wiley.

Charleton, S. (2002). Measurement of cognitive states in testing and evaluation. In S. G. Charleton & T. C. O'Brien (Eds.), *Handbook of human factors and evaluation* (pp. 97-126).

Crosby, M., Auernheimer, B., Aschwanden, C., & Ikehara, C. (2001). Physiological data feedback for application in distance education. *PUI 2001 Proceedings,* Orlando FL.

Crosby, M., & Chin, D. (1997). Evaluating multi-user interfaces (EMI). In M. Smith, G. Salvendy, & R. Koubek (Eds.), *Design of computing systems: Social and ergonomic considerations: Vol. 21-B* (pp. 675-678). Amsterdam: Elsevier Science.

Crosby, M., & Chin, D. (1999). Investigating user comprehension of complex multi-user interfaces. In H. J. Bullinger & J. Ziegler (Eds.), *Human-computer interaction: ergonomics and user interfaces: Vol. 1* (pp. 856-860). London: Lawrence Erlbaum Associates.

Crosby, M., & Iding, M. (1997a). A comparison of two individual differences measures and performance on a multimedia tutor for learning physics, *Computers and Education, Pergamon, 29*(23), 127-136.

Crosby, M., & Iding, M. (1997b). The influence of cognitive styles on the effectiveness of a multimedia tutor. *Computer Assisted Language Learning, 10*(4), 375-386. Lisse: Swets & Zeitlinger.

Crosby, M., Iding, M., & Chin, D. (2003, January 3-6). Research on task complexity as a foundation for augmented cognition. *Proceedings of HICSS 36,* Kona, Hawaii (pp. 1-9).

Crosby, M. E., Iding, M. K., & Chin, D. N. (2001). Visual search and background complexity: Does the forest hide the trees? In M. Bauer, P. J. Gmytrasiewicz, & J. Vassileva, (Eds.), *User modeling 2001* (pp. 225-227). Berlin; Heidelberg; New York: Springer-Verlag.

Crosby, M., & Ikehara, C. (2004, April 12-16). Continuous identity authentication using multi-modal physiological sensors. *Proceedings of the International Society for Optical Engineering (SPIE), Defense and Security Symposium,* Orlando, FL.

Copyright © 2006, Idea Group Inc. Copying or distributing in print or electronic forms without written permission of Idea Group Inc. is prohibited.

Crosby, M., Ikehara, C., & Chin, D. (2002, August 8-10). Measures of real-time assessment to use in adaptive augmentation. *Proceedings of the 24th Annual Meeting of the Cognitive Science Society,* VA.

Crosby, M., & Sophian, C. (2003). Processing spatial configurations in computer interfaces. In J. Hyona, R. Radach, & H. Deubel (Eds.), *The mind's eye: Cognitive and applied aspects of eye movement research* (pp. 517-530). Amsterdam: North Holland-Elsevier Science.

Crosby, M., & Stelovsky, J. (1995). From multimedia instruction to multimedia evaluation. *Journal of Educational Multimedia and Hypermedia, 4*(2/3), 147-162.

Crosby, M., Stelovsky, J., & Ashworth, D. (1994). Hypermedia as a facilitator for retention: A case study using Kanji City. *Computer Assisted Language Learning, 7*(1), 3-13. Lisse: Swets & Zeitlinger.

Crosby, M, Stelovsky, J., & Ashworth, D. (1996). Predicting language proficiency based on the use of multimedia interfaces for transcription tasks. *Computer Assisted Language Learning, 9*(2-3), 251-262. Lisse: Swets & Zeitlilnger.

Dryer, D. C. (1993). *Multi-dimensional and discriminant function analyses of affective state data.* Unpublished manuscript, Stanford University.

Dumais, S., & Wright, A. (1986). Reference by name vs. location in a computer filing system. *Proceedings of the Human Factors Society* (pp. 824-929).

Egan, D. (1988). Individual differences in human-computer interaction. In M. Helender (Ed.), *Handbook of human-computer interaction* (pp. 543-580). New York: Elsevier (North Holland).

Egan, D., & Gomez, L. (1985). Assaying, isolating, and accommodating differences. In R. Dillon (Ed.), *Individual differences in cognition* (pp. 174-216). Academic Press.

Ekstrom, R., French, J., & Harman, H. (1976). *Manual for kit of factor-referenced cognitive tests.* Princeton, NJ: Educational Testing Service.

Friedhoff, R., & Benzon, W. (1989). *The second computer revolution: Visualization.* New York: Harry W. Abrams.

Gomez, L., Egan, D., Wheeler, E., Sharma, D., & Gruchaz, A. (1983). How interface design determines who has difficulty learning to use a text editor. *Proceedings of Computer-Human Interaction (CHI) '83* (pp. 176-181).

Howard, D. L., & Crosby, M. (1993). Snapshots from the eye: Towards strategies for viewing bibliographic citations. In G. Salvendy & M. Smith (Eds.), *Advances in human factors/ergonomics: Human-computer interaction: Software and hardware interfaces: Vol. 19-B* (pp. 488-493). Amsterdam: Elsevier Science.

Ikehara, C., Chin, D., & Crosby, M. (2004, January 5-8). Modeling and implementing an adaptive human-computer interface using passive biosensors. *Proceedings of the 37th Annual Hawaii International Conference on System Sciences (HICSS 37),* Kona, Hawaii.

Ikehara, C., & Crosby, M. (2003, Jan. 3-6). User identification based on the analysis of the forces applied by a user to a computer mouse. *Proceedings of HICSS 36,* Kona, Hawaii.

Copyright © 2006, Idea Group Inc. Copying or distributing in print or electronic forms without written permission of Idea Group Inc. is prohibited.

Ikehara, C., & Crosby, M. (2005, January 6-9). Assessing cognitive load with physiological sensors. *Proceedings of the 38th Annual Hawaii International Conference on System Sciences (HICSS 38),* Kona, Hawaii.

Lange Küüttner, C. (1998). *Perceptual and Motor Skills,* 86 (3 Pt 2), 1299 310.

Mayer, R., & Sims, V. (1994). The role of spatial ability in learning with computer generated animations. *Journal of Educational Psychology, 86*(3), 389-401.

Nordbotten, J., & Crosby, M. (1997). Individual user differences in data model comprehension. In M. Smith, G. Salvendy, & R. Koubek (Eds.), *Design of computing systems: Social and ergonomic considerations: Vol. 21-B* (pp. 663-670). Amsterdam: Elsevier Science

Nordbotten, J., & Crosby, M. (1999). The effect of graphic style on data model interpretation. *Information Systems Journal, 9,* 139 - 155.

Picard, R. (1999). Affective computing for HCI. In H. J. Bullinger & J. Ziegler (Eds.), *Human-computer interaction: Ergonomics and user interfaces* (pp. 829-833). London: Lawrence Erlbaum Associates.

Quiroga, L., & Crosby, M. (2004). Information filtering. In W. S. Bainbridge (Ed.), *Berkshire encyclopedia of human-computer interaction* (pp. 351-355). National Science Foundation; Berkshire Publishing Group LLC: Great Barrington, MA.

Sheldon, E. (2001). *Virtual agent interactions.* Doctoral dissertation, University of Central Florida, Orlando.

Sein, M., & Bostrom, R. (1989). Individual differences and conceptual models in training novice users. *Human-Computer Interaction, 4,* 197-229.

Sein, M., Olfman, L., Bostrom, R., & Savis, S. (1993). The importance of visualization ability in predicting learning success. *International Journal of Man-Machine Studies, 39*(4), 599-620.

Sophian, C., & Crosby, M. (1999). A picture is worth more than two lines. In H. J. Bullinger & J. Ziegler (Eds.), *Human-computer interaction: Ergonomics and user interfaces: Vol. 1* (pp. 376-380). London: Lawrence Erlbaum Associates.

St. John, M., Kobus, D. A., & Morrison, J. G. (2003). *DARPA augmented cognition technical integration experiment* (Rep. No. TR-1905, Contract Number: N66001-99-D-0050, Pub.) Pacific Science and Engineering Group Inc. Retrieved from http://handle.dtic.mil/100.2/ADA420147

Vincente, K., Hayes, B., & Williges, R. (1987). Assaying and isolating individual differences in searching a hierarchical file system. *Human Factors, 29,* 647-668.

Copyright © 2006, Idea Group Inc. Copying or distributing in print or electronic forms without written permission of Idea Group Inc. is prohibited.

Section III

Human Factors

Copyright © 2006, Idea Group Inc. Copying or distributing in print or electronic forms without written permission of Idea Group Inc. is prohibited.

Chapter IX

Perceptual Multimedia:
A Cognitive Style Perspective

Gheorghita Ghinea, Brunel University, UK

Sherry Y. Chen, Brunel University, UK

Abstract

In this chapter, we describe the results of empirical studies which examined the effect of cognitive style on the perceived quality of distributed multimedia. We use two dimensions of Cognitive Style Analysis, Field Dependent/Independent and Verbaliser/ Visualiser, and the Quality of Perception metric to characterise the human perceptual experience. This is a metric which takes into account multimedia's infotainment (combined informational and entertainment) nature, and comprises not only a human's subjective level of enjoyment with regards to multimedia content quality, but also his/ her ability to analyse, synthesise and assimilate the informational content of such presentations. Results show that multimedia content and dynamism are strong factors influencing perceptual quality.

Copyright © 2006, Idea Group Inc. Copying or distributing in print or electronic forms without written permission of Idea Group Inc. is prohibited.

Introduction

Multimedia has been identified as a potential method of improving the learning process in particular and the user computing experience in general. Its use encourages user interaction, thus ensuring that users cannot become a passive participant of the learning experience (Neo & Neo, 2004). Not only does the use of multimedia in applications increase interaction levels, but increases interest, motivation, and retention of information (Demetriadis, Triantafilou, & Pombortis, 2003). Moreover, the fact that students seem to prefer the use of multimedia for teaching to the standard teacher-student paradigm as it is more user-centred (Zwyno, 2003), comes as no surprise. The potential of multimedia has, to date, not been fully realised. Users perceive, process, and organise information in individualistic ways, yet current multimedia applications routinely fail to take this into consideration (Stash, Cristea, & De Bra, 2004).

We therefore believe that the effectiveness of a multimedia presentation would be hindered if it did not include the user experience in terms of enjoyment and information assimilation. Key to this is the issue of the quality of the multimedia presentation. Quality, in our perspective, has two main facets in a distributed multimedia environment: *of service* and *of perception*. The former, Quality of Service (QoS), illustrates the technical side of computer networking and represents the performance properties that the underlying network is able to provide. The latter, Quality of Perception (QoP), characterises the perceptual experience of the user when interacting with multimedia applications. While the quality delivered by communication networks has traditionally been measured using QoS metrics, we believe that, as users are "consumers" of multimedia applications, it is their opinions about the quality of multimedia material visualised which ultimately measures the success (or indeed, failure) of such applications to deliver desktop instruction material. When this delivery is done over Wide Area Networks such as the World Wide Web ("the Web"), transmission of multimedia data has to accommodate not only user subjective preferences, but also fluctuating networking environments.

The other concern is whether distributed multimedia presentations can accommodate individual differences. Previous studies indicate that users with different characteristics have different perceptions of multimedia presentation (Chen & Angelides, 2003). In particular, different cognitive style groups benefit from different types of multimedia presentation. Therefore, empirical evaluation that examines the impact of cognitive styles becomes paramount because such evaluations can provide concrete prescriptions for developing learner-centred systems that can match the particular needs of each cognitive style group. While QoP has been investigated in the context of distributed multimedia quality (Ghinea & Thomas, 2005), the study did not take into account the possible effect of users' cognitive styles on their QoP.

In this chapter, we present the results of studies which looked at how multimedia content is perceived by different cognitive style groups. Accordingly, the chapter begins by building a theoretical background to present previous work in the area of subjective distributed multimedia quality and to discuss the influence of cognitive style on user perception of multimedia presentations. It then describes and discusses the findings of our empirical studies. The chapter ends with conclusions being drawn, highlighting the value of integrating QoP considerations with users' cognitive styles in the delivery of distributed multimedia presentations.

Copyright © 2006, Idea Group Inc. Copying or distributing in print or electronic forms without written permission of Idea Group Inc. is prohibited.

Theoretical Background

Quality of Service

The networking foundation on which current distributed multimedia applications are built either do not specify QoS parameters (also known as best effort service) or specify them in terms of traffic engineering parameters such as delay, jitter, and loss or error rates. However, these parameters do not convey application-specific needs such as the influence of clip content and informational load on the user multimedia experience.

Furthermore, traditional approaches of providing QoS to multimedia applications have focused on ways and means of ensuring and managing different technical parameters such as delay, jitter, and packet loss over unreliable networks. To a multimedia user, however, these parameters have little immediate meaning or impact. Although (s)he might be slightly annoyed at the lack of synchronisation between audio and video streams, it is highly unlikely that (s)he will notice, for instance, the loss of a video frame out of the 25 which could be transmitted during a second of footage, especially if the multimedia video in question is one in which the difference between successive frames is small. Moreover, in a distributed setting, the underlying communication system will not be able to provide an optimum QoS due to two competing factors, multimedia data sizes and network bandwidth. This results in phenomena such as congestion, packet loss, and errors. However, little work has been reported on the relationship between the network provided QoS and the satisfaction and perception of the user.

While the QoS impacts upon the perceived multimedia quality in distributed systems, previous work examining the influence of varying QoS on user perceptions of quality has almost totally neglected multimedia's *infotainment* quality (i.e., a mixture of both of informational as well as entertainment content), and has concentrated primarily on the perceived entertainment value of presentations displayed with varying QoS parameters.

Accordingly, previous work has studied the impact of varying clip frame rates on the user's enjoyment of multimedia applications (Apteker, Fisher, Kisimov, & Neishlos, 1995; Fukuda, Wakamiya, Murata, & Miyahara, 1997), and it has been shown that the dependency between human satisfaction and the required bandwidth of multimedia clips is non-linear. Consequently, a small change in human receptivity leads to a much larger relative variation of the required bandwidth. From a different perspective, Wijesekera and Srivastava (1996) and Wijesekera, Srivastava, Nerode, and Foresti (1999) have examined the effect that random media losses have on the user-perceived quality. Their work showed that missing a few media units will not be negatively perceived by a user, as long as too many such units are not missed consecutively and that this occurrence is infrequent. Moreover, because of the bursty nature of human speech (i.e. talk periods interspersed with intervals of silence), audio loss is tolerated quite well by humans as it results merely in silence elimination (21% audio loss did not provoke user discontent (Wijesekera et al., 1999). However, viewer discontent for aggregate video losses increases gradually with the amount of losses, while for other types of losses and synchronisation defects, there is an initial sharp rise in viewer annoyance that afterwards plateaus out.

Copyright © 2006, Idea Group Inc. Copying or distributing in print or electronic forms without written permission of Idea Group Inc. is prohibited.

Further work has been undertaken by Steinmetz (1996) who explored the bounds within which lip synchronisation can fluctuate without undue annoyance on the viewer's part, while the establishment of metrics for subjective assessment of teleconferencing applications was explored in Watson and Sasse (1996). Indeed, the correlation between a user's subjective ratings of differing-quality multimedia presentation and physiological indicators has been studied by Wilson and Sasse (2000). However, research has largely ignored the influence that the user's psychological factors have on the perceived quality of distributed multimedia.

The focus of our research has been the enhancement of the traditional view of QoS with a user-level defined QoP. This is a measure that encompasses not only a user's satisfaction with multimedia clips, but also his/her ability to perceive, synthesise, and analyse the informational content of such presentations. As such, we have investigated the interaction between QoP and QoS and its implications from both a user perspective as well as from a networking angle.

Cognitive Styles

Cognitive style is an individual's characteristic and consistent approach to organising and processing information. Riding and Rayner (1998) defined cognitive style as "an individual preferred and habitual approach to organising and representing information" (p.25). Among a variety of cognitive styles, Field Dependence/Independence and Visualiser/Verbaliser are related to perceptual multimedia. The former concerns how users process and organise information, whereas the latter emphasises on how users perceive the presentation of information.

Field Dependence/Independence is related to the "degree to which a learner's perception or comprehension of information is affected by the surrounding perceptual or contextual field" (Jonassen & Grabowski, 1993, p. 87). Field-dependent people tend to perceive objects as a whole, whereas Field -independent people focus more on individual parts

Table 1. The differences between field-dependent and field-independent learners (Adapted from Jonassen & Grabowski, 1993; Riding & Rayner, 1998)

Field Dependent Learners	Field Independent Learners
They are externally directed and are easily influenced by salient features.	They are internally directed and process information with their own structure.
They experience surroundings in a relatively global fashion and struggle with individual elements.	They experience surroundings analytically and are good with problems that require taking elements out of their whole context.
They are more likely to accept ideas as presented.	They are more likely to accept ideas only strengthened through analysis.

Copyright © 2006, Idea Group Inc. Copying or distributing in print or electronic forms without written permission of Idea Group Inc. is prohibited.

of the object. Field-dependent individuals rely more on external references; by contrast, field-independent individuals rely more on internal references (Witkin, Moore, Goodenough, & Cox, 1977). The differences between field-independent and field-dependent users are summarized in Table 1.

The main difference between visualiser/verbaliser focuses on a preference for learning with words versus pictures. A visualiser would prefer to receive information via graphics, pictures, and images, whereas a verbaliser would prefer to process information in the form of words, either written or spoken (Jonassen & Grabowski, 1993). In addition, visualisers prefer to process information by seeing, and they will learn most easily through visual and verbal presentations, rather than through an exclusively verbal medium. Moreover, their visual memory is much stronger than their verbal. On the other hand, verbalisers prefer to process information through words, and find they learn most easily by listening and talking (Laing, 2001). Their differences are summarised in Table 2.

These two dimensions of cognitive styles have been investigated by several works in learning environments. For example, a study by Chuang (1999) produced four-courseware versions: animation+text, animation+voice, animation+text+voice, and free choice. The result showed that field-independent subjects in the animation+text+voice group or in the free choice group scored significantly higher than those did in the animation+text group or in the animation+voice group. No significant presentation effect was found for the field-dependent subjects. Furthermore, Riding and Douglas (1993), with 15-16-year-old students, found that the computer-presentation of material on motorcar braking systems in a Text-plus-Picture format facilitated the learning by visualisers compared with the same content in a Text-plus-Text version. They further found that in the recall task in the Text-plus-Picture condition, 50% of the visualisers used illustrations as part of their answers, compared to only 12% of the verbalisers. Generally, visualisers learn best from pictorial presentations, while verbalisers learn best from verbal presentations. However, paucity of study investigates the relationship between the use patterns of these two dimensions of cognitive styles in multimedia systems in general, and specifically in distributed multimedia systems, where quality fluctuations can occur owing to dynamically varying network conditions. As the QoP metric is one which has an integrated view of user-perceived multimedia quality in such distributed systems, it is of particular interest to investigate the impact of cognitive styles on QoS-mediated QoP, as it will help in achieving a better understanding of the factors involved in such

Table 2. The differences between visualisers and verbalisers (Adapted from Jonassen & Grabowski, 1993; Riding & Rayner, 1998)

Visualisers	Verbalisers
Think concretely	Think abstractly
Have high imagery ability and vivid daydreams	Have low imagery ability
Like illustrations, diagrams, and charts	Like reading text or listening
Prefer to be shown how to do something	Prefer to read about how to do something
Are more subjective about what they are learning	Are more objective about what they are learning

Copyright © 2006, Idea Group Inc. Copying or distributing in print or electronic forms without written permission of Idea Group Inc. is prohibited.

environments (distance learning and CSCW, to name but two) and ultimately help in the elaboration of robust user models which could be used to develop applications that meet with individual needs.

Methodology Design

Overview

This chapter reports the results of studies which investigated how different multimedia content is perceived by different cognitive style dimensions. Perceived multimedia quality was examined using the QoP measure, the only such metric that takes into account multimedia's infotainment duality.

Measuring QoP

As previously mentioned, QoP has two components: an information analysis, synthesis, and assimilation part (henceforth denoted by *QoP-IA*) and a subjective level of enjoyment (henceforth denoted by *QoP-LOE*). To understand QoP in the context of our work, it is important to explain how both these components were defined and measured.

Measuring Information Assimilation (QoP-IA)

In our approach, QoP-IA was expressed as a percentage measure, which reflected a user's level of information assimilated from visualised multimedia content. Thus, after watching a particular multimedia clip, the user was asked a standard number of questions (10, in our case) which examined information being conveyed in the clip just seen, and QoP-IA was calculated as being the proportion of correct answers that users gave to these questions. All such questions asked must, of course, have definite answers, for example: (from the Rugby video clip used in our experiments) "What teams are playing?" had an unambiguous answer (England and New Zealand) which had been presented in the multimedia clip, and it was therefore possible to determine if a participant had answered this correctly or not.

Thus, by calculating the percentage of correctly-absorbed information from different information sources, it was possible to determine from which information sources participants absorbed the most information. Using this data, it is possible to determine and compare, over a range of multimedia content, potential differences that might exist in QoP-IA.

Copyright © 2006, Idea Group Inc. Copying or distributing in print or electronic forms without written permission of Idea Group Inc. is prohibited.

Measuring Subjective Level of Enjoyment (QoP-LOE)

The subjective level of enjoyment (QoP-LOE) experienced by a user when watching a multimedia presentation, was polled by asking users to express, on a scale of 1-6, how much they enjoyed the presentation (with scores of 1 and 6 respectively representing "no" and "absolute" user satisfaction with the multimedia video presentation).

In keeping with the methodology followed by Apteker et al (1995), users were instructed not to let personal bias towards the subject matter in the clip or production-related preferences (for instance, the way in which movie cuts had been made) influence their enjoyment quality rating of a clip. Instead, they were asked to judge a clip's enjoyment quality by the degree to which they, the users, felt that they would be satisfied with a general purpose multimedia service of such quality. Users were told that factors which should influence their quality rating of a clip included clarity and acceptability of audio signals, lip synchronisation during speech, and the general relationship between visual and auditory message components. This information was also subsequently used to determine whether ability to assimilate information has any relation to user level of enjoyment, the second essential constituent (beside information analysis, synthesis, and assimilation) of QoP.

Participants

This chapter brings together the results of two empirical studies conducted at Brunel University's School of Information Systems, Computing, and Mathematics. In the first study, participants' cognitive styles were categorised according to the Field Dependent/Independent dimension, while in the second, participants' cognitive styles were categorised using the Verbaliser/Visualiser dimension.

Experiment 1

Sixty-six subjects participated in this study. Despite the fact that the participants volunteered to take part in the experiment, they were extremely evenly distributed in terms of cognitive styles, including 22 field-independent users, 22 intermediate users, and 22 field-dependent users. In terms of gender, there were 34 male users and 32 female users.

Experiment 2

This study involved 71 participants, which turned out to be quite evenly distributed in terms of cognitive styles, including 23 field-independent users, 25 intermediate users, and 23 field- dependent users. Moreover, participant breakdown according to gender was also quite evenly matched (37 males and 34 females). For both studies, all participating users were inexperienced in the content domain of the multimedia video clips visualised as part of our experiments, which will be described next.

Copyright © 2006, Idea Group Inc. Copying or distributing in print or electronic forms without written permission of Idea Group Inc. is prohibited.

Research Instruments

Video Clips

A total of 12 video clips were used in our study. The multimedia clips were visualised under a Microsoft Internet Explorer browser with a Microsoft Media player plug-in, with users subsequently filling in a Web-based questionnaire to evaluate QoP for each clip.

These 12 clips had been used in previous QoP experiments (Ghinea & Thomas, 1998), and were between 30-44 seconds long and digitised in MPEG-1 format. The subject matter

Table 3. Video categories used in experiments

VIDEO CATEGORY	Dynamic	Audio	Video	Text
1 - Action Movie	Strong	Medium	Strong	Weak/None
2 - Animated Clip	Medium	Medium	Strong	Weak/None
3 - Band Clip	Medium	Strong	Mediu	Weak/None
4 - Chorus Clip	Weak	Strong	Mediu	Weak/None
5 - Commercial/Ad Clip	Medium	Strong	Strong	Medium
6 - Cooking Clip	Weak	Strong	Strong	Weak/None
7 - Documentary Clip	Medium	Strong	Strong	Weak/None
8 - News Clip	Weak	Strong	Strong	Medium
9 - Pop Music Clip	Medium	Strong	Strong	Strong
10 – Rugby Clip	Strong	Medium	Strong	Medium
11 - Snooker Clip	Weak	Medium	Mediu	Strong
12 - Weather Forecast Clip	Weak	Strong	Strong	Strong

Figure 1. Snapshots of clips used in experiments

Copyright © 2006, Idea Group Inc. Copying or distributing in print or electronic forms without written permission of Idea Group Inc. is prohibited.

they portrayed was varied (as detailed in Table 3 and Figure 1) and taken from selected television programmes, thereby reflecting informational and entertainment sources that average users might encounter in their everyday lives. Thus, six of the clips (2, 5, 6, 7, 8, and 12 in Table 1) comprised predominantly informational content, with the remainder of the clips being viewed mainly for entertainment purposes. Also varied was the dynamism of the clips (i.e., the rate of change between the frames of the clip), which ranged from a relatively static news clip to a highly dynamic space action movie. Table 3 also describes the importance, within the context of each clip, of the audio, video, and textual components as purveyors of information, as previously established through user tests (Ghinea & Thomas, 1998).

Cognitive Style Analysis

The cognitive style dimensions investigated in this study include Field Dependence/ Independence and Verbaliser/Visualiser. A number of instruments have been developed to measure these two dimensions. Riding's (1991) Cognitive Style Analysis (CSA) was applied to identify each participant's cognitive styles in this study, because the CSA offers computerised administration and scoring. In addition, the CSA can offer various English versions, including Australasian, North American, and United Kingdom contexts.

The CSA uses two sub-tests to identify Field Dependence/Independence. The first presents items containing pairs of complex geometrical figures that the individual is required to judge as either the same or different. The second presents items each comprising a simple geometrical shape, such as a square or a triangle, and a complex geometrical figure, as in the GEFT, and the individual is asked to indicate whether or not the simple shape is contained in a complex one by pressing one of two marked response keys (Riding & Grimley, 1999).. The first sub-test is a task requiring field-dependent capacity. Conversely, the second sub-test requires the disembedding capacity associated with field-independence.

The CSA uses two types of statement to measure the Verbal-Imagery dimension and asks participants to judge whether the statements are true or false. The first type of statement contains information about conceptual categories while the second describes the appearance of items. There are 48 statements in total covering both types of statement. Each type of statement has an equal number of true statements and false statements. It is assumed that visualisers respond more quickly to the appearance statements, because the objects can be readily represented as mental pictures and the information for the comparison can be obtained directly and rapidly from these images. In the case of the conceptual category items, it is assumed that verbalisers have a shorter response time because the semantic conceptual category membership is verbally abstract in nature and cannot be represented in visual form. The computer records the response time to each statement and calculates the Verbal-Visualiser Ratio. A low ratio corresponds to a verbaliser and a high ratio to a visualiser, with the intermediate position being described as biomodal.

Copyright © 2006, Idea Group Inc. Copying or distributing in print or electronic forms without written permission of Idea Group Inc. is prohibited.

This study followed Riding's recommendation for the measurements of Field Dependence/Independence and Verbaliser/Visualiser. In terms of Field Dependence/Independence, Riding's (1991) recommendations are that scores below 1.03 denote field-dependent individuals; scores of 1.36 and above denote field-independent individuals; students scoring between 1.03 and 1.35 are classed as Intermediate. Regarding the measurement of Verbaliser/Visualiser, the recommendations are scores below 0.98 denote verbalisers; scores of 1.09 and above denote visualisers; students scoring between 0.98 and 1.09 are classed as biomodal.

Procedure

The experiment consisted of several steps. Initially, the CSA was used to classify users' cognitive styles as Field Dependent /Intermediate/Field Independent (Experiment 1) or Verbaliser/Biomodal/Visualiser (Experiment 2). Subjects then viewed the 12 multimedia video clips. In order to counteract any order effects, the order in which clips were visualised was varied randomly for each participant. After the users had seen each clip once, the window was closed, and they had to answer a number of questions about the video clip they had just seen. The actual number of such questions depended on the video clip, and varied between 10 and 12. After the user had answered the set of questions pertaining to a particular video clip and the responses had been duly recorded, (s)he was asked to rate the enjoyment quality of the clip that had just been seen on a Likert scale of 1 - 6 (with scores of 1 and 6 representing the worst and, respectively, best perceived qualities possible). The user then went on and watched the next clip.

Users were instructed not to let personal bias towards the subject matter in the clip or production-related preferences (for instance, the way in which movie cuts had been made) influence their enjoyment quality rating of a clip. Instead, they were asked to judge a clip's enjoyment quality by the degree to which they, the users, felt that they would be satisfied with a general purpose multimedia service of such quality. Users were told that factors which should influence their quality rating of a clip included clarity and acceptability of audio signals, lip synchronisation during speech, and the general relationship between visual and auditory message components.

Data Analyses

In this study, the independent variables include the participants' cognitive styles, as well as clip categories and their degree of dynamism. The dependent variables were the two components of Quality of Perception: the level of understanding (QoP-IA, expressed as a percentage measure describing the proportion of questions that the user had correctly answered for each clip) as well as the level of enjoyment (QoP-LOE, expressed on a six-point Likert scale). Data were analysed with the Statistical Package for the Social Sciences (SPSS) for Windows version (release 9.0). An ANalysis Of VAriance (ANOVA), suitable to test the significant differences of three or more categories, and t-test, suitable to identify the differences between two categories (Stephen & Hornby, 1997), were applied to analyse the participants' responses. A significance level of $p < 0.05$ was adopted for the study.

Copyright © 2006, Idea Group Inc. Copying or distributing in print or electronic forms without written permission of Idea Group Inc. is prohibited.

Discussion of Results

Subject Content

Experiment 1

Our analysis has highlighted that the subject content (i.e., particular clip category) has a statistically significant impact on the QoP-IA level of participants (p=.0000). This confirms previous results (Ghinea & Thomas, 1998) and we extended our analysis to include the impact of users' cognitive styles. As depicted in Table 4, the clip on which participants performed best varied according to the users' cognitive style. Accordingly, intermediate and field-independent users had the highest level of understanding for the Snooker video clip. However, field- dependent users perform better in the Documentary clip. As Table 1 shows, the most discriminating feature between the two clips is the fact that the latter clip does not contain any textual description. The difference in the understanding corresponding to the different cognitive styles is thus probably due to the fact that field-dependent learners are "influenced by salient features" (Jonassen & Grabowski, 1993, p.88). Without the distraction of text description, field-dependent users could concentrate their learning on video clips, so they could have better performance. On the other hand, irrespective of the type of cognitive style, all users performed worst in the highly dynamic Rugby sports action clip, in which, as Table 4 shows, all media components were medium-strong purveyors of information. This finding seems to imply that users have difficulty concentrating on multiple, different sources of information, and that the dynamism of the clip is also a contributing factor to participants' level of understanding, as was confirmed by further analysis, presented in the section titled "Degree of Dynamism"..

The specific multimedia clip type also influences the QoP-LOE, namely the level of enjoyment experienced by users (p=.0000). As Table 5 shows, although the Documentary and Rugby video clips predominate in the "Most Enjoyed" and "Least Enjoyed" categories, only for field-dependent users does the choice of most/least enjoyed clip coincide with the clips on which the level of understanding is highest, respectively lowest (see Table 4, in this regard). These results are in line with those of Fullerton (2000) and Ford and Chen (2001), which showed that field- dependent users performed better in a learning environment matching their preferences; conversely, their performance will be reduced in a mismatched condition. Moreover, our results also show that the Forecast

Table 4. Cognitive styles and QoP-IA in clip categories

	Field Dependent	Intermediate	Field Independent
Best Performance	Documentary	Snooker	
	63.84%	62.93%	63.46%
Worst Performance	Rugby		
	33.24%	37.5%	34.82%

Copyright © 2006, Idea Group Inc. Copying or distributing in print or electronic forms without written permission of Idea Group Inc. is prohibited.

Table 5. Cognitive styles and QoP-LOE of clip categories

	Field Dependent	Intermediate	Field Independent
Most Enjoyed	Documentary		Forecast
	3.18	2.86	3.02
Least Enjoyed	Rugby	Band	Rugby
	1.39	1.93	1.61

video clip was the one which field-independent users enjoyed most. It is probable that the wealth of detail present in this clip (information was conveyed through all three channels, video, audio, and text) is what makes the clip appealing to this particular type of users, who concentrated primarily on procedural details when processing information in a learning context (Pask, 1976, 1979). This is in contrast to the "Most Enjoyed" clip (i.e. Documentary) for the other two categories of users, in which information was only conveyed through the video and audio streams.

Experiment 2

Our results indicate that clip categories, as given by their specific multimedia content matter, significantly influence participants' components of QoP-IA. This shows that the information assimilation scores are significantly influenced by the content being visualised; moreover, this observation is valid irrespective of the particular cognitive style of the participant. However, closer analysis reveals that different cognitive style groups have different favourite clips. Pop Music, which displays information using multiple channels, including video, audio, and text, is the favourite clip, from an information assimilation point of view, for biomodals who combine the characteristics of both verbalisers and visualisers and are particularly adept at receiving information from either textual descriptions or graphic presentations. However, we did obtain some significant results that contradict those of previous research (Laing, 2001; Riding & Watts, 1997) (Tables 6 and 7). Although the Documentary clip does not display any text description, it is the clip in which, on average, verbalisers obtain the highest QoP-IA ($F=10.592$, $df=5,40$, $p=.000$). On the other hand, visualisers perform better in the Snooker clip, which, though static, includes information conveyed through video, audio, and text ($F=14.8451$, $df=6,36$, $p=.000$).

However, irrespective of cognitive style, we found that the Rugby clip was the one in which participants obtained the lowest QoP-IA scores ($F=32.743$, $df=15,72$, $p=.000$). Although this clip is similar in some respects to others studied by us (such as the Snooker clip, which also has an abundance of information being portrayed through video, audio, and textual means), its main distinguishing feature is its high dynamism; there is considerable temporal variability due to the high inter-frame differences specific to clips featuring action sports. We therefore assume that the reason why participants scored so lowly in terms of QoP-IA on this clip is precisely because of its high dynamism, a hypothesis that shall be further explored in the section titled "Degree of Dynamism".

Copyright © 2006, Idea Group Inc. Copying or distributing in print or electronic forms without written permission of Idea Group Inc. is prohibited.

Table 6. Favourite clips

	Verbaliser	Biomodal	Visualiser
Score	62.84%	62.98%	65.52%
	Documentary	Snooker	
Enjoyment	4.17	3.94	3.91
	Documentary	Pop Music	Documentary

Table 7. Least favourite clip

	Verbaliser	Biomodal	Visualiser
Score	35.59%	35.74%	34.49%
		Rugby	
Enjoyment	2.59	2.78	2.81
		Rugby	

Enjoyment will also influence users' performance, especially for verbalisers, who perform better and enjoy more the Documentary clip and performed worse and enjoyed less the Rugby clip. It is consistent with the results of previous research (Chen, 2002), which highlight that positive user perceptions of the learning environment can enhance their performance; conversely, negative attitudes will tend to hinder learning performance.

Degree of Dynamism

Multimedia clip dynamism was found to significantly impact upon participants' QoP-IA levels in our study (Figure 2). The correct one is that the level of significance was found to be p=.000 for field-dependent users and p=.001 for intermediate and field-independent users. Clip dynamism is given by Table 3, where the terms strong, medium, and weak were coded with the values of 3, 2, and 1, respectively. All users performed worst in the clips with strong dynamism. In particular, field- dependent users do not perform as well as field-independent and intermediate users. As suggested by previous works (Chen, 2002; Chen & Macredie, 2002), field-dependent users' performance was hindered in situations where they need to extract cues by themselves. Thus, in multimedia clips with strong dynamism that provided too many cues, field-dependent users might find it difficult to select relevant cues.

The dynamism of the visualised clips also influenced the level of enjoyment experienced by participants (p=.000). If a per-cognitive style analysis is pursued, we find that the level of enjoyment is influenced by the dynamism of the multimedia clip for both field-independent (p=.004) and field-dependent (p=.000) users. As shown in Figure 3, both field-independent and field-dependent users experienced higher levels of enjoyment from the clips with medium dynamism, while strongly dynamic clips were liked least of

Copyright © 2006, Idea Group Inc. Copying or distributing in print or electronic forms without written permission of Idea Group Inc. is prohibited.

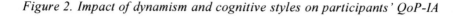

Figure 2. Impact of dynamism and cognitive styles on participants' QoP-IA

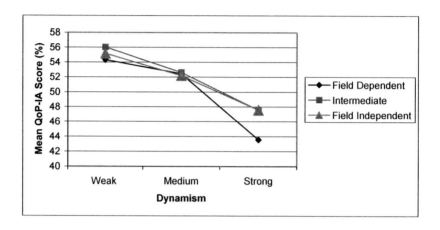

all. However, dynamism does not seem to be a factor influencing multimedia clip enjoyment of intermediate users. One possible interpretation is that individuals possessing an Intermediate cognitive style employ a more versatile repertoire of information-seeking strategies. Versatile users, who have acquired the skill to move back and forth between different information-seeking strategies, are more capable of adapting themselves to suit the subject content presented by the multimedia video clips. This finding is consistent with the views of previous work, namely that a versatile strategy can be better equipped for multimedia learning technology (Chen & Ford, 1998; Paterson, 1996).

Analysis of the results obtained from the experiments shows that the degree of clip dynamism significantly impacts upon the QoP-IA component of QoP, irrespective of the user's cognitive style. The analysis has highlighted, moreover, the fact that the highest QoP-IA scores are obtained for clips which have a low degree of dynamism. Conversely, multimedia clips which have a high degree of dynamism have a negative impact on the user assimilation of the informational content being conveyed by the respective clips (Figure 4). Thus, clips which have relatively small inter-frame variability will facilitate higher QoP-IA scores: An object which might appear for only 0.5 seconds in a highly dynamic clip is less easily remembered than is the case when it appears for one second in a clip which is less dynamic.

As far as the QoP-LOE component is concerned, analysis of our results reveals that while dynamism is a significant factor in the case of verbalisers and visualisers, this is not true of biomodals (Figure 5). As suggested by previous research (Riding & Rayner, 1998), biomodals can tailor learning strategies to the specific learning environments so the features of learning environments have no significant effects on their enjoyment. For verbalisers and visualisers, however, it was found that clips of medium dynamism had the highest levels of QoP-LOE, which suggests that such users do not find enjoyable clips which are static (or, conversely, highly dynamic). While the user may feel somewhat overwhelmed by a fast-paced clip, (s)he might possibly feel uninterested by a static clip with almost repetitive frame displays; it should come as no surprise, then, that such users

Copyright © 2006, Idea Group Inc. Copying or distributing in print or electronic forms without written permission of Idea Group Inc. is prohibited.

Figure 3. Impact of dynamism and cognitive styles on participants' QoP-LOE

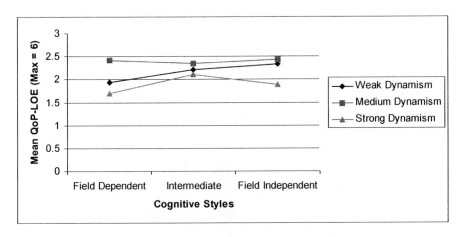

Figure 4. Cognitive styles and clip dynamism impact on QoP-IA

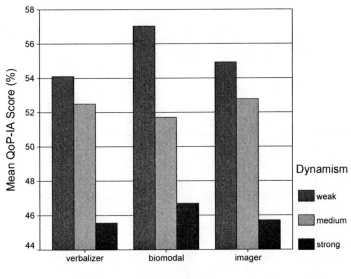

Copyright © 2006, Idea Group Inc. Copying or distributing in print or electronic forms without written permission of Idea Group Inc. is prohibited.

Figure 5: Cognitive styles and clip dynamism impact on QoP-LOE

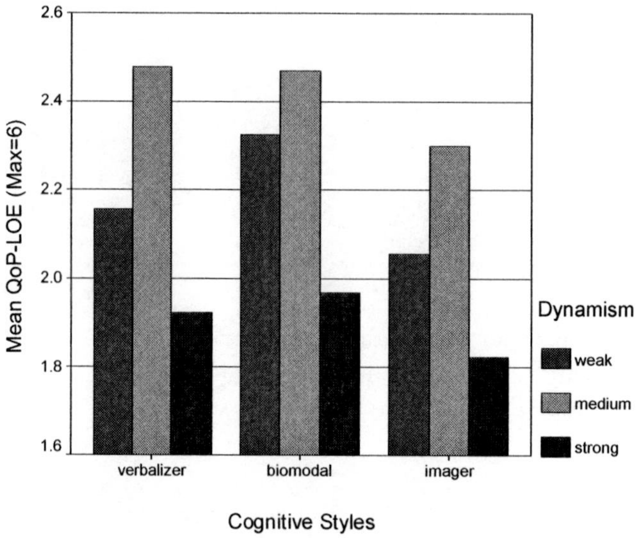

prefer clips of medium dynamism, where they do not feel overwhelmed, but neither are they bored by the presentation of the subject matter concerned.

Summary

This chapter has presented the results of two studies that looked at the impact of cognitive styles and multimedia content on users' subjective Quality of Perception. The latter is a term which encompasses not only a user's enjoyment, but also his/her level of understanding of visualised multimedia content. Each of the studies used a different dimension for categorising participants' cognitive style – thus, the first study used the Field Dependent/ Independent categorisation, whilst the second employed the Verbaliser/ Visualiser dimension.

Our results reveal that multimedia video clip dynamism is an important factor impacting, irrespective of the particular dimension of cognitive style being employed, upon participants' QoP-IA levels. A similar conclusion as regards QoP-LOE can only be made, however, if the Field Dependent/Independent dimension is used. If one uses the Verbaliser/Visualiser dimension to classify cognitive style, clip dynamism has no significant effects on Biomodals, which, displaying characteristics of both Verbalisers and Visualisers, have adaptable preferences of accessing information and enjoy receiving information from multiple channels.

Copyright © 2006, Idea Group Inc. Copying or distributing in print or electronic forms without written permission of Idea Group Inc. is prohibited.

However, what our results do highlight, independent of the cognitive style taxonomy being used, is that multimedia content does have a significant influence on the user experience, as measured by the QoP metric. This would imply that in order to deliver an enhanced multimedia infotainment experience, multimedia content providers should focus on relatively static multimedia video and take into consideration the appropriateness of the subject matter in order to aid in the uptake and proliferation of distributed multimedia.

References

Apteker, R. T., Fisher, J. A., Kisimov, V. S., & Neishlos, H. (1995). Video acceptability and frame rate. *IEEE Multimedia, 2*(3), 32-40.

Chen, S. Y. (2002). A cognitive model for non-linear learning in hypermedia programmes. *British Journal of Educational Technology, 33*(4), 453-464.

Chen, S. Y., & Angelides, M. C. (2003). Customisation of Internet multimedia information systems design through user modelling. In S. Nansi (Ed.), *Architectural issues of Web-enabled electronic business* (pp. 241-255). Hershey, PA: Idea Group Publishing.

Chen, S. Y., & Ford, N. J. (1998). Modelling user navigation behaviours in a hypermedia-based learning system: An individual differences approach. *International Journal of Knowledge Organization, 25*(3), 67-78.

Chen, S. Y., & Macredie, R. D. (2002). Cognitive styles and hypermedia navigation: Development of a learning model. *Journal of the American Society for Information Science and Technology, 53*(1), 3-15.

Chuang, Y. R. (1999). *Teaching in a multimedia computer environment: A study of effects of learning style, gender, and math achievement.* Retrieved December 21, 2001, from http://imej.wfu.edu./articles/1999/1/10/index/asp

Demetriadis, S., Triantafilou, E., & Pombortis, A. (2003, June 30-July 2). A phenomenographic study of students' attitudes toward the use of multiple media for learning. *ACM SIGCSE Bulletin, Proceedings of the 8th Annual Conference on Innovation and Technology in Computer Science Education,* Thessaloniki, Greece (pp. 183-187).

Ford, N., & Chen, S. Y. (2001). Matching/mismatching revisited: An empirical study of learning and teaching styles. *British Journal of Educational Technology, 32*(1), 5-22.

Fukuda, K., Wakamiya, N., Murata, M., & Miyahara, H. (1997). QoS mapping between user's preference and bandwidth control for video transport. *Proceedings of the 5th International Workshop on QoS (IWQoS)* (pp. 291-301).

Fullerton, K. (2000). *The interactive effects of field dependence-independence and Internet document manipulation style on student achievement from computer-based instruction.* Doctoral dissertation, University of Pittsburgh.

Copyright © 2006, Idea Group Inc. Copying or distributing in print or electronic forms without written permission of Idea Group Inc. is prohibited.

Ghinea, G., & Thomas, J. P. (2005). Quality of perception: User quality of service in multimedia presentations. *IEEE Transactions on Multimedia, 7*(4), 786-789.

Jonassen, D. H., & Grabowski, B. (1993). *Individual differences and instruction.* New York: Allen & Bacon.

Laing, M. (2001). Teaching learning and learning teaching: An introduction to learning styles. *New Frontiers in Education, 31*(4), 463-475.

Neo, T., & Neo, M. (2004). Classroom innovation: Engaging students in interactive multimedia learning. *Campus-Wide Information Systems, 21*(3), 118-124.

Pask, G. (1976). Styles and strategies of learning. *British Journal of Educational Psychology, 46,* 128-148.

Pask, G. (1979). *Final report of S.S.R.C. Research Programme HR 2708.* Richmond (Surrey): System Research Ltd.

Paterson, P. (1996). *The influence of learning strategy in a computer mediated learning environment.* Paper presented at ALT-Conference '96. Retrieved November 19, 1997, from http://www.warwick.ac.uk/alt-/alt-96/papers.html

Riding, R. J. (1991). *Cognitive styles analysis.* Birmingham: Learning and Training Technology.

Riding, R. J., & Douglas, G. (1993). The effect of cognitive style and mode of presentation on learning performance. *British Journal of Educational Psychology, 63,* 297-307.

Riding, R. J., & Grimley, M. (1999). Cognitive style and learning from multimedia materials in 11 year old children. *British Journal of Educational Technology, 30*(2), 43-56.

Riding, R. J., & Rayner, S. G. (1998). *Cognitive styles and learning strategies.* London: David Fulton Publisher.

Riding, R. J., & Watts, M. (1997). The effect of cognitive style on the preferred format of instructional material. *Educational Psychology, 17*(1 & 2), 179-183.

Stash, N., Cristea, A., & De Bra, P. (2004, May). Authoring of learning styles in adaptive hypermedia: Problems and solutions. *Proceedings of the Thirteenth International World Wide Web Conference,* New York, (pp 114-123).

Steinmetz, R. (1996). Human perception of jitter and media synchronisation. *IEEE Journal on Selected Areas in Communications, 14*(1), 61-72.

Watson, A., & Sasse, M. A. (1996). Evaluating audio and video quality in low cost multimedia conferencing systems. *Interacting with Computers, 8*(3), 255-275.

Wijesekera, D., & Srivastava, J. (1996). Quality of service (QoS) metrics for continuous media. *Multimedia Tools and Applications, 3*(1), 127-136.

Wijesekera, D., Srivastava, J., Nerode, A., & Foresti, M. (1999). Experimental evaluation of loss perception in continuous media. *Multimedia Systems, 7*(6), 486-499.

Wilson, G. M., & Sasse, M. A. (2000, December). Investigating the impact of audio degradations on users: Subjective vs. objective assessment methods. *Proceedings of OZCHI2000,* Sydney (pp. 135-142).

Copyright © 2006, Idea Group Inc. Copying or distributing in print or electronic forms without written permission of Idea Group Inc. is prohibited.

Witkin, H. A., Moore, C. A., Goodenough, D. R., & Cox., P. W. (1977). Field-dependent and field independent cognitive styles and their educational implications. *Review of Educational Research, 47,* 1-64.

Zwyno, M. S. (2003, November). Student learning styles, Web use patterns, and attitudes towards hypermedia-enhanced instruction. *Proceedings of the 33rd ASEE/IEEE Frontiers in Education Conference* (pp. 1-6).

Copyright © 2006, Idea Group Inc. Copying or distributing in print or electronic forms without written permission of Idea Group Inc. is prohibited.

Chapter X

Expert-Novice Differences and Adaptive Multimedia

Slava Kalyuga, University of New South Wales, Australia

Abstract

This chapter provides an overview of theoretical frameworks and empirical evidence for the design of adaptive multimedia that is tailored to individual levels of user expertise to optimize cognitive resources available for learning. Recent studies indicate that multimedia design principles that benefit low-knowledge users may disadvantage more experienced ones due to increase in cognitive load required for integration of presented information with available knowledge base. The major implication for multimedia design is the need to tailor instructional formats to individual levels of expertise. The suggested adaptive procedure is based on empirically-established interactions between levels of user proficiency and formats of multimedia presentations (the expertise reversal effect), and on real-time monitoring of users' expertise using rapid cognitive diagnostic methods.

Copyright © 2006, Idea Group Inc. Copying or distributing in print or electronic forms without written permission of Idea Group Inc. is prohibited.

Introduction

When the same digital multimedia materials are presented to novices and experts in a domain, these two categories of users would perceive the presentation rather differently. Imagine a sufficiently complex electrical wiring diagram displayed on a computer screen. Experienced electrical engineers, even if they have not seen this particular diagram before, may immediately recognize familiar functional parts and see the diagram as a representation of a whole device or system. They would be able to rapidly figure out how the circuit operates and what it is used for. On the other hand, first-year electrical trade apprentices may at best see only a set of familiar elements (resistors, switches, etc.) without understanding why all these elements are connected in this particular way. (Let us leave out laypersons that would see a collection of lines, circles, rectangles, and other strange symbols). Obviously, the apprentices need detailed explanations of the circuit operation to understand the diagram.

Assume the above categories of users are provided with well-developed comprehensive multimedia instructions (for example, constructed according to multimedia design principles of Mayer, 2001, 2005). The instructions may have the diagram animated and include appropriately-placed (synchronized with animations) narrated auditory explanations. Such multimedia materials would certainly help the novice trainees to understand the operation of the circuit. However, when the same instructions are presented to the experienced engineers, they may find them rather annoying, especially if it is not possible to avoid the instructional details. If post-instruction test tasks are used to assess the understanding of the operation of the circuit, we could find that experts who were presented with the diagram-only format without any explanations outperform the experts who studied the multimedia instructional message. In fact, such relative declines in experts' learning were observed on many occasions (see the following section for some references). The effect was related to changes in domain-specific knowledge base that people acquire as they become more experienced in the domain, and to the role this knowledge plays in guiding their cognitive performance.

According to a large number of studies in cognitive psychology conducted over many decades, our knowledge base is the most important factor influencing our perception of incoming information. Some recent evidence of "early vision" that is independent of knowledge structures (e.g., see Pylyshyn, 1999, for an overview) would not change the general significance of the knowledge effect, especially during complex cognitive activities. According to another impressive set of cognitive studies initiated by investigations of the nature of chess expertise (de Groot, 1966), the knowledge base is a single most important factor determining expert-novice differences in cognitive performance (see Chi, Glaser, & Farr, 1988, for an overview). Finally, studies of information processing limitations imposed by the structure of human cognitive architecture (originated by Miller, 1956), and the ways we deal with these limitations by reorganizing our knowledge base and using multiple (e.g., visual and auditory) modalities, provide another theoretical framework and empirical evidence to complete the picture of expert-novice differences in perceiving multimedia messages.

This chapter starts with a brief overview of major theoretical issues and empirical evidence that are essential for understanding multimedia design implications of expert-

Copyright © 2006, Idea Group Inc. Copying or distributing in print or electronic forms without written permission of Idea Group Inc. is prohibited.

novice differences. The most important consequence of the reviewed studies is that the design of effective and cognitively-efficient multimedia environments needs to be tailored to changing levels of user expertise in a domain in order to optimize cognitive resources available for understanding multimedia messages. Empirically-established interactions between levels of user expertise and different formats of multimedia presentations are described. Also, a brief overview of recently- developed technical adaptation solutions based on user modeling in the adaptive hypermedia field is provided. Finally, the chapter discusses a possible adaptive methodology that is based on real-time monitoring of users' proficiency in a domain by using rapid cognitive diagnostic methods for capturing authentic domain-specific knowledge structures involved in processing presented information. This diagnostic approach may have the potential for developing more rapid and sensitive knowledge-tracing techniques than traditional tests. It could be used to increase the accuracy of information about levels of knowledge and expertise stored in an individual user model. To illustrate the approach, the rapid diagnostic method has been applied (in a preliminary pilot study) as a means of tailoring instructions to levels of learner expertise in a simple adaptive computer-based tutor in kinematics.

Expert-Novice Differences in Processing Multimedia Information

Two major components of our cognitive architecture that are directly related to processing multimedia information are working memory and long-term memory. Working memory provides temporary storage and transformation of verbal and pictorial information that is currently in the focus of our attention (e.g., constructing and updating mental representations of a current situation or task). If too many elements of information are processed simultaneously in working memory, its capacity may become overloaded (Baddeley, 1986; Miller, 1956). Processing limitations of working memory and associated cognitive load represent a major factor influencing the effectiveness of learning (Sweller, 1999; Sweller, van Merrienboer, & Paas, 1998). Presenting related elements of information (e.g., pictures and related words) in alternative modalities (visual and auditory) may reduce cognitive load by employing two relatively independent sub-systems of working memory responsible for dealing with visual and auditory information.

In order to overcome limitations of working memory and reduce associated cognitive load, organized domain-specific knowledge structures in long-term memory allow treating multiple elements of information as a single higher-level element (Chi, Glaser, & Rees, 1982). Such structures also allow experts to rapidly classify problem situations and retrieve appropriate schematic procedures for handling these situations instead of employing cognitively demanding and inefficient search-based strategies that novices usually use. For example, studies of problem solving in physics by individuals with different levels of expertise demonstrated that experts approached the problems in terms of the basic principles of physics, while novices heavily depended on surface features mentioned in each specific task (Chi, Feltovich, & Glaser, 1981). In addition, experts are also able to bypass working memory limitations by having their knowledge structures

Copyright © 2006, Idea Group Inc. Copying or distributing in print or electronic forms without written permission of Idea Group Inc. is prohibited.

in long-term memory highly automated due to extensive practice. To emphasize the decisive role of long-term memory in expert performance, Ericsson and Kintsch (1995) proposed the theory of long-term working memory (LTWM). According to this theory, long-term memory structures associated with components of working memory create a LTWM structure that is capable of holding virtually unlimited amounts of information.

Recent studies of the *expertise reversal effect* (see Kalyuga, 2005; Kalyuga, Ayres, Chandler, & Sweller, 2003, for an overview) have demonstrated that information or a learning procedure that is beneficial for novice learners may become redundant for more knowledgeable learners. The expertise reversal effect can be related to research on aptitude-treatment interactions (e.g., Cronbach & Snow, 1977; Shute, 1992) that occur when different instructional treatments result in different learning outcomes depending on student aptitudes (knowledge, skills, learning styles, personality characteristics, etc.). In the expertise reversal effect, prior knowledge is the aptitude of interest. The effect can be explained by assuming that for more knowledgeable learners, the redundant material or instructional guidance overloads working memory relative to information without redundancy because resources are required for cross-referencing presented and previously-learned information. Accordingly, cognitive efficiency of multimedia presentations is relative to levels of user proficiency in a domain. Using appropriate procedures and removing redundant information at each level of user expertise, thus minimizing interfering cognitive load, is necessary for optimizing cognitive resources when designing multimedia presentations.

For example, in a set of studies conducted with technical apprentices of a manufacturing company (Kalyuga, Chandler, & Sweller, 2000), detailed auditory explanations of procedures for using specific types of diagrammatic representations (cutting speed nomograms) that were presented simultaneously with animated diagrams were cognitively optimal multimedia instructional formats for novice trainees. However, at higher levels of expertise achieved after a series of intensive training sessions, when cognitive activities of the same users were based on well-learned schematic procedures, presenting a slightly different type of nomograms with detailed auditory explanations was suboptimal. Explanations designed to support construction of schematic knowledge structures that had already been acquired by trainees were redundant and inefficient.

According to cognitive theories of multimedia learning (Mayer, 2001, 2005; Sweller, 1999), when text and pictures are not synchronized in space (located separately) or time (presented after or before each other), the integration process may increase cognitive load due to cross-referencing different representations. Physically integrating verbal and pictorial representations may eliminate this split-attention effect (Mayer & Gallini, 1990; Sweller, Chandler, Tierney, & Cooper, 1990). Therefore, a cognitively-optimal design of multimedia presentations for novice users usually requires eliminating situations when attention is split between multiple complementing information representations (e.g., on-screen text and diagrams) by embedding sections of textual explanations directly into the diagram in close proximity to relevant components of the diagram. Alternatively, dual-modality formats should be used with segments of narrated text presented simultaneously with the diagram (or relevant animation frames). Also, providing detailed instructional guidance by using plenty of fully worked-out examples at the initial stages of learning is required for novice learners (Sweller, et al., 1998). On the other

Copyright © 2006, Idea Group Inc. Copying or distributing in print or electronic forms without written permission of Idea Group Inc. is prohibited.

hand, recently established cognitive load effects in multimedia design for more advanced learners suggest eliminating non-essential redundant representations in multimedia formats and gradually reducing levels of guidance by increasing the relative share of problem-based and exploratory environments as levels of user proficiency in the domain increase (Kalyuga, 2005).

Thus, studies of expert-novice differences have demonstrated that organized schema-based knowledge structures in long-term memory are the most critical factor influencing proficient performance. These cognitive constructs effectively reduce or eliminate severe processing limitations of our cognitive system and fundamentally alter charac-teristics of our performance. They guide allocation of attentional resources and signifi-cantly influence our perception of multimedia materials. Non-optimal multimedia formats may overload limited attentional capacity of working memory. As a consequence, multimedia presentations which include information that is essential and appropriate for novices, may need to be re-designed by eliminating redundant information for more expert individuals in order to optimize cognitive resources.

An important implication of these findings is that multimedia needs to be tailored to levels of user expertise in a domain. To be able to dynamically select multimedia formats optimal for individual users, it is necessary not only to understand cognitive mechanisms that influence efficiency of multimedia information presentations, but also to have suitable methods for collecting information about user levels of proficiency in a domain suitable for real-time applications.

User Modelling in Adaptive Hypermedia Environments

Hypermedia systems add navigation support to traditional linear multimedia environ-ments. This capability provides appropriate levels of user interactivity and user control implemented as an organized network of hyperlinks that allow nonlinear access to graphics, sound, animation, and other multimedia elements. Adaptive hypermedia environments accommodate user characteristics (knowledge, interests, goals, etc.) into an explicit user model and then use this model to adapt interactions with each user to her or his characteristics and needs, for example, by providing adaptive content selection and presentation, or suggesting a set of most relevant links to proceed (see Brusilovsky, 2001; De Bra & Calvi, 1998; Kobsa, 2001, for comprehensive overviews of the field). Adapting the content modality to an individual user (selecting the most relevant modes of presentation from text, narration, animation, video, etc.) is an important part of adaptive presentation techniques based on the user-modeling technology.

User models (student or learner models in learning systems) represent the key component of an adaptive hypermedia system. These models are multi-dimensional constructs that may include many different user characteristics in addition to subject matter knowledge, for example, level of computer literacy, experience in using specific software applications, learning styles, background, preferences, goals, interests, and so forth. User models are

Copyright © 2006, Idea Group Inc. Copying or distributing in print or electronic forms without written permission of Idea Group Inc. is prohibited.

usually constructed by using traditional testing and survey methods, or recording the history of user interactions with the system (e.g., browsing behaviour or navigation trace) to determine users' knowledge and experience, background, interests, preferences, learning styles, and other characteristics. These models are regularly updated as users work their way through the environment. User models are utilized by the system to individualize components of the content and user activities (the domain model) according to a specified adaptive methodology (the adaptation model).

For example, AHA (Adaptive Hypermedia Architecture) system (De Bra & Calvi, 1998) includes an engine that maintains a user model based on knowledge of the concepts involved in a domain. The model is generated as the user reads pages and takes tests. Depending on the user's knowledge, different fragments of the learning material are presented. The user is guided towards more appropriate pages that contain information most relevant at that time by hiding, removing, or disabling less appropriate links, or by providing adaptive link annotations (e.g., by using a specific color scheme for desired, undesired, neutral, or external links). ELM-ART and InterBook adaptive hypermedia learning environments (Brusilovsky, Eklund, & Schwarz, 1998) use history-based, knowledge-based, and prerequisite-based adaptive annotations of links to suggest a best path through a learning space. Adaptive navigation support adjusts the links accessible to a particular learner using such techniques as direct guidance, adaptive link sorting, adaptive link hiding, removal, or disabling. Adaptive tables of contents, known and required concepts, adaptive content pages, and adaptive messages about the educational status of a page (e.g., warning that the page is not yet ready to be learned) are provided.

Levels of user domain expertise are usually represented by the knowledge component of traditional user models. Because domain-specific knowledge is a major factor that directly influences learning processes, it is usually included in most student models. However, the way it is modeled and the levels of granularity of the models vary considerably. In most cases, they are rather coarse-grained representations using a few numeric or categorical values (e.g., high, intermediate, low levels, or no knowledge; or just Booleans yes or no) for a few concepts. Even systems that allow many values (e.g., percentage values from 0 to 100) use only a few discrete levels in the actual adaptation process (De Bra & Calvi, 1998). Initial information about user knowledge is usually obtained from tests at the beginning of the first session or is set as default values. Thereafter, the system updates the level of knowledge in the user model based on direct assessment tests or history of student actions (e.g., number of reattempts during task solutions, number of requests for help, etc.). The adaptation model then uses the updated knowledge levels to adjust multimedia presentations for individual users.

The accuracy of information in user models is one of the defining factors that influence quality of adaptive environments. An important direction of improvement of user models for adaptive hypermedia (and multimedia) environments is constructing richer and more diagnostically-informative models that capture the nature and levels of user proficiency more precisely. Using traditional (mostly multiple-choice) tests and tracing sequences of mouse clicks provide rather limited sources of diagnostic information. Analyses of student solutions to presented problems usually deal with final answers to those problems without considering details of how those answers were actually obtained. The

Copyright © 2006, Idea Group Inc. Copying or distributing in print or electronic forms without written permission of Idea Group Inc. is prohibited.

data available from tracing user interactions with the system are usually imprecise, incomplete, and uncertain. Applying modern artificial intelligence approaches and methods (e.g., machine learning, Bayesian inference networks, neural networks, etc.) could help increase the precision of adaptive technologies. For example, intelligent solution analyses could diagnose missing or defective components of knowledge and skill, and provide learners with more accurate feedback and support. On the other hand, quality of adaptive environments could also be improved by developing new cognitive diagnostic techniques to replace traditional assessment methods used in constructing user models. The following sections describe a possible implementation of this approach.

Rapid Diagnostic Method for Tailoring Multimedia to Levels of User Expertise

The research on expertise emphasizes the importance of diagnosing domain-specific organized knowledge structures when evaluating levels of proficiency. Traditional methods of knowledge assessment are usually lengthy and limited in their ability to rapidly diagnose different levels of knowledge acquisition. They are not suitable for real-time, on-line adaptation of multimedia formats to dynamically changing levels of expertise. Available methods of cognitive diagnosis used in cognitive laboratory studies (e.g., concurrent and retrospective reporting, observations, etc.) are also unfit for real-time monitoring of user performance in adaptive digital multimedia environments because they are very time-consuming. Therefore, no appropriate, cognitively-oriented diagnostic methods are available to be used in adaptive procedures for user-tailored multimedia environments.

The content of users' knowledge base could not be accessed directly. Usually, we are able to obtain some evidence of that knowledge from results of various cognitive activities (e.g., solving test problem) and make probabilistic inferences about possible underlying cognitive constructs. This evidence could be inadequate in many situations. For example, students' answers to a series of test problems would not tell us if those problems were solved by using a novice-like search approaches or an expert-like method based on knowledge of appropriate solution procedures (or, in the latter case, what level of knowledge was applied). We could do better in cognitive diagnosis if we were able to rapidly register immediate traces of individuals' use of their knowledge structures while they approach a problem or situation. The diagnostic power of this method could approach that of concurrent reporting or think-aloud diagnostic techniques; however, it could work on a considerably shorter time scale. The rapid tracing of currently-activated knowledge structures essentially means accessing and monitoring content of working memory or, more accurately LTWM, since we are diagnosing knowledge-based cognitive performance. Therefore, to evaluate user levels of expertise in real-time, we may need to rapidly diagnose the content of LTWM during complex cognitive activities. With this approach, LTWM characteristics are used to determine relevant components of knowledge base held in long-term memory.

Copyright © 2006, Idea Group Inc. Copying or distributing in print or electronic forms without written permission of Idea Group Inc. is prohibited.

Overall, it is possible to suggest that multimedia presentations could be more cognitively efficient if they continuously and dynamically tailor formats of information presentation to changing levels of user proficiency in a domain. Appropriately-constructed rapid diagnostic methods based on immediate traces of the content of long-term working memory while users approach a task or solve a problem, could be used to dynamically monitor levels of expertise in the domain. Rapid cognitive diagnosis in combination with principles for optimizing cognitive load derived from the expertise reversal effect could provide effective adaptive procedures.

Developing appropriate rapid cognitive diagnostic techniques is the key task in implementing this adaptive multimedia methodology. As mentioned previously, long-term memory structures define the characteristics of our performance during knowledge-based cognitive activities. If a person is facing a task in a familiar domain, and her or his immediate approach to this task is based on available knowledge structures, these structures will be rapidly activated and brought into the person's working memory. A corresponding LTWM structure will be created. These LTWM structures are durable and interference-proof to allow sufficient time for a practically-usable diagnostic procedure. There is no need to capture the immediate content of memory strictly within a split-second of corresponding cognitive operations. The available time could be sufficient for recording or otherwise registering task responses in a suitable format. The general idea of this diagnostic approach is to determine the highest level of organized knowledge structures (if any) that a person is capable of retrieving and applying rapidly to a task or situation she or he encounters.

The approach has been realized as the *first-step diagnostic method*. Learners were presented with a task for a limited time and asked to indicate *their* first step towards solution (Kalyuga & Sweller, 2004). The first step would involve different cognitive operations for individuals with different levels of expertise in a domain. An expert may immediately provide the final answer; a less knowledgeable person may indicate the very first operation according to a detailed step-by-step solution procedure; and a novice may start some search process, for example, using a trial-and-error technique. Therefore, different first-step responses would reflect different levels of acquisition of corresponding knowledge structures. Skipping some intermediate procedural operations when showing the first subjectively-significant solution step would indicate a higher level of proficiency: the expert may have corresponding operations automated or well-learned to be able to perform them mentally without exceeding working memory capacity.

The first-step method was used (both in paper- and computer-based formats) to diagnose secondary and high-school students' knowledge of procedures for solving linear algebraic equations, simple coordinate geometry tasks, and arithmetic word problems (Kalyuga & Sweller, 2004; Kalyuga, in press), and for testing reading comprehension skills (Kalyuga, 2006) Experimental results indicated significant correlations (.72 - .92) between performance on these tasks and traditional measures of knowledge that required complete solutions of corresponding tasks. Test times were reduced by factors of up to 4.9 in comparison with traditional test times. The first-step diagnostic method was not only less time-consuming but also more sensitive to underlying knowledge structures than traditional tests.

Copyright © 2006, Idea Group Inc. Copying or distributing in print or electronic forms without written permission of Idea Group Inc. is prohibited.

Similar first-step diagnostic tasks could be used to determine optimal instructional procedures for individuals with different levels of expertise in a domain. For example, based on the expertise reversal effect, we know that presenting novice learners with worked examples is superior to presenting them with problems to solve, but that more knowledgeable learners should be presented with problems rather than worked examples (Kalyuga, Chandler, Tuovinen, & Sweller, 2001). We do not know at what point the switch from examples to problems should occur because we have not had a suitable diagnostic instrument to provide us with levels of expertise. The suggested diagnostic technique is just such an instrument.

In a preliminary experiment using coordinate geometry, we were able to use the rapid test to successfully predict which students should be presented with worked examples and which should be presented with problems (Kalyuga & Sweller, 2004). Similarly, the rapid diagnostic test can be used to predict whether learners should be presented with information in integrated or dual-modality format (novices) or in non-redundant diagrammatic format (more expert learners). In other words, the rapid test can determine the point at which information should no longer be presented in integrated format (e.g., textual explanations embedded into a diagram or presented as auditory narrations) but rather be presented as a single diagram without any textual (on-screen or auditory) explanations.

Figure 1. A snapshot of the multimedia instructional format for the cutting speed nomogram (adapted from Kalyuga, Chandler, & Sweller (2000); copyright © 2000 by the American Psychological Association, Inc)

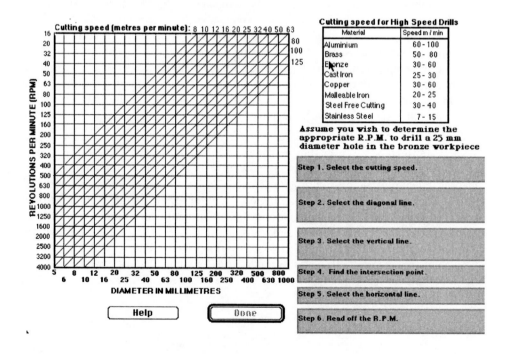

Copyright © 2006, Idea Group Inc. Copying or distributing in print or electronic forms without written permission of Idea Group Inc. is prohibited.

For example, when training apprentices of manufacturing companies in reading different cutting speed nomograms (Figure 1) used to determine the appropriate number of revolutions per minute to run a specific type of cutting machines (Kalyuga, et al., 2000), we observed that replacing visual texts with corresponding auditory explanations was beneficial for novice learners (modality effect). When a novice trainee clicked on a particular step heading button, an auditory narration of an explanation of this step was delivered through headphones instead of being displayed as an identical visual text next to the diagram. However, when learners became much more experienced in using different nomograms, the best way to present them a new type of nomograms was to display just a diagram without any explanations at all (an example of the expertise reversal effect).

The rapid test that may allow us to switch instructional formats at the most appropriate time for an individual trainee can be based on regularly presenting trainees partially completed procedures (with different degrees of completeness) and asking them to rapidly indicate their next step towards solution. At the lowest level of completeness, no task information is indicated on the diagram. At the next level, only some initial data in the task statement is highlighted along the axes or in the table. At the following levels, more lines and their intersection points are shown. In this way, levels of expertise can be rapidly determined. Accordingly, less expert participants, as determined by the rapid test, should be presented with comprehensive auditory explanations. In contrast, more expert participants, for whom the auditory explanations might be redundant, would perform better with a diagram and limited or no explanations.

The rapid testing method was applied for real-time adaptation of instructional procedures (worked examples and problem-based instruction) to current levels of individual learners' knowledge in the domain of linear algebra equations using a simple adaptive computer-based tutor (Kalyuga & Sweller, 2004). The aim of the experiment was to demonstrate that the rapid test could be effectively used in a computer-based training environment for adapting instruction to changing levels of learners' knowledge of solution procedures. The rapid test was used for initial selection of the appropriate levels of instructional materials according to levels of learners' preliminary knowledge, as well as for monitoring learners' progress during instruction and real-time selection of the most appropriate instructional formats. For learners with lower levels of expertise, based on the rapid test, additional worked-example information was provided. For learners with higher levels of expertise, less worked-example information and more problem-solving exercises were provided.

The learner-adapted procedure was compared to an equivalent procedure without real-time adaptation of instruction to levels of learner knowledge. Learning was enhanced by adapting instruction to learner levels of expertise (effect size 0.46 for relative knowledge gains due to instruction). This study provided evidence for the usability of the rapid test, although in a relatively simple and not media-rich domain. Similar rapid diagnosis-based approaches could be used in other more complex environments for initial selection of the appropriate formats of multimedia materials according to levels of users' preliminary knowledge in the domain, monitoring their progress during training, and real-time selection of the most appropriate multimedia formats to build a fully learner-adapted presentation.

Copyright © 2006, Idea Group Inc. Copying or distributing in print or electronic forms without written permission of Idea Group Inc. is prohibited.

Studying the Usability of
the Adaptive Procedure

Rapid diagnostic tasks in many complex domains require responses that cannot always be specified precisely in advance. For example, for many tasks in physics or engineering, drawing graphical representations is essential, and these drawings are usually very subjective. In paper-based formats, the above first-step method could still be applied in all these situations. However, when indicating a first-step response requires graphical representations, recording and analyzing students' responses in computer-based multimedia environments may be technically challenging. In such situations, an alternative rapid diagnostic approach could be based on users' rapid verifications of suggested solution steps.

With the *rapid verification method*, after studying a task for a limited time, users are presented with a series of possible (both correct and incorrect) solution steps reflecting various stages of the solution procedure, and are asked to rapidly verify the suggested steps (for example, by pressing corresponding keys on the computer keyboard). For example, for the vector addition motion task *A crane is moving horizontally at 3 m/s. A load is being lifted at 1 m/s. What is the velocity of the load relative to the ground?,* each solution verification window may include a diagrammatic and/or numerical representation of a possible (correct or incorrect) solution step and buttons "Right", "Wrong", and "Don't know" for students to click on (see Figure 2 for an example of a suggested incorrect step). The "Don't know" button was included as the third answer option in order to reduce a possible guessing effect.

Figure 2. A suggested incorrect solution step for a rapid verification diagnostic task

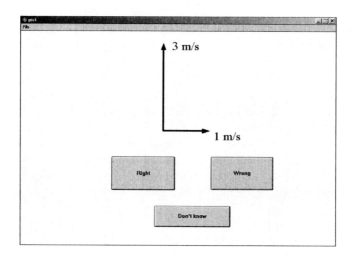

Copyright © 2006, Idea Group Inc. Copying or distributing in print or electronic forms without written permission of Idea Group Inc. is prohibited.

The usability of the rapid verification method as a means of real-time adaptation of multimedia (diagrams with on-screen textual explanations) presentations to current levels of users' expertise in solving vector addition motion problems was pilot-tested with a limited sample of 16 Grade 11 high-school students in the school's computer lab. Using an adaptive computer-based tutor, the learner-adapted procedure was compared to an equivalent instruction without real-time adaptation to the level of learner expertise. The training packages were designed using Authorware Professional and included an initial rapid diagnostic test, an adaptive training session for the experimental group and a non-adaptive version for the control group, and a final rapid diagnostic test (similar to the initial test with re-worded tasks).

To limit the task domain to relatively simple classes of problems, a restricted range of angle values between vectors were used in the tasks: 0° (the same direction of movements), 90° (perpendicular vectors), 180° (opposite directions of movements), 60°, and 120°. When 60° or 120° angles are used, only equal velocity values for both vectors were allowed. The diagnostic items in this restricted domain included one diagnostic task statement for each angle value, followed by a series of five possible (both correct and incorrect) solution steps for rapid verification with gradually-increasing levels of graphical and numerical solution details provided to students. For example, the first step indicated only directions of vectors; the second step also showed their numerical values; the third step indicated the direction of the resulting vector; the fourth step included the numerical expression for calculating the value of the resulting vector; and the fifth step indicated only the final answer. Each task statement was presented to students for around 15 seconds that were sufficient for reading the statement. Instead of technically restricting response times to several seconds (what could have forcefully interrupted some genuine responses), students were "coached" in responding fast during pre-test exercises with a sample of tasks from a different area.

Users' prior knowledge is an important factor contributing to individual differences in the effect of instruction based on text and visual displays (Schnotz, 2002). For learners with lower levels of expertise, based on the rapid diagnostic assessment, additional pictorial and textual information was provided. For learners with higher levels of expertise, redundant representations were eliminated. The adaptive training session was based on a series of faded worked examples or completion tasks (Renkl & Atkinson, 2003; Van Merriënboer, 1990), each followed by a problem-solving practice. According to this approach, novices learn most effectively when instructed using fully worked out examples. As levels of learners' knowledge in the domain increase, parts of worked examples could be gradually omitted, thus increasing a relative share of problem-solving practice in instruction.

In the learner-adapted group, learners were allocated to appropriate stages of the instructional procedure corresponding to the performance break-down points that were determined by the outcomes of the initial rapid diagnostic test. Appropriate fully and partially worked-out examples were presented, each followed by a problem-solving exercise. Depending on the outcome of the rapid diagnostic probes during instruction, each learner was allowed to proceed to the next stage of the training session or was required to repeat the same stage and then take the rapid test again. At each subsequent stage of the training session, a lower level of instructional guidance was provided to learners by eliminating increasingly more explanations of initial procedural steps in faded

Copyright © 2006, Idea Group Inc. Copying or distributing in print or electronic forms without written permission of Idea Group Inc. is prohibited.

examples, and a higher level of the rapid test task was used at the end of the stage. How long each learner stayed at each stage depended on her or his performance on rapid diagnostic tasks during the session. Thus, according to the number of levels of acquisition of the solution skill, there were five levels of adaptation for each of the five types of tasks (corresponding to different angle values).

In contrast, in the non-learner-adapted group, all learners went through all the stages of the training session regardless of their performance on the initial rapid test. Each learner had to study all worked examples, perform all problem exercises, and undertake all rapid diagnostic tasks (however, the outcomes of these tests were not used for selecting the subsequent instructional materials).

The user-adapted group indicated better knowledge gains (differences between the sum of the test scores for the final rapid test and sum of the test scores for the initial rapid test). Ninety-five percent confidence intervals were 19.2 ± 4.5, for the learner-adapted group and 12.7 ± 4.1, for the non-adapted group; effect size = .52 (a medium size effect). Training session time was reduced by factor 1.5. Ninety-five percent confidence intervals were (in *sec*) 1960 ± 222, for the learner-adapted format and 2797 ± 408, for the non-adapted format; effect size = .73. The higher knowledge gains for the learner-adapted format in comparison with the non-adapted format, together with significantly reduced training time, provided preliminary evidence that the suggested rapid technique for diagnosing learner levels of expertise can be used to individualize instructional procedures in multimedia environments.

Towards Cognitively Efficient Adaptive Multimedia

Higher levels of expertise in a domain are characterized not only by rapid and effective performance due to well-organized knowledge base in long-term memory. In most cases, expert performance is also relatively effortless: it does not require much cognitive resources and does not exceed available working memory capacity. Even if an expert and a novice both successfully solve a task, the cognitive cost of their performance could be markedly different. On the other hand, if the same level of cognitive load is involved in solving a problem, experts are expected to achieve higher performance results. Therefore, measuring levels of cognitive load in addition to performance test results may provide better indicators of expertise. The cognitively optimal status of multimedia presentations themselves could also be verified by directly monitoring cognitive load during user-tailored sessions.

There are several relatively simple and practically-usable methods of measuring cognitive load (see Paas, Tuovinen, Tabbers, & van Gerven, 2003, for an overview). Simple ratings of subjective mental effort during cognitive performance can be an effective and reliable method of evaluating cognitive load. This technique is easy to apply and does not intrude on primary task performance (Paas & van Merriënboer, 1994). Likert-type scales have been used with participants required to estimate how easy or difficult

Copyright © 2006, Idea Group Inc. Copying or distributing in print or electronic forms without written permission of Idea Group Inc. is prohibited.

materials were to understand by choosing a response option or a number on the scale ranging from *extremely easy* (corresponding to the score 1) to *extremely difficult* (corresponding to the maximum difficulty score). For example, for a seven-point scale, the options are: *extremely easy, very easy, easy, neither easy nor difficult, difficult, very difficult, extremely difficult.*

Although cognitive studies indicate that subjective ratings of task difficulty have been successfully used to measure cognitive load in many realistic and experimental educational settings (Paas et al., 2003), there could be potential problems with this method, for example, due to varying subjective interpretations of the rating scale. Alternative methods based on more objective indicators have also been developed. For example, dual-task paradigms use performance on a simple secondary task as an indicator of cognitive load associated with performance on a primary task of interest. In digital multimedia environments, effectiveness of dual-task techniques for measurement of cognitive load was investigated by Brünken, Plass, and Leutner (2003, 2004) and Brünken, Steinbacher, Plass, and Leutner (2002). In these studies, the secondary task required users to react (e.g., press a key on the computer keyboard) as soon as possible to a color change of a letter displayed in a small frame above the main task frame. Reaction time in the secondary monitoring task was used as a measure of cognitive load associated with the primary multimedia task.

Adaptive multimedia environments could be more efficient if rapid diagnostic tests of proficiency are combined with measures of cognitive load. There are different methods, both quantitative and qualitative, that could be used to combine measures of performance and cognitive load. Paas and van Merriënboer (1993) defined a quantitative integrated indicator of cognitive efficiency as a difference between standardized z-scores of performance and standardized z-scores of subjective ratings of mental effort. This indicator has been successfully used in cognitive load research since then. Camp, Paas, Rikers, and van Merriënboer (2001) and Salden, Paas, Broers, and van Merriënboer (2004) used this approach for the dynamic selection of digital multimedia learning tasks in air traffic control training. The method can also be used in a simple qualitative manner. For example, different levels of expertise could be associated with combinations of certain ranges of values of performance indicators and ratings of cognitive load in a matrix-like representation (e,g,. the top level of expertise may require achieving the top performance scores combined with difficulty ratings not lower than "easy", etc.).

The efficiency-based approach to optimizing cognitive load was tested using a simple adaptive algebra tutor (Kalyuga & Sweller, 2005). Cognitive efficiency (E) was defined as the level of performance P (measured by the rapid first-step diagnostic method) divided by the rating of mental effort R: $E = P/R$. In contrast to the approach using z-scores for performance and mental effort ratings, that can be calculated only when means and standard deviations are available for the whole sample, this definition could be used in real-time for each individual user. It is also consistent with the general understanding of efficiency as performance levels relative to the cognitive resources required. Critical values of cognitive efficiency required for achieving a sufficient level of expertise could be defined, for example, as $E_{cr} = P_{max}/R_{max}$, where P_{max} is the maximum performance score for the task and R_{max} is the maximum difficulty (or mental effort) rating score. With this definition, if someone invests maximum mental effort in a task but does not display

Copyright © 2006, Idea Group Inc. Copying or distributing in print or electronic forms without written permission of Idea Group Inc. is prohibited.

the maximum level of the task performance, his or her cognitive performance is not regarded as efficient. The developed adaptive training program, which incorporated this approach for the learning task selection procedure, demonstrated its superiority to non-adaptive procedure based on the same learning materials.

The cognitive efficiency-based approach requires further fine-tuning by exploring various criteria for achieving the levels of performance efficiency required for changing formats of multimedia presentations and designing less intrusive non-verbal formats of subjective rating scales. The approach could also be extended further by combining multiple measures of cognitive load and performance, for example, measures of learning task effort, test task effort, and test performance indicators (a multi-dimensional approach, see Tuovinen & Paas, 2004). The influence of user motivational characteristics on the relation between mental effort and performance could also be taken into account (see Paas, Tuovinen, van Merriënboer, & Darabi, 2005).

Summary

Digital multimedia can demonstrate its potential by integrating our current knowledge of cognitive processes involved in human information processing with technological capabilities that allow using the most appropriate content and presentation formats, at the most appropriate time, and in the most cognitively efficient way for each individual user, thus providing truly user-centred multimedia environments (Shute & Towle, 2003). The recent research studies in cognitive load issues in multimedia learning have demonstrated that many suggested multimedia presentation techniques that were highly effective with less knowledgeable learners could lose their effectiveness and even have negative consequences when used with more experienced learners. An important implication of these findings is that multimedia presentation techniques and formats need to be tailored to levels of user experience in a domain. To be able to dynamically select multimedia formats optimal for individuals with different levels of domain-specific proficiency, it is important to understand cognitive mechanisms that influence efficiency of multimedia learning for individual users and to have simple rapid diagnostic measures of expertise in a domain suitable for real-time on-line evaluation of users' proficiency.

Recent advances in our knowledge of human cognitive architecture, learning processes, and the nature of expertise have created foundations for understanding changes in mechanisms of processing multimedia information that occur with development of expertise in a domain. Also, a rapid diagnostic approach to cognitive assessment of levels of user expertise has been developed and tested in several domains. Finally, prototypes of adaptive procedures using rapid diagnostic techniques were pilot-tested in preliminary studies with adaptive computer-based training packages. Such techniques can be used to improve accuracy of information in user models for adaptive multimedia and hypermedia systems, or as stand-alone diagnostic methods. All these developments have created prerequisites for the design of comprehensive cognitive theory-based adaptive multimedia environments that would be dynamically tailored to changing levels of user expertise.

Copyright © 2006, Idea Group Inc. Copying or distributing in print or electronic forms without written permission of Idea Group Inc. is prohibited.

An important advantage of the suggested method for the design of user-adapted multimedia environments is its relative simplicity. It could be implemented with most common multimedia authoring tools and does not require the complex computational modeling and high-level programming expertise that are essential for developing sophisticated intelligent tutoring systems. However, even such a relatively simple adaptive procedure for multimedia environments have the potential to enhance performance outcomes, increase levels of competence for each user and, at the same time, reduce training and diagnostic assessment time, thus resulting in corresponding direct and indirect financial benefits.

References

Baddeley, A. D. (1986). *Working memory*. New York: Oxford University Press.

Brünken, R., Plass, J., & Leutner, D. (2003). Direct measurement of cognitive load in multimedia learning. *Educational Psychologist, 38,* 53-61.

Brünken, R., Plass, J. L., & Leutner, D. (2004). Assessment of cognitive load in multimedia learning with dual-task methodology: Auditory load and modality effects. *Instructional Science, 32,* 115-132.

Brünken, R., Steinbacher, S., Plass, J., & Leutner, D. (2002). Assessment of cognitive load in multimedia learning using dual-task methodology. *Experimental Psychology, 49,* 109-119.

Brusilovsky, P. (2001). Adaptive hypermedia. *User Modeling and User-Adapted Interaction, 11,* 87-110.

Brusilovsky, P., Eklund, J., & Schwarz, E. (1998). Web-based education for all: A tool for developing adaptive courseware. *Computer Networks and ISDN Systems, 30,* 291-300.

Camp, G., Paas, F., Rikers, R., & van Merriënboer, J. J. G. (2001). Dynamic problem selection in air traffic control training: A comparison between performance, mental effort, and mental efficiency. *Computers in Human Behavior, 17,* 575-595.

Chi, M. T. H., Feltovich, P., & Glaser, R. (1981). Categorisation and representation of physics problems by experts and novices. *Cognitive Science, 5,* 121-152.

Chi, M. T. H., Glaser, R., & Farr, M. J. (Eds.). (1988). *The nature of expertise*. Hillsdale, NJ: Erlbaum.

Chi, M., Glaser, R., & Rees, E. (1982). Expertise in problem solving. In R. Sternberg (Ed.), *Advances in the psychology of human intelligence* (pp. 7-75). Hillsdale, NJ: Erlbaum.

Cronbach, L. J., & Snow, R. E. (1977). *Aptitudes and instructional methods: A handbook for research on interaction*. New York: Irvington Publishers.

De Bra, P., & Calvi, L. (1998). AHA! An open Adaptive Hypermedia Architecture. *The New Review of Hypermedia and Multimedia, 4,* 115-139.

Copyright © 2006, Idea Group Inc. Copying or distributing in print or electronic forms without written permission of Idea Group Inc. is prohibited.

De Groot, A. D. (1966). Perception and memory versus thought: Some old ideas and recent findings. In B. Kleinmuntz (Ed.), *Problem solving: Research, method, and theory* (pp. 19-50). New York: Wiley.

Ericsson, K. A., & Kintsch, W. (1995). Long-term working memory. *Psychological Review, 102*, 211-245.

Kalyuga, S. (2005). Prior knowledge principle in multimedia learning. In R. Mayer (Ed.), *Cambridge handbook of multimedia learning*. New York: Cambridge University Press.

Kalyuga, S. (in press). Rapid assessment of learners' knowledge structures. *Learning & Instruction*.

Kalyuga, S. (2006). Rapid assessment of learners' proficiency: A cognitive load approach. *Educational Psychology, 26,* 613-627.

Kalyuga, S., Ayres, P., Chandler, P., & Sweller, J. (2003). Expertise reversal effect. *Educational Psychologist, 38,* 23-31.

Kalyuga, S., Chandler, P., Tuovinen, J., & Sweller, J. (2001). When problem solving is superior to studying worked examples. *Journal of Educational Psychology, 93,* 579-588.

Kalyuga, S., Chandler, P., & Sweller, J. (2000). Incorporating learner experience into the design of multimedia instruction. *Journal of Educational Psychology, 92,* 126-136

Kalyuga, S., & Sweller, J. (2004). Measuring knowledge to optimize cognitive load factors during instruction. *Journal of Educational Psychology, 96,* 558-568.

Kalyuga, S., & Sweller, J. (2005). Rapid dynamic assessment of expertise to improve the efficiency of adaptive e-learning. *Educational Technology, Research and Development, 53,* 83-93.

Kobsa, A. (2001). Generic user modelling systems. *User Modeling and User-Adapted Interaction, 11,* 49-63.

Mayer, R. E. (2001). *Multimedia learning.* Cambridge, MA: Cambridge University Press.

Mayer, R. E. (Ed.) (2005). *Cambridge handbook of multimedia learning.* New York: Cambridge University Press.

Mayer, R., & Gallini, J. (1990). When is an illustration worth ten thousand words? *Journal of Educational Psychology, 82,* 715-726.

Miller, G. A. (1956). The magical number seven, plus or minus two: Some limits on our capacity for processing information. *Psychological Review, 63,* 81-97.

Paas, F., Tuovinen, J., Tabbers, H., & van Gerven, P. (2003). Cognitive load measurement as a means to advance cognitive load theory. *Educational Psychologist, 38,* 63-71.

Paas, F., Tuovinen, J. E., van Merriënboer, J. J. G., & Darabi, A. A. (2005). A motivational perspective on the relation between mental effort and performance: Optimizing learners' involvement in instructional conditions. *Educational Technology, Research & Development, 53,* 25-34.

Copyright © 2006, Idea Group Inc. Copying or distributing in print or electronic forms without written permission of Idea Group Inc. is prohibited.

Paas, F., & van Merriënboer, J. J. G. (1993). The efficiency of instructional conditions: An approach to combine mental-effort and performance measures. *Human Factors, 35*, 737-743.

Paas, F., & van Merriënboer, J. J. G. (1994). Instructional control of cognitive load in the training of complex cognitive tasks. *Educational Psychology Review, 6*, 351-371.

Pylyshyn, Z. (1999). Is vision continuous with cognition? The case for cognitive impenetrability of visual perception. *Behavioral and Brain Sciences, 22*, 341-423.

Renkl, A., & Atkinson, R. K. (2003). Structuring the transition from example study to problem solving in cognitive skills acquisition: A cognitive load perspective. *Educational Psychologist, 38*, 15-22.

Salden, R. J. C. M., Paas, F., Broers, N. J., & van Merriënboer, J. J. G. (2004). Mental effort and performance as determinants for the dynamic selection of learning tasks in air traffic control training. *Instructional Science, 32*, 153-172.

Schnotz, W. (2002). Towards an integrated view of learning from text and visual displays. *Educational Psychology Review, 14*, 101-120.

Shute, V. J. (1992). Aptitude-treatment interactions and cognitive skill diagnosis. In J. W. Regian & V. J. Shute (Eds.), *Cognitive approaches to automated instruction* (pp. 15-47). Hillsdale, NJ: Lawrence Erlbaum Associates.

Shute, V., & Towle, B. (2003). Adaptive e-learning. *Educational Psychologist, 38*, 105-114.

Sweller, J. (1999). *Instructional design in technical areas.* Melbourne: Australian Council for Educational Research.

Sweller, J., Chandler, P., Tierney, P., & Cooper, M. (1990). Cognitive load and selective attention as factors in the structuring of technical material. *Journal of Experimental Psychology: General, 119*, 176-192.

Sweller, J., Van Merriënboer, J., & Paas, F. (1998). Cognitive architecture and instructional design. *Educational Psychology Review, 10*, 251-296.

Tuovinen, J. E., & Paas, F. (2004). Exploring multidimensional approaches to the efficiency of instructional conditions. *Instructional Science, 32*, 133-152.

van Merriënboer, J. J. G. (1990). Strategies for programming instruction in high school: Program completion vs. program generation. *Journal of Educational Computing Research, 6*, 265-287.

Copyright © 2006, Idea Group Inc. Copying or distributing in print or electronic forms without written permission of Idea Group Inc. is prohibited.

Chapter XI

Issues of Hand Preference in Computer Presented Information and Virtual Realities

Adam Tilinger, University of Veszprém, Hungary

Cecilia Sik-Lányi, University of Veszprém, Hungary

Abstract

This chapter presents the differences between left- and right-handed persons in the aspect of computer-presented information and virtual realities. It introduces five test scenarios and their results addressing this question. We showed that there are moderate differences between groups preferring different hands. The different needs of left- and right-handed people may play an important role in user-friendly interface and virtual environment design, since about a tenth of the population is left-handed. This could help to undo the difficulties that the left-handed and ambidextrous routinely encounter in their daily lives.

Copyright © 2006, Idea Group Inc. Copying or distributing in print or electronic forms without written permission of Idea Group Inc. is prohibited.

Introduction

Hand preference plays an important role in using tools and utensils. While, for example, left-handed persons can apply scissors specially designed for them much more efficiently than ordinary ones and vice versa; in multimedia design and virtual environment architecture, the developer is able to suit both groups by producing software with adjustable content according to the user's hand preference. Endeavouring to create interfaces that are more and more user-friendly, developers are aspiring to take into account individual demands. Hand preference differences, due to differences in the cortical processing of the perceived information, is accompanied with different thinking features; thus an obvious characteristic probably defining these needs is whether the user prefers to use his left or right hand.

The objective of the present chapter is to show the differences of left- and right-handed people in the aspect of computer-presented information and virtual realities.

Background

Arguably, 2 to 30 percent of any human population is left-handed or ambidextrous, with most estimates hovering around 10%, depending upon the criteria used to assess handedness (Holder, 1992). Generally, males are three times more likely to be left-handed than females. Statistically, one twin of a pair has a 20% chance to be left-handed. We can well observe the differences between left- and right-handers' motion and in the use of utensils in our everyday lives. Due to the countless experiments made in this topic, we can describe left- and right-handedness quite well. It is the consequence of the asymmetry of the human brain, which can be hardly observed at first glance. The asymmetry is in the functionality of the two hemispheres and slightly in the structure (Atkinson, Atkinson, Smith, Bem, Nolen-Hoeksema, & Smith, 1996). We know that motor nerves cross each other as they leave the brain; thus the left hemisphere controls the right side of the body including the right hand, and similarly the right hemisphere controls the left side of the body and the left hand. In the case of the optic nerve, the crossing works like this: If we look straight ahead, the stimuli which are on the left of the fixation point lead on to the right hemisphere, while the ones on the right side of the fixation point lead on to the left hemisphere from both eyes. The input of audition is crossed as well, but the incoming voices are represented on the side of the perceptive ear as well. There is no crossing in olfaction. The left hemisphere controls the written and spoken language and mathematical computations. The right hemisphere can slightly understand language; its most important abilities are spatial and pattern senses.

The two hemispheres have different characteristics and are responsible for different functionalities. While, for example, the left hemisphere is responsible for logical, rational, digital and algebraic thinking, for self-awareness, and a sense of time; the right hemisphere is responsible for creativity, musicality, for geometrical and analogue thinking, and the sense of humour. The two halves of the human brain are also differentiating

Copyright © 2006, Idea Group Inc. Copying or distributing in print or electronic forms without written permission of Idea Group Inc. is prohibited.

among the tasks they have: In an oversimplified form, one might state that the processing of verbal information is done in the left side of the brain. On the other hand, the right hemisphere of the brain is normally responsible for perceiving pictorial information. It is usually better in processing images and is responsible for the left-hand motoric skills (Annett, 1985; Hellige, 1993). Experiments on people with splitted corpus callosum (prohibiting communication of the two hemispheres) showed that the vision of the left hemisphere is different from that of the right. One can perceive the depth on a cube drawn by the left hand, while the right hand is unable to draw it. Geschwind repeated these tests on healthy participants. The left hemisphere created rhapsodic, anxious figures while those of the right hemisphere were much more harmonic and complex. It is also well-known that the left-handed possess an advantage in sports where the athlete has to throw a ball, in tennis, or in fencing. This is because the right hemisphere, which is more advanced in depth perception, sets out the responses in those situations.

In the case of most people, when the left hemisphere is dominant, they are right-handed.

The asymmetries of the human body, for example, cerebral asymmetry (Geschwind, Miller, DeCarli, & Carmelli, 2002) or manual asymmetry (Purves, White, & Andrews, 1994), are also strongly correlated with handedness. The loss of genetic determination of the left and right cerebral hemispheres in the non-right-handed results in the loss of a right-hand, left-hemisphere-biasing genetic influence, a "right-shift" genotype, thus decreased cerebral asymmetry (Annett, 1985).

Why the human population is dominantly right-handed is yet unknown. The most commonly accepted theory of handedness is the following. Since both speaking and handiwork require fine motor skills, having one hemisphere of the brain do both would be more efficient than having it divided up. And since in most people the left side of the brain controls speaking, right-handedness would prevail. It also predicts that left-handed people would have a reversed brain division of labour. Lastly, since other primates do not have a spoken language (at least of the type we have), there would be no stimulus for right-handed preference among them, and that is true—there is no right-hand preference among them. However, this does not explain why the left hemisphere would always be the one controlling language. Why not for 50% of the population the left and for 50% the right? While 95% of right-handers do indeed use the left side of the brain for speaking, it is more variable for left-handers. Some do use the right for linguistic skills, some use the left hemisphere, and others use both. On the balance, it appears that this theory could well explain some left-handedness, but it has too many gaps to explain all left-handedness.

Handedness runs in families, although even when both parents are left-handed, there is only a 26% chance of their child being left-handed. Thus, it is clear that genetics is not the only cause.

Left-handers and the ambidextrous routinely encounter difficulties in their daily lives that right-handers do not. Many of the more serious problems may be avoided or overcome with a little basic understanding and a few simple coping strategies. Up to now, many utensils (scissors, can openers) and working tools are also made especially for left-handers to undo the difficulties they must face in their daily lives. This also includes hardware design; left-handed mouse, special left-handed keyboard, and other devices are accessible nowadays. However, this is not the case in software ergonomics.

Copyright © 2006, Idea Group Inc. Copying or distributing in print or electronic forms without written permission of Idea Group Inc. is prohibited.

As hand preference is deduced from the different dominance of brain hemispheres, it is possible that there are dissimilarities in abstraction, reflection, or perception between left- and right-handed people. This could lead to the need of an alternative user interface or one specially designed to suit both groups. This also applies to virtual environment architecture. Especially in virtual realities and multimedia applications, the designer can utilise the advantage of the variability of the environment or interface. To design appropriate interfaces and virtual worlds, we need to know the different needs of left- and right-handers.

There are many recommendations on design guidelines of multimedia and virtual realities (Fencott & Isdale, 2001; Hix, Swan II, Gabbard, McGee, Durbin, & King, 1999; Wilson, Eastgate, & D'Cruz, 2002). However, none of these deals with the differences between left- and right-handers. Should the design of a computer display or a virtual environment adapt to the differences between users, or can we live with a single – right-handed – design? This question might be even more important in educational multimedia applications, as modern pedagogic theory favours to permit left-handed children to stay left-handed and tries to provide them with utensils optimised for their use.

Both sides can be pleased in case of multimedia and virtual realities either by designing layouts and environments that suit both left- and right-handed people or by using the variability property of these software. Either way, to design appropriate worlds, we need to know the different needs of left- and right-handers.

Differences of Left- and Right-Handed Persons Regarding Computer-Presented Information and Virtual Realities

Thus, to analyse the different behaviour and needs of left- and right-handed people, three multimedia applications, focusing on computer-presented information, and four virtual reality test scenarios were developed at the University of Veszprém.

Computer-Presented Information

Experiments were conducted to test the hypothesis that left- and right-handed people, due to differences in the cortical processing of the perceived information, will respond differently if a signal is presented in the left or right peripheral visual zone. The fundamental idea to test was the following: Is there any difference in cognitive reactions between left- and right-handed observers if a special task pops up on the left or the right side of the screen while working with a normal visual task in the middle of the computer screen? Is there any difference in experiencing this special task by right- and by left-handed people?

Two kinds of tasks were used during these tests, tasks where the meaning of a sign had to be recognised (i.e. identification of letters - such identification exercise should be

Copyright © 2006, Idea Group Inc. Copying or distributing in print or electronic forms without written permission of Idea Group Inc. is prohibited.

performed by the left hemisphere of the brain) and where pictorial information was used (most likely to be processed primarily by the right hemisphere of the brain).

All three independent experiments were conducted using a PC with a 19" screen diameter monitor. The angular distance between the task seen in the middle of the screen and on the right or left side was approximately 18°.

In the first experiment, a printed text was displayed in the middle of the screen, and as the normal visual task, the participant had to seek a particular character in this text by hitting the space bar when it was found. After a random duration (this period was recorded along with the number of hits), the textbox disappeared and reappeared either on the left or the right side of the screen. Following this, the observer had to signal that he/she found the new textbox on the side of the screen by pressing the F key for left or the J key for right (the F and J key were selected because the participant did not have to look down to the keyboard, since these keys are easy to find due to the small bulge on them). Then the participant had to continue the search for the particular letter from the third paragraph on and signal every hit of the letter by pressing the space bar. The idea behind this experiment was that if there is any difference in processing printed text information by the two halves of the brain (Caramazza, 1996), this would manifest itself in the speed of finding the particular letter in the reappeared text.

During the second experiment, random three-digit numbers appeared in the middle of the screen. The primary task of the participant was to signal a given property (parity or imparity) (Deheane, 1998; Dehaene & Cohen, 1997) of that particular number by pressing the space bar. At the same time, on both sides of the screen the same images (digital photographs) were displayed (Figure 1.). After a random period a small part of the image, either on the left or right side of the screen, changed and the participant had to respond when he/she realised this by pressing the F or the J key according to which of the two images has been altered. We investigated the reaction time differences of the two groups.

The third test experiment was quite similar to the previous one. In this test, geometrical primitives (like circles, rectangles, etc.) were presented in the middle of the screen instead of numbers, and the task was to signal if this plane figure was of a previously-given shape by hitting space. The special task was the same as in the previous experiment (signalling

Figure 1. Screenshot of the second multimedia experiment

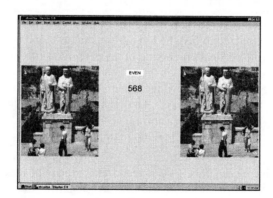

Copyright © 2006, Idea Group Inc. Copying or distributing in print or electronic forms without written permission of Idea Group Inc. is prohibited.

the change of images on the left and right side of the screen) but with horizontally-mirrored image pairs. Besides the reaction time differences of the two groups, another question was, comparing this and the previous experiment, that whether there are any differences in the reaction time between mathematical tasks (processed primarily by the left side of the brain) and pictorial ones (where the primary processing is supposed to be done by the right side of the brain).

Results

Results of 18 participants were analysed, six left-handed, six right-handed and six ambidextrous (Sik-Lányi, 2003). All of them were university students (averagely 20 years old). For each test person, we collected 20 individual results with all three experiments.

Student's test was performed to compare the average results of all the left-handed participants with those of all the right-handed. This was followed by a multiple analysis of variance. The averages were regarded not to be different if, for a significance level of 0.05, the two averages could not be regarded as different. We found that significant differences between the responses given by left- and right-handed could only be found in the second experiment when changing the image on the right side of the screen. However, right-handed participants signal earlier for the text search task if it appears on the right side than left-handed do, and left-handed signal in average earlier if the change appears on the left-side. A possible explanation could lie in the motoric skills (reappearing of the text on the right side was to be signalled with the right hand and vice versa), although the results of the second experiment contradict this assumption (the reaction time of right-handed participants with their left hand was shorter than with their right hand).

Left-handed subjects realise the change on the right side of the screen faster than that on the left side if the task is parity or imparity signalling. A possible explanation could be that left-handed people are apt to mix up numbers. It is well known that they easily mix signs being mirror images for each other.

In all three experiments, the reaction time of left-handed people was slightly shorter. However, comparing the results of the left- and right-handed participants, we can state that there is no significant difference (apart from the one mentioned case).

Virtual Realities

To analyse behavioural differences of left- and right-handed in virtual realities, four non-immersive virtual environments were developed. All tests were concluded on PCs with simple monitors as display. Three of the scenarios were environments where the user could freely walk around, observed through 1st person perspective (the perspective was selected because of the better immersion, although there are observations that inexperienced users prefer 3rd person perspective (Shopland, Lewis, Brown, & Dattani-Pitt, 2004)). Throughout the tests, participants could operate the mouse and/or keyboard using the hand with which they felt most comfortable.

Copyright © 2006, Idea Group Inc. Copying or distributing in print or electronic forms without written permission of Idea Group Inc. is prohibited.

Virtual Gallery

In the first virtual reality scenario, we investigated the differences between the motions of left- and right-handers in a virtual environment that can be independently toured. The aim of this experiment was to detect differences in walk patterns. We designed a considerably symmetrical gallery room exhibiting landscape paintings of Dutch artists and a sculpture in the middle (Figure 2). This room has no prominent places to draw attention and bias results by compensating differences. The test persons are given unlimited time to explore the gallery while a relaxing classical music is played. We investigated the differences between paths of left- and right-handed users.

After starting the program, the user has to calibrate the screen resolution and colour quality to suit the hardware configuration. Although windowed mode is supported, it is worth running the software on full screen for better immersion (every test person in our result database used the full screen mode). The user is allowed to move freely in the gallery using the mouse and/or the keyboard. Besides the cursor control arrow keys, the test person can use the WASD keys for movement control, which is widespread for example in computer games. During the simulation, hitting the Esc key reveals the control panel. Here we can start or stop the recording of the user movement, define the user properties, playback a previously- recorded motion, save the current recordings to a text file, and stop the simulation. While recording the software stores the exact location of the avatar in the three-dimensional world and the view direction in every second. We can use this for further investigations. The output text file also stores information about the actual test person. These properties are: username for data distinguishing, age, gender, hand dominance, and additional information (for example, if a left-handed user is accustomed to use his/her right hand).

Results

Two studies were concluded, a pilot study with 20 test participants and a repeated study with 45 participants.

Figure 2. Screenshot of the gallery

Copyright © 2006, Idea Group Inc. Copying or distributing in print or electronic forms without written permission of Idea Group Inc. is prohibited.

Figure 3. The division of the gallery into six segments

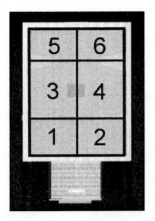

Some prominent characteristics were easy to observe after a relatively small number of test persons (20 participants, 6 of them left-handed, 12 right-handed and 2 ambidextrous) only by looking at the paths. Left-handed users tended to walk around the gallery turning left at the entrance, where right-handed mostly walked in zigzags turning right at the beginning. This difference was then proven in a pilot study by K-means clustering.

To apply K-means clustering (Guralnik & Karypis, 2001), we needed to simplify the paths considerably. We divided the gallery into six sectors and numbered them (Figure 3), then replaced the path with the sequence of these numbers in the order that the user entered the correspondent parts. This resulted in an ordered sixth (we eliminated multiple visits in sectors), thus we got a point in a six-dimensional space. We used K-means clustering in this space to verify differences between left- and right-handed. We divided the set of the six dimensional points into two clusters with the algorithm. For initial group centroids, we arbitrarily selected two points of the set. The applied distance metric was the Euclidean distance. We used the algorithm 20 times with different random initial centroids. Two of 20 applications were not interpretable because of the inaccurate selection of initial group centroids. All the points belonging to left-handed test persons were classified to the same cluster in 13 of the 18 interpretable cases. In three cases, two extreme paths were determined in a distinct cluster (these two test persons didn't started the recording of their motion at the beginning of the simulation).

The tested two ambidextrous users differ greatly. One of them moved like an average left-handed person, while the other moved like a right-handed person. Only 3 of 12 (two female and one male) right-handed test persons behaved themselves like the average left-handed person, the others walked in zigzags, turning either left or right at the entrance.

In a second test with larger number of participants - 25 left- and 20 right-handed - these tendencies seemed to dissolve. Table 1 presents the parameters of the sample. It also shows some primary characteristics, differences, and similarities between left- and right-handed. More left-handed walked in zigzags than in the first test, and even some right-handed appeared who walked around in circles. An interesting result is that the ratio of

Copyright © 2006, Idea Group Inc. Copying or distributing in print or electronic forms without written permission of Idea Group Inc. is prohibited.

Table1. The sample and primary characteristics

		Left-hander	Right-hander
N		25	20
Female		12 (48%)	10 (50%)
Male		13 (52%)	10 (50%)
Age	Average	19.4	22.4
	Deviation	4.9	4.7
Type of walk pattern	Circle	10 (40 %)	5 (25 %)
	Zigzag	12 (48 %)	14 (70 %)
	Other	3 (12 %)	1 (5 %)
Turning direction at start	Left	18 (72 %)	14 (70 %)
	Right	7 (28 %)	6 (30 %)
Time spent in gallery (in seconds)	Average	148	184.1
	Median	135	142
	Deviation	66.5	139.7

Table 2. Position in the gallery per time segment

		Hand pref.	Time segment					
			1	2	3	4	5	6
Position in gallery (in room segments)	Modus	Left	1	3	3	4	5	6
		Right	1	2	4	3	5	6
	Median	Left	1	3	4	4	5	4
		Right	1	2	4	4	5	5
	D	Left	1.12	1.21	1.23	1.21	1.05	2.01
		Right	0.47	1.21	0.79	1.24	1.24	1.72
Z position in gallery (1=near, 3=far)	Modus	Left	1	2	2	2	3	3
		Right	1	1	2	2	3	3
	Median	Left	1	2	2	2	3	2
		Right	1	1	2	2	3	3
	D	Left	0.47	0.69	0.56	0.58	0.59	0.95
		Right	0	0.61	0.47	0.49	0.61	0.92
X position in gallery (1=left, 2=right)	Modus	Left	1	1	1	2	1	2
		Right	1	2	1	1	2	2
	Median	Left	1	1	1	2	1	2
		Right	1	2	1	1	2	2
	D	Left	0.46	0.49	0.49	0.48	0.51	0.44
		Right	0.47	0.51	0.51	0.51	0.51	0.46
Position by modus of X/Z values (in room segments)		Left	1	3	3	4	5	6
		Right	1	2	3	3	6	6

Copyright © 2006, Idea Group Inc. Copying or distributing in print or electronic forms without written permission of Idea Group Inc. is prohibited.

turning left or right at the beginning is the same in the left- and right-handed test groups. As the room was designed to be symmetrical and without any areas of interest that may cause consistent rotation, we expected major differences in this respect (as it was seen in the pilot study). Our assumption is that this may be influenced by cultural customs. For example, all test participants write from left to right, making left an obvious starting direction.

Due to the nature of these results, K-means clustering does not lead to left- and right-handed classes. Although the Welch test shows that the values of five of the six dimensions have different expectation values in the two groups at 90% confidence, the groups are not sharply divided as in the first test.

For further investigations, we used the six dimension vectors as time-line arrays with six time-segments (these are the relevant time-segments). Table 2 shows the modus, median, and standard deviation of expected room segment number in the first six time segments. To avoid the inaccuracy caused by the numbering of room segments, we also analysed the movement through the X and Z dimensions separately. The modus, median, and standard deviation of these data is also shown in Table 2. The last rows of this table present the most possible sojourn of left- and right-handers in the different time segments derived from the modus of position in the two separate dimensions. These results are slightly different from the direct modus of room segment numbers per time segments. As Table 2 shows, there are moderate differences between the expectable positions of left- and right-handed persons, but because of the relatively high standard deviation we cannot conclude sharp distinctness. This loss of significant difference is due to the different combinations of walk pattern types and turning directions at the entrance.

A likewise important characteristic is when the test persons enter the different room segments. The mean, median, modus, and standard deviation values of order of first appearance in different room segments are presented in Table 3. The deviation in this case is also relatively high, and the differences are slight. Segment 2 (first right) is visited before segment 3 (middle left) by right-handed people and vice versa by those who prefer their left hand. This also consistently follows from the higher ratio of zigzag walk patterns of right-handed people.

Table 3. First appearances in room segments

First appearance in a room segment (in time orders)		Room segment					
		1	2	3	4	5	6
Mean	Left-hander	2.04	3.46	2.88	3.96	4.08	4.78
	Right-hander	1.42	2.56	3.05	3.75	4.12	5
Median	Left-hander	1	2	3	4	4	5
	Right-hander	1	2	3	4	4	5
Modus	Left-hander	1	2	2	4	5	6
	Right-hander	1	2	3	3	5	6
Deviation	Left-hander	2.01	2.34	1.01	1.02	1.1	1.45
	Right-hander	0.96	1.95	0.83	0.93	0.86	1.12

Copyright © 2006, Idea Group Inc. Copying or distributing in print or electronic forms without written permission of Idea Group Inc. is prohibited.

Labyrinths

The following scenarios presented virtual environments where the users had to complete a given task, namely finding the exit of different labyrinths. This scenario consists of three mazes of two kinds. The first two are symmetrically designed and the test person only has information about the direction of the exit from the entrance. No other aids like map or compass are presented (not to bias the results, since

left-handed people are known to have stronger spatial skills). We investigated the differences of paths and time required to escape the labyrinth. The third maze is different from the first two. This asymmetrical labyrinth has different-coloured lines painted on the floor. Every third crossing is a checkpoint, where the user is presented with the right way to the next checkpoint by the colours he/she should follow (the colours of lines to follow are written above these checkpoints). To complicate the task, the words are written in different colours than their meaning, but the participants have to follow the meaning of the words. For example, at the first crossing the word "red" is written in blue, "green" in red and "blue" in green (Figure 4). At the beginning of this task, the participants were tested to confirm if they could recognise the colours used. This task reduces the differences between the paths of left- and right-handed to a minimum (only in a few cases did the test persons turn to the wrong direction). Thus we analysed only escape times and mistakes.

The simulation starts after adjusting the video settings (screen resolution and colour quality), the parameters of the test person (username, age, gender, hand preference, additional information), and the recording options (time-span between recorded positions for providing smoother paths when needed). The controls were reduced to the keyboard to slightly compensate the differences between more experienced and beginner users, which still persist. As the test person navigates through the labyrinths, his/her position is recorded at the time-span set at the beginning of the simulation. These and the parameters of the test persons are saved to a text file for further analysis.

Figure 4. Screenshot of the third labyrinth

Copyright © 2006, Idea Group Inc. Copying or distributing in print or electronic forms without written permission of Idea Group Inc. is prohibited.

Results

The statistically-tested sample consists of 10 left- and 10 right-handers. The walk paths in this evaluation were also simplified, in this case to a sequence of $+1$ and -1 according to which direction the users took at crossings. $+1$ means a right turn, -1 means a left turn. Our observations focused on bypassing the first three obstacles in the smaller maze and the first four in the more complex maze. The overall turn directions were also analysed. In the smaller labyrinth, a significant difference can be observed at the first two obstacles. Left-handers tend to turn left, where right-handed people prefer to evade to the right. This was also confirmed by Student's test. Though the third obstacle was mainly bypassed to the right also in the left-handed test group (thus no significant difference in the two samples), this may be due to the design of the maze. In overall turn directions, significant difference can be detected again.

In the more complex labyrinth, the left-handers' performance is similar to that in the simple maze: They prefer to turn left at crossings. However, more and more right-handers take the left at bypassing obstacles. Thus no significant differences can be shown at the first four obstacles. This turning-left tendency of right-handed people was also observed in the Virtual Gallery scenario. However, in overall turn directions, significant differences can be proved by Student's test (right-handed persons turn right still significantly more often). We can see that this scenario showed more immense difference between left- and right-handers like the Virtual Gallery scenario. This might be on the account of the relatively small sample but also of the limited freedom of movement. Where in the gallery, the test persons could freely move from one room segment to the other, in the labyrinth in practice they only had to choose left or right at the crossings.

The results of the third labyrinth, the escape times (averagely 58.11 seconds by left- and 72.36 seconds by right-handed), and the mistakes made (0 of 10 by left- and 4 of 10 by right-handed) are worth mentioning but might not be relevant due to the small number of test persons.

Fire-Alarm Simulation

The third virtual reality scenario is also a virtual environment where the test person has to find an exit. For providing a more natural environment than a labyrinth, this virtual world simulates an office on fire where the user has to reach one of the three exits as fast as possible. Besides the characteristics of the paths, we also analysed other results like escape time, number of steps, number of entering dead-ends, and specific rooms.

After a short description of the task and the explanation of controls, a map is shown to the user (also to test the assumption of left-handed people having stronger spatial skills). This shows the wall arrangement, the position of furniture and other objects, the initial position, and the location of the three exits. However, it does not show where the fire will occur. After the test person has made a mental note of this map, by hitting the Space key he/she could directly enter the virtual environment. The user can navigate through the office with the mouse and/or the keyboard. Besides the cursor control arrow keys, the test person can use the WASD keys for movement control. There are bursts of flames

Copyright © 2006, Idea Group Inc. Copying or distributing in print or electronic forms without written permission of Idea Group Inc. is prohibited.

Figure 5. Screenshot of the fire-alarm simulation

placed here and there in the rooms that cannot be passed across (Figure 5). The placement of the flames is fixed, the same at every simulation, defining a way to the exit with some branches. Only one of the three exits is accessible; the way to the other two is blocked by the fire. The software stores the position of the avatar in the vertical plane only while moving. This path and the escape time are saved in the output text file. This kind of data storage enables a more easy analysis of other characteristics of motion in virtual environments, namely the number of steps and average step duration.

Results

A pilot study with seven left- and seven right-handers showed notable differences in characteristics. Where we could observe immense dissimilarity in average escape time (214.29 seconds by left- and 111.86 seconds by right-handed persons) the differences between average total step number is more moderate (87.86 by left- and 62.43 by right-handed persons). This means more hesitation by left-handed people, which was confirmed by subjective observation. The number of entering dead-ends was also moderately higher in the left-handed group (1.43, where in the right-hander group it was only 1.29). There were other qualitative differences between left- and right-handed paths. The simulated office building has one dead-end room easy to access by turning right at every opportunity. As we expected, twice as many right-handers entered this room as left-handers. To reach the exit there are obligatory passable rooms. In these rooms, we placed obstacles and observed which direction they get bypassed. These showed surprisingly slight differences, and in the case of one obstacle, left-handed persons even turned more often to the right then did right-handed persons.

The sample of the second test for analysing quantitative measurements consisted of 31 left- and 12 right-handed. The average, median, and standard deviation values of escape times, total steps made, and average step time are presented for both groups in Table 4. It shows that despite the considerable difference between average values, the expecta-

Copyright © 2006, Idea Group Inc. Copying or distributing in print or electronic forms without written permission of Idea Group Inc. is prohibited.

Table 4. Performance in the fire-alarm simulation

		Average	Median	Deviation
Escape time (in seconds)	Left-hander	146.7	83	132.8
	Right-hander	86.3	82	56
Total steps made	Left-hander	82.4	54	76.6
	Right-hander	50.1	50.5	21.5
Step duration (in seconds)	Left-hander	1.73	1.53	0.59
	Right-hander	1.61	1.34	0.56

tion values of both escape time and total steps made are similar in left- and right-handed samples. It also shows a compensation of differences presented in the first test although the expectation value of step duration is still 15% larger in the left-handed group. An interesting result is that left- handed users, despite being known to have better spatial skills, took longer to complete the task. It is more interesting compared to the results of the labyrinth scenario where left-handed individuals were faster. A possible cause for this may lay in the design of the environment. The path to the exit in the labyrinth is more straightaway than in the fire-alarm simulation. Thus a zigzag-like walk strategy is more effective in the virtual office building, resulting a shorter escape time. Moreover, right-handed people are accounted to be better at logical thought in a higher stress condition. This may also be a factor.

Virtual Environment Equipment

The possibility that left- and right-handed persons move differently in virtual realities leads to the need of diverse environmental layout designs and different equipment of the virtual world. Thus the last scenario handles the differences and similarities of virtual rooms furnished by left- and right-handed people. We analysed the differences between spatial positioning of the different furniture by the left- and right-handed and also touched upon the correlated positioning of some furniture.

A special software was designed to easily furnish given rooms (Figure 6), thus available to experiment with participants who are not professional virtual reality designers. At the beginning of the simulation, the user has to fill in a questionnaire on his or her age, gender, and hand preference. Eight virtual rooms have to be furnished by the test persons. The user has to place all the given furniture to get to the next room. The different objects can be moved in the horizontal plane and rotated around a vertical axis. The positioning method and the motion velocities are adjustable for providing a comfortable interface (none of the test persons had reported that the test became tiresome even after 45 minutes; moreover, most of them enjoyed the task). The positions of the doors and

Copyright © 2006, Idea Group Inc. Copying or distributing in print or electronic forms without written permission of Idea Group Inc. is prohibited.

Figure 6. Virtual environment equipment scenario

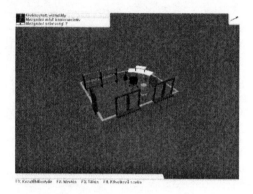

Figure 7. Four example layouts arranged by left-handed

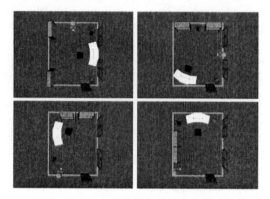

windows were given, but amongst the eight rooms were similar ones with different window and door layout. Although the fixed location of doors and windows eliminates some of the furniture layouts, the resulting arrangements showed that it still leaves many possibilities (Figure 7). The output text files contain data of the position and direction of the different furniture and the user properties.

Results

Figure 8. Distribution of the positions of an object placed by left- and right-handed. The two samples show great similarities.

The statistical sample consisted of 36 left- and 21 right-handers. They had to furnish eight rooms with 6 objects in the first room, 5 in the second, 6 in the third, 7 in the fourth, 6 in the fifth, 5 in the sixth, 4 in the seventh, and 5 eighth. This furniture could be moved in two dimensions. By analysing the spatial position, we handled the two free dimensions

Copyright © 2006, Idea Group Inc. Copying or distributing in print or electronic forms without written permission of Idea Group Inc. is prohibited.

separately. First we applied F-test on the deviation. All the standard deviations of the left-handed and right-handed samples are pair-wise equal with a 98% confidence except one (one co-ordinate of a desk in a considerably smaller bedroom). After that a Student's test showed only in four cases equal expectation values in both co-ordinates of objects with at least 80% confidence. Despite that, the distribution of the two samples showed great similarities (Figure 8). This was even proved in most cases by Wilcoxon test at 95% confidence. Therefore, for all the slight differences between the two groups, hand preference is not a primary factor of general furniture position in virtual environments.

Besides the spatial position of the objects in the room, we investigated eight cases of correlated position of special furniture. Four of these were bed and night-table pairs, one desk and waste-basket, one washbasin and waste-basket, and one couch and floor-lamp. The last pair was a bench and table pair (in this room, there was only one bench for a larger table; we investigated the positioning of this). Our expectations were that the users would place the objects like night-table or waste-basket where they can reach it with their dominant hand. Surprisingly, it was not the case. The percentage of night-tables on the right and on the left side was hovering around 50% in both the left- and right-handed groups (with overall percentages of night-tables on the right side of 56.9% in the left- and 52.8% in the right-handed sample). Only half of the right-handed placed the waste-baskets beside the desk or the washbasin, where this was the case in two-third parts of the left-handed group. The remaining users placed them, again in a remarkably similar percentage, on the right side (46.2% and 45.5% in the desks and 61.5% and 60% in the washbasins case). Difference could only be observed in the position of the bench related to the table (Table 5.). The relevance of this difference is low, though, because the test persons could rotate the room freely, thus directions as north-south and east-west bear less meaning.

Figure 8. Distribution of the positions of an object placed by left- and right-handed. The two samples show great similarities

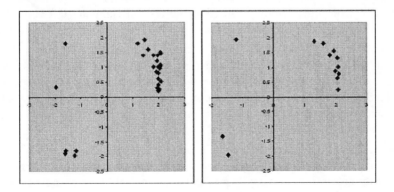

Copyright © 2006, Idea Group Inc. Copying or distributing in print or electronic forms without written permission of Idea Group Inc. is prohibited.

Table 5. Bench position related to desk position

	North	South	East	West	North-east
Left-hander (36)	27.78%	13.89%	22.22%	30.56%	5.56%
Right-hander (16)	12.50%	31.25%	18.75%	18.75%	18.75%

Future Trends

The future of user-friendly design is possibly in fulfilling individual users' needs. One of the questions we must answer is how to suit virtual environments to the needs of left- or right-handed. Future virtual worlds may have layouts for left-handed or automatically generate it from given contents.

The tests in this chapter were applying non-immersive virtual environments. In immersive virtual realities, where the users could behave in a more natural manner, the differences might be more intensive. Also the influence of the degree of freedom is worth analysing in more detail. Another question is how younger, left-handed people, and children differ in behaviour in virtual environments; whether the experienced left-handed were accustomed to move in virtual realities designed for right-handed. Cultural customs, like whether they drive on the left- or the right side of the road, may also have influence on the differences.

Summary

Left- and right-handed people differ from each other. Most utensils designed for right-handed persons are hard to use by left-handed people. In computer applications and virtual realities, we have the advantage of the variability of the interface or the environment. To design ergonomic virtual worlds for both right- and left-handed people, we need to know if there are dissimilarities in their needs.

Concerning the perception of information presented on the computer screen, there is no significant difference between left- and right-handed. The difference between the processing of information presented on the left or on the right side of a screen seems to be smaller than other motoric skill differences. If one is accustomed to observe at a given location one type of information, it is more important to keep that location constant than to adjust it to different observers.

In the present chapter, we showed that there are slight differences in behaviour in virtual environments between left- and right-handers. The lack of immense difference may be counted to the cultural or everyday customs. It is also possible that because in these non-

Copyright © 2006, Idea Group Inc. Copying or distributing in print or electronic forms without written permission of Idea Group Inc. is prohibited.

immersive virtual environments every interaction is mostly controlled by one hand (with the mouse or keyboard), these differences are compensating. Even virtual realities with two-handed controls apply the two hands specialised (for example one hand for moving with the keyboard and one for orienting with the mouse) thus decreasing preference of one side. Other features, like the limitation of freedom in the environment, can enhance or compensate the differences.

References

Annett, M. (1985). *Left, right, hand, and brain: The right shift theory*. London: Erlbaum.

Atkinson, R. L., Atkinson, R. C., Smith, E. E., Bem, D. J., Nolen-Hoeksema, S., & Smith, C. D. (1996). *Hilgard's introduction to psychology* (12th ed.). Fort Worth, TX: Harcourt Brace College Publishers.

Caramazza, A. (1996). Pictures, words, and the brain. *Nature, 383*, 234-235.

Deheane, S. (1998). The number sense: How the mind creates mathematics. *Nature, 391*, 856.

Dehaene, S., & Cohen, L. (1997). Cerebral pathways for calculation: Double dissociation between rote verbal and quantitative knowledge of arithmetic. *Cortex, 33*, 219-250.

Fencott, C., & Isdale, J. (2001). Design issues for virtual environments. *Proceedings of Workshop on Structured Design of Virtual Environments at Web3D 2001*. Paderborn: Shaker Verlag.

Geschwind, D. H., Miller, B. L., DeCarli, C., & Carmelli, D. (2002). Heritability of lobar brain volumes in twins supports genetic models of cerebral laterality and handedness. *Proceedings of the National Academy of Sciences of the United States of America, 99*(5), 3176-3181.

Guralnik, V., & Karypis, G. (2001, August 26). *A scalable algorithm for clustering protein sequences*. Workshop on Data Mining in Bioinformatics, San Francisco, CA (pp. 73-80).

Hellige, J. B. (1993). *Hemispheric asymmetry: What's right and what's left*. Cambridge, MA: Harvard University Press, 168-206.

Hix, D., Swan II, J. E., Gabbard, J. L., McGee, M., Durbin, J., & King, T. (1999). User-centered design and evaluation of a real-time battlefield visualisation virtual environment. *Proceedings of the Virtual Reality '99, IEEE* (pp. 96-103).

Holder, M. K. (1992). *Hand preference questionnaires: One gets what one asks for*. Master's thesis, Department of Anthropology, Rutgers University, New Brunswick, NJ.

Purves, D., White, L. E., & Andrews, T. J. (1994). Manual asymmetry and handedness. *Proceedings of the National Academy of Sciences of the United States of America, 91* (pp. 5030-5032).

Copyright © 2006, Idea Group Inc. Copying or distributing in print or electronic forms without written permission of Idea Group Inc. is prohibited.

Shopland, N., Lewis, J., Brown, D. J., & Dattani-Pitt, K. (2004). *Design and evaluation of flexible travel training environment for use in a supported employment setting.* Paper presented at The 5th International Conference on Disability, Virtual Reality, and Associated Technologies 2004, 69-76.

Sik-Lányi, C. (2003). Optimization of computer-presented information for left-handed observers. *PsychNology 2003, 1*(2). Retrieved from http://www.psychnology.org/article205.htm

Wilson, J. R., Eastgate, R., & D'Cruz, M. (2002). Structured development of virtual environments. In K. Stanney (Ed.), *Virtual environment handbook* (pp. 353-378). Lawrence Erlbaum.

Copyright © 2006, Idea Group Inc. Copying or distributing in print or electronic forms without written permission of Idea Group Inc. is prohibited.

Section IV

Multimedia Communication and Adaptation

Copyright © 2006, Idea Group Inc. Copying or distributing in print or electronic forms without written permission of Idea Group Inc. is prohibited.

Chapter XII

Incorporating User Perception in Adaptive Video Streaming Systems

Nicola Cranley, University College Dublin, Ireland

Liam Murphy, University College Dublin, Ireland

Abstract

There is an increasing demand for streaming video applications over both the fixed Internet and wireless IP networks. The fluctuating bandwidth and time-varying delays of best-effort networks makes providing good quality streaming a challenge. Many adaptive video delivery mechanisms have been proposed over recent years; however, most do not explicitly consider user-perceived quality when making adaptations, nor do they define what quality is. This chapter describes research that proposes that an optimal adaptation trajectory through the set of possible encodings exists, and indicates how to adapt transmission in response to changes in network conditions in order to maximize user-perceived quality.

Copyright © 2006, Idea Group Inc. Copying or distributing in print or electronic forms without written permission of Idea Group Inc. is prohibited.

Introduction

Best-effort IP networks are unreliable and unpredictable, particularly in a wireless environment. There can be many factors that affect the quality of a transmission, such as delay, jitter, and loss. Congested network conditions result in lost video packets, which, as a consequence, produce poor quality video. Further, there are strict delay constraints imposed by streamed multimedia traffic. If a video packet does not arrive before its playout time, the packet is effectively lost. Packet losses have a particularly devastating effect on the smooth continuous playout of a video sequence due to inter-frame dependencies. A slightly degraded quality but uncorrupted video stream is less irritating to the user than a randomly-corrupted stream. However, rapidly fluctuating quality should also be avoided as the human vision system adapts to a specific quality after a few seconds, and it becomes annoying if the viewer has to adjust to a varying quality over short time scales (Ghinea, Thomas, & Fish, 1999). Controlled video quality adaptation is needed to reduce the negative effects of congestion on the stream while providing the highest possible level of service and quality. For example, consider a user watching some video clip; when the network is congested, the video server must reduce the transmitted bitrate to overcome the negative effects of congestion. In order to reduce the bitrate of the video stream, the quality of the video stream must be reduced by sacrificing some aspect of the video quality. There are a number of ways in which the quality can be adapted; for example, the image resolution (i.e. the amount of detail in the video image), the frame rate (i.e. the continuity of motion), or a combination of both can be adapted. The choice of which aspect of the video quality should depend on how the quality reduction will be perceived.

In the past few years, there has been much work on *video quality adaptation* and *video quality evaluation*. In general, video quality adaptation indicates how the bit rate of the video should be adjusted in response to changing network conditions. However, this is not addressed in terms of video quality, as for a given bit rate budget there are many ways in which the video quality can be adapted. Video quality evaluation measures the quality of video as perceived by the users, but current evaluation approaches are not designed for adaptive video streaming transmissions.

This chapter will firstly provide a generalized overview of adaptive multimedia systems and describe recent systems that use end-user perception as part of the adaptation process. Many of these adaptive systems rely on objective metrics to calculate the user-perceived quality. Several objective metrics of video quality have been developed, but they are limited and not satisfactory in quantifying human perception. Further, it can be argued that to date, objective metrics were not designed to assess the quality of an adapting video stream. As a case study, the discussion will focus on recent research that demonstrates how user-perceived quality can be used as part of the adaptation process for multimedia. In this work, the concept of an Optimal Adaptation Trajectory (OAT) has been proposed. The OAT indicates how to adapt multimedia in response to changes in network conditions to maximize user-perceived quality. Finally experimental subjective testing results are presented that demonstrate the dynamic nature of user-perception with adapting multimedia. The results illustrate that using a two-dimensional adaptation strategy based on the OAT out-performs one-dimensional adaptation schemes, giving better short-term and long-term user-perceived quality.

Copyright © 2006, Idea Group Inc. Copying or distributing in print or electronic forms without written permission of Idea Group Inc. is prohibited.

Review of Adaptive Multimedia Systems

Given the seriousness of congestion on the smooth continuous play-out of multimedia, there is a strong need for adaptation. The primary goals of adapting multimedia are to ensure graceful quality adaptation, maintain a smooth continuous play-out and maximize the user-perceived quality. Multimedia servers should be able to intelligently adapt the video quality to match the available resources in the network. There are a number of key features that need to be considered in the development of an adaptive streaming system (Wang & Schulzrinne, 1999) such as feedback to relay the state of the network between client and server, the frequency of this feedback, the adaptation algorithm used, the sensitivity of the algorithm to feedback, and the resulting user-perceived quality.

However, the most important thing is how the system reacts, how it adapts to congestion, and the perceived quality that results from this adaptation.

Adaptation Techniques

Broadly speaking, adaptation techniques attempt to reduce network congestion by matching the rate of the multimedia stream to the available network bandwidth. Without some sort of rate control, any data transmitted exceeding the available bandwidth would be discarded, lost, or corrupted in the network. Adaptation techniques can be classified into the following generalized categories: rate control, rate shaping, and rate adaptive encoding (Figure 1). Each of these techniques adapts the transmitted video stream to match the available resources in the network by either adapting the rate at which packets are sent or adjusting the quality of the delivered video (Wu, Hou, Zhu, Lee, Chiang, Zhang, & Chao, 2000, 2002). These are briefly described in the following sections.

Figure 1. Adaptation techniques

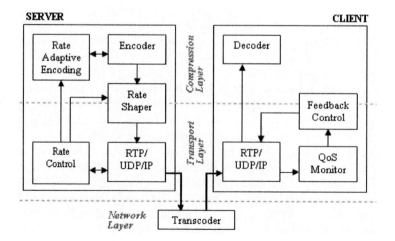

Copyright © 2006, Idea Group Inc. Copying or distributing in print or electronic forms without written permission of Idea Group Inc. is prohibited.

Rate Control

Rate control is the most commonly-used mechanism employed in adaptive multimedia systems. Rate control can be implemented either at the server, the client, or a hybrid scheme whereby the client and server cooperate to achieve rate control.

- **Sender-based rate control:** On receipt of feedback from the client, the server adapts the transmission rate of the multimedia stream being transmitted in order to minimize the levels of packet loss at the client by matching the transmission rate of the multimedia stream to the available network bandwidth. Without any rate control, the data transmitted exceeding the available bandwidth would be discarded in the network.

- **Receiver-based rate control:** The clients control the receiving rate of video streams by adding/dropping layers. In layered multicast, the video sequence is compressed into multiple layers: a base layer and one or more enhancement layers. The base layer can be independently decoded and provides basic video quality; the enhancement layers can only be decoded together with the base layer, and they enhance the quality of the base layer.

- **Hybrid rate control:** This consists of rate control at both the sender and receiver. The hybrid rate control is targeted at multicast video and is applicable to both layered video and non-layered video. Typically, clients regulate the receiving rate of video streams by adding or dropping layers while the sender also adjusts the transmission rate of each layer based on feedback information from the receivers.

Unlike server-based schemes, the server uses multiple layers, and the rate of each layer may vary due to the hybrid approach of adapting both at the server and receiver.

Rate Shaping

Rate shaping is a technique to adapt the rate of compressed video bit-streams to meet some target bit rate by acting as a filter (or interface) between the compression layer and the transport layer. There are a number of filters that can be used to achieve rate shaping.

- **Frame-dropping filter:** This filter distinguishes between the different frame types in a video stream (i.e., I-, P- and B-frames). The frame-dropping filter is used to reduce the data rate of a video stream by discarding frames according to their relative importance. For example, B-frames are preferentially dropped, followed by P-frames and finally I-frames.

- **Frequency filter:** This filter performs filtering operations on the compression layer, for example, by discarding DCT coefficients at higher frequencies or reducing the color depth.

Copyright © 2006, Idea Group Inc. Copying or distributing in print or electronic forms without written permission of Idea Group Inc. is prohibited.

- **Re-quantization filter:** Re-quantizes the DCT coefficients. The filter extracts and de-quantizes the DCT coefficients from the compressed video stream then re-quantizes the coefficients with a larger quantization step which results in a reduced bitrate and reduced quality.

Rate Adaptive Encoding

Rate adaptive encoding performs adaptation by adjusting the encoding parameters which in turn adapts the output bit rate. However, adaptive encoding is constrained by the capabilities of the encoder and the compression scheme used. There are a number of encoding parameters that can be adapted in rate adaptive encoding, such as dynamically adapting the quantization parameter, frame rate, and/or the spatial resolution.

Discussion

The key questions that arise when developing or designing adaptation algorithms are how the system adapts and the perceived quality at the receiver.

There are a number of common components in each of the different adaptation techniques described. Many adaptation algorithms have a strong dependency on the choice of control parameters used within the adaptation process. For example, in a server-based rate control system, upon receipt of feedback the server either increases its transmission rate by α or decreases its rate by β. If the rate of α is chosen to be too large, the increased transmission rate could push the system into causing congestion, which can in turn cause the client to experience loss and poor perceived quality. However, if α is too small, the server will be very slow to make use of the extra available bandwidth and send a higher bit rate video stream. Thus, the algorithm is heavily dependent on the value of the control parameters, α and β, which drive the adaptation.

Even more problematic is translating rate into real video encoding parameters. Consider a simple system where the server is delivering video at 150kbps, and based on feedback, the algorithm indicates that the transmission rate should be increased to 160kps. The question that remains is: How should the extra 10kps be achieved, how can the video stream be adjusted to achieve this rate? This is further complicated by the limitations of the encoder to adapt the video. Layer-based schemes are equally problematic since there is no firm definition of what constitutes a base layer and each of the enhancement layers.

The most important issue that is often omitted in the design of adaptation algorithms is user-perception. User-perception should be incorporated into the adaptation algorithms, since it is the user who is the primary entity affected by adaptation, and should therefore be given priority in the adaptation decision-making process. For example, if a video clip is being streamed at a particular encoding configuration and the system needs to degrade the quality being delivered, how this adaptation occurs should be dictated by the users' perception. The way to degrade should be such as to have the least negative impact on the users' perception. There needs to be some sort of understanding of video quality and the perception of the video quality in order for adaptation to occur in an achievable and intelligent manner.

Copyright © 2006, Idea Group Inc. Copying or distributing in print or electronic forms without written permission of Idea Group Inc. is prohibited.

Review of Objective Metrics

The main goal of objective metrics is to measure the perceived quality of a given image or video. Sophisticated objective metrics incorporate perceptual quality measures by considering the properties of the *Human Visual System* (HVS) in order to determine the visibility of distortions and thus the perceived quality. However, given that there are many factors that affect how users perceive quality, such as video content, viewing distance, display size, resolution, brightness, contrast, sharpness/fidelity, and colour, many objective metrics have limited success in calculating the perceived quality accurately for a diverse range of testing conditions and content characteristics. Several objective metrics of video quality have been proposed (Hekstra, 2002; van den Branden Lambrecht, 1996; Watson, Hu, & McGowan, 2000; Winkler, 1999), but they are limited and not satisfactory in quantifying human perception (Masry & Hemami, 2002; Yu & Wu, 2000).

In this section two key objective metrics, the *Peak Signal to Noise Ratio* (PSNR) and the *Video Quality Metric* (VQM) are reviewed. These two metrics have been widely applied to many applications and adaptation algorithms to assess video quality.

Peak Signal to Noise Ratio (PSNR)

The most commonly-used objective metric of video quality assessment is the *Peak Signal to Noise Ratio* (PSNR). The advantage of PSNR is that it is very easy to compute. However, PSNR does not match well to the characteristics of HVS. The main problem with using PSNR values as a quality assessment method is that even though two images are different, the visibility of this difference is not considered. The PSNR metric does not take the visual masking phenomenon or any aspects of the HVS into consideration, that is, every single errored pixel contributes to the decrease of the PSNR, even if this error is not perceived. For example, consider an image where the pixel values have been altered slightly over the entire image and an image where there is a concentrated distortion in a small part of the image both will result in the PSNR value however, one will be more perceptible to the user than the other. It is accepted that the PSNR does not match well to the characteristics of the HVS (Girod, 1993; van den Branden Lambrecht & Verscheure, 1996).

Video Quality Metric (VQM)

The ITU-T has recently accepted the Video Quality Metric (VQM) from the National Telecommunications and Information Administration (NTIA) as a recommended objective video quality metric that correlates adequately to human perception in ITU-T J.148 (2003) and ITU-T J.149 (2004). The *Video Quality Metric* (VQM) provides a means of objectively evaluating video quality. The system compares an original video clip and a processed video clip and reports a Video Quality Metric (VQM) that correlates to the perception of a typical end user. The VQM objective metrics are claimed to provide close

Copyright © 2006, Idea Group Inc. Copying or distributing in print or electronic forms without written permission of Idea Group Inc. is prohibited.

approximations to the overall quality impressions, or mean opinion scores (Wolf & Pinson, 1999). The quality measurement process includes sampling of the original and processed video streams, calibration of the original and processed video streams, extraction of perception-based features, computation of video quality parameters, and finally calculation using various VQM models.

Using Objective Metrics for Multimedia Adaptation

Given the wide range of video quality metrics developed, the *Video Quality Experts Group* (VQEG) was formed in 1997 with the task of collecting reliable subjective ratings for a defined set of test sequences and to evaluate the performance of various objective video quality metrics (VQEG, 2005). In 2000, the VQEG performed a major study of various objective metrics on behalf of the ITU to compare the performances of various objective metrics against subjective testing in terms of prediction accuracy, prediction monotonicity, and prediction consistency. The results of the VQEG study found that no objective metric is able to fully replace subjective testing, but even more surprisingly, that no objective metric performed statistically better than the PSNR metric.

The main difficultly with video quality metrics is that even though they give an indication of the video quality, they do not indicate how the video quality should be adapted in an adaptive system. Furthermore, many of these objective metrics require a comparison between the reference clip and the degraded video clip in order to calculate the video quality. This comparison is often done on a frame-by-frame basis and therefore requires both the reference and degraded clips to have the same frame rate. The more sophisticated metrics proposed are extremely computationally intense and are unsuitable for use in a real-time adaptive system. Given the limitations of objective metrics, it has been recognized that user-perception needs to be incorporated in adaptation algorithms for streamed multimedia. There are emerging adaptive streaming systems being developed that address this issue (Muntean, Perry, & Murphy, 2004; Wang, Chang, & Loui, 2004).

Optimum Adaptation Trajectories (OATs)

This section will focus on an approach that incorporates user-perception into adaptation algorithms for video streaming. This work proposes that there is an optimal way in which multimedia transmissions should be adapted in response to network conditions to maximize the user-perceived quality (Cranley, Murphy, & Perry, 2003). This is based on the hypothesis that within the set of different ways to achieve a target bit rate, there exists an encoding configuration that maximizes the user-perceived quality. If a particular multimedia file has n independent encoding configurations, then there exists an adaptation space with n dimensions. When adapting the transmission from some point within that space to meet a new target bit rate, the adaptive server should select the encoding configuration that maximizes the user-perceived quality for that given bit rate. When the

Copyright © 2006, Idea Group Inc. Copying or distributing in print or electronic forms without written permission of Idea Group Inc. is prohibited.

transmission is adjusted across its full range, the locus of these selected encoding configurations should yield an ***Optimum Adaptation Trajectory*** (OAT) within that adaptation space.

This approach is applicable to any type of multimedia content. The work presented here focuses for concreteness on the adaptation of MPEG-4 video streams within a finite two-dimensional adaptation space defined by the range of the chosen encoding configurations. Each encoding configuration consists of a combination of frame rate and resolution and is denoted as [Frame rate $_{FPS}$, Resolution $_R$]. These encoding variables were chosen as they most closely map to the spatial and temporal complexities of the video content. The example shown in Figure 2(a) indicates that, when degrading the quality from an encoding configuration of 25fps and 100% resolution or [25_{FPS}, 100_R], there are a number of possibilities such as reducing the frame rate only, [X_{FPS}, 100_R], reducing the resolution only, [25_{FPS}, Y_R], or reducing a combination of both parameters, [U_{FPS}, V_R]. Each of these possibilities lies within a zone of ***Equal Average Bit Rate*** (EABR). The clips falling within a particular zone of EABR have different, but similar bit rates. For example, the bit rates corresponding to the encoding points [17_{FPS}, 100_R], [25_{FPS}, 79_R] and [25_{FPS}, 63_R] were 85, 88, and 82 kbps, respectively. To compare clips of exactly the same bit rate would require a target bit rate to be specified, and then the encoder would use proprietary means to achieve this bit rate by compromising the quality of the encoding in an unknown manner. Using zones of EABR effectively quantizes the bit rate of different video sequences with different encoding configurations. The boundaries of these zones of EABR are represented as linear contours for simplicity, since their actual shape is irrelevant for this scheme.

The OAT indicates how the quality should be adapted (upgraded or downgraded) so as to maximize the user-perceived quality. The OAT may be dependent on the characteristics of the content. There is a content space in which all types of video content exist in terms of ***spatial and temporal complexity*** (or *detail* and *action*). Every type of video content within this space can be expanded to an adaptation space as shown in Figure 2(b). Adaptation space consists of all possible dimensions of adaptation for the content. It can be implemented as part of an adaptive streaming server or adaptive encoder.

Figure 2(a). Adaptation possibilities   Figure 2(b). Adaptation space

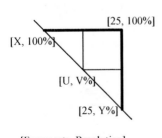

[Frame rate, Resolution]

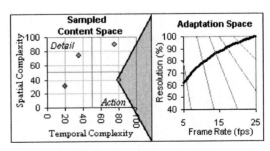

— Optimum Adaptation Trajectory (OAT)
------ Zones of Equal Average Bit Rate (EABR)

Copyright © 2006, Idea Group Inc. Copying or distributing in print or electronic forms without written permission of Idea Group Inc. is prohibited.

OAT Discovery

User perception of video quality may vary with the content type; for example, viewers may perceive action clips differently from slow-moving clips. Thus, there may exist a different OAT for different types of content based on their spatial and temporal characteristics. In order to characterize content in terms of its spatial and temporal complexity, a spatial-temporal grid was constructed, as shown in Figure 3(a). The spatial and temporal perceptual information of the content was determined using the metrics Spatial Information (SI) and Temporal Information (TI) (ITU-T P.910, 1999).

Eight different content types were selected based on their SI and TI values in order to cover as much of the Spatial-Temporal grid as possible. These test sequences were acquired from the VQEG. Each test sequence was then expanded to form an adaptation space, as shown in Figure 3(b). During the preparation of the test sequences for the subjective testing, the encoding method used was the "*most accurate*", that is, no target bit rate was specified, and the encoder followed the supplied encoding parameters as closely as possible regardless of the resulting bit rate.

The subjective testing consisted of two independent testers performing identical test procedures and using identical test sequences on subjects. Subjects were eliminated if the subject was either knowledgeable about video quality assessment or had any visual impairments. Testing was conducted in two phases. Phase One considered four test sequences, one taken from each quadrant of the SI-TI grid. To facilitate subjective testing and reduce the number of test cases, adaptation space was sampled using a logarithmic scale to reflect Weber's Law of Just Noticeable Difference (JND). Phase Two considered four different test sequences with similar SI-TI values to those used for Phase One. However, this time, the adaptation space was sampled using a linear scale. The main objective of having two different test phases was to verify and validate the results from

Figure 3(a). Spatial-temporal grid sampled with four content types for phase one of testing; (b) logarithmically-sampled adaptation space for content type C1

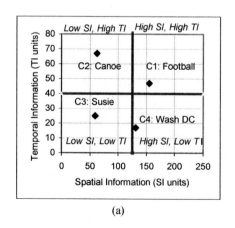

(a)

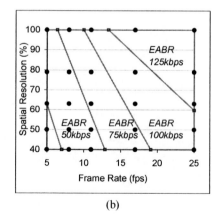

(b)

Copyright © 2006, Idea Group Inc. Copying or distributing in print or electronic forms without written permission of Idea Group Inc. is prohibited.

Phase One. In addition, by using different encoding scales, it could be verified that the OAT was similar in shape regardless of whether a linear or logarithmic scale was used, and regardless of the encoding points tested.

There are a number of different subjective testing methodologies that are proposed by the ITU-T, including the *Absolute Category Rating* (ACR), the *Degraded Category Rating* (DCR), and *Pair Comparison* (PC) methods. The DCR method uses a five-point impairment scale whilst the ACR method uses a five-point quality grading scale, or alternatively a *Continuous Quality Scale* (CQS) (ITU-T P.910, 1999). However, by using such grading scales, it is criticized that different subjects may interpret the associated grading scale in different ways and use the grading scale in a non-uniform fashion (Watson, 1998). To overcome these difficulties in the grading procedure, the *Forced Choice* methodology is often employed. In the forced choice method, the subject is presented with a number of spatial or temporal alternatives in each trial. The subject is forced to choose the location or interval in which their preferred stimulus occurred. Using the forced choice method, the bias is binary, which simplifies the rating procedure and allows for reliability, verification, and validation of the results. The subjective tests consisted of a subject watching every combination of pairs of clips from each EABR zone for each content type and making a forced choice of the preferred encoding configuration. Intra-reliability and inter-reliability of a subject were factored into the test procedure by including repetition of the same test sequence presentation.

The diagram in Figure 4 shows the subjective test results obtained for a particular content type. The diagram consists of a grid of circular encoding points where the frame rate is on the x-axis and the resolution is on the y-axis. Through these encoding points are diagonal grey lines denoting the zones of EABR, ranging from 100kbps to 25kbps. The encoding points marked with a percentage preference value are those points that were tested within a zone of EABR. For example, in EABR-100kbps, there were two encoding configurations tested, $[17_{FPS}, 100_R]$ and $[25_{FPS}, 79_R]$. Seventy percent of the subjects preferred encoding configuration $[17_{FPS}, 100_R]$, while the remaining 30% preferred encoding configuration $[25_{FPS}, 79_R]$. However, in the left-most zone of EABR, the preferred encoding configuration is $[5_{FPS}, 63_R]$. In this zone of EABR there are three encoding configurations, but since the frame rate is the same, the preferred encoding configuration is that with the highest resolution, $[5_{FPS}, 63_R]$.

The *Path of Maximum Preference* is the path through the zones of EABR joining the encoding configurations with the maximum user preference. Weighted points were then used to obtain the *Optimal Adaptation Perception (OAP)* points. The weighted points were interpolated as the sum of the product of preference with encoding configuration. For example, 70% of subjects preferred encoding $[17_{FPS}, 100_R]$ and 30% preferred encoding point $[25_{FPS}, 79_R]$. The weighted vector of these two encoding configurations is $[70\%(17_{FPS})+30\%(25_{FPS}), 70\%(100_R)+30\%(79_R)]$ which equals OAP point $[19.4_{FPS}, 93.7_R]$. The *Weighted Path of Preference* is the path joining the OAPs. There are two possible paths which can be used to represent the OAT: the path of maximum user preference, and the weighted path of preference. It seems likely that by using the weighted path of preference, the system can satisfy more users by providing a smooth graceful quality adaptation trajectory. Using the same subjective testing methodology, the OAPs in each zone of EABR were compared against the maximum preferred encoding and all other encoding configurations. In all cases, the interpolated OAP did not have a statistically-

Copyright © 2006, Idea Group Inc. Copying or distributing in print or electronic forms without written permission of Idea Group Inc. is prohibited.

Figure 4. Subjective test results for content type, C3

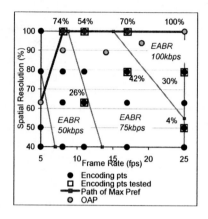

significant preference from the maximum preferred encoding indicating that this simple weighted vector approach is acceptable. It was also observed that there was a higher incidence of forced choices when the maximum preferred encoding and the OAP were close together.

Figure 5 shows the paths of maximum preference and weighted paths of preference for the four content types used during Phase One of testing. It can be clearly seen from the paths of maximum user preference that when there is high action (C1 and C2), the resolution is less dominant regardless of whether the clip has high spatial characteristics or not. This implies that the user is more sensitive to continuous motion when there is high temporal information in the video content. Intuitively this makes sense as when

Figure 5. Path of maximum user preference and weighted path of preference for four different content types

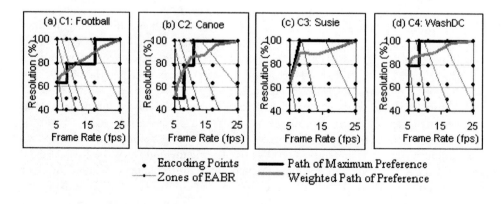

Copyright © 2006, Idea Group Inc. Copying or distributing in print or electronic forms without written permission of Idea Group Inc. is prohibited.

there is high action in a scene; often the scene changes are too fast for the user to be able to assimilate the scene detail. Conversely, when the scene has low temporal requirements (C3 and C4), the resolution becomes more dominant regardless of the spatial characteristics.

Objective metrics were investigated to determine whether they yielded an OAT that correlated to that discovered using subjective testing. The results showed that there is a significant difference between the adaptation trajectories yielded using objective metrics and subjective testing techniques. This suggests that *measuring quality* and *adapting quality based on this measurement* are different tasks.

Oats in Practice

In this section, how user-perception is affected by adapting video quality is investigated. In particular, the user-perceived quality is compared when video quality is varied by adapting the frame rate only, the resolution only, or adapting both the frame rate and the resolution using the OAT. Streaming multimedia over best-effort networks is becoming an increasingly important source of revenue. A content provider is unlikely to have the resources to provide real-time adaptive encoding for each unicast request and, as such, reserves this for "live" multicast sessions only. Typically, pre-encoded content is transmitted by unicast streams where the client chooses the connection that most closely matches their requirements. For such unicast sessions, the adaptive streaming server can employ several techniques to adapt the pre-encoded content to match the clients' resources. In such adaptive streaming systems, two techniques that are most commonly used are frame dropping and stream switching. The OAT shows how to stream the video in order to maximize the user's perceived quality in a two-dimensional adaptation space defined by frame rate and resolution (Figure 6). Adaptive frame rate can be achieved by frame dropping, while adapting spatial resolution can be achieved using track or stream switching.

All adaptation algorithms behave in an A-Increase/B-Decrease manner where A and B are the methods of change and can be either Additive, Multiplicative, Proportional, Incremental, or Decremented (Figure 7). When there is no congestion, the server increases its transmission rate either additively (AI), proportionally (PI), or multiplicatively (MI), and similarly when there is congestion, it decreases its transmission rate either additively (AD), proportionally (PD), or multiplicatively (MD). There are many ways to adapt video quality, for example:

- **Additive Increase/Multiplicative Decrease** (AIMD) (Chiu & Jain, 1989)
- **Additive Increase/Additive Decrease** (AIAD),
- **Additive Increase/Proportional Decrease** (AIPD) (Venkitaraman, Kim, Lee, Lu, & Bharghavan, 1999),
- **Multiplicative Increase/Multiplicative Decrease** (MIMD) (Turletti & Huitema, 1996).

Copyright © 2006, Idea Group Inc. Copying or distributing in print or electronic forms without written permission of Idea Group Inc. is prohibited.

Figure 6. One-dimensional versus two-dimensional adaptation

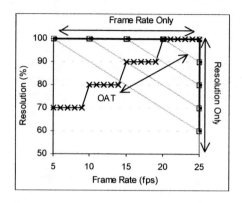

Figure 7. AIMD and MIMD

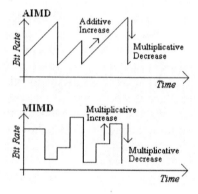

In general, all rate-control algorithms exhibit some form of AI and AD behavior, although the majority of adaptation algorithms are AIMD (Feamster, Bansal, & Balakrishnan, 2001). Thus the perception of adapting video quality is assessed in three different test cases. The first test assesses user perception when quality is adapted up in an AI manner, while the second assesses perception when quality is degraded down in an AD manner. Finally, the third assesses quality adapting in an Additive Increase/Multiplicative Decrease (AIMD) manner.

Test Methodology

The Forced Choice methodology is suitable for clips lasting not longer than 15 seconds. For video clips lasting longer than this duration, there are *recency* and *forgiveness* effects by the subject, which are a big factor when the subject must grade the overall

Copyright © 2006, Idea Group Inc. Copying or distributing in print or electronic forms without written permission of Idea Group Inc. is prohibited.

quality of a video sequence. For example, the subject may forget and/or forgive random appearances of content-dependent artifacts when they are making their overall grade of the video sequence. To test clips of a longer duration, a different test methodology to the forced choice method needs to be applied to overcome the forgiveness and recency effects and to ensure the subject can make an accurate judgement.

The *Single Stimulus Continuous Quality Evaluation* (SSCQE) methodology is intended for the presentation of sequences lasting several minutes (ITU-R BT.500-7, 1997). Continuous evaluation is performed using a slider scale on the screen to record the subjects' responses without introducing too much interference or distraction, and provides a trace of the overall quality of the sequence (Pinson & Wolf, 2003). A reference clip was played out at the beginning of the test so that the subjects were aware of the highest quality sequence. The three varying quality sequences were then presented in random order to each subject in turn. As each sequence was played out, the subject continuously rated the quality of the sequence using the slider. When the slider is moved, the quality grade of the slider is captured and related to the playout time of the media. The *Mean Opinion Score* (MOS) and standard deviation are calculated at each media time instant. In this case, each media time instant corresponds to one second of media. The MOS and standard deviation is calculated for each clip segment.

The test sequence chosen for this experiment contains a wide range of spatial and temporal complexity. The test sequence contains periods of high temporal complexity that are generally bursty containing many scene changes. In this test sequence, periods of high temporal complexity are generally followed by periods of relatively low temporal complexity but high spatial complexity consisting of detailed scenes such as facial close-ups and panoramic views. This test sequence contains a broad diversity of complexity and is typical of film trailers. The test sequence was divided into segments of 15 seconds duration, and each segment was encoded at various combinations of spatial resolution and frame rate. These video segments were then pieced together seamlessly to produce three varying bit rate versions of the test sequence. It was necessary to control and align each adaptation in each of the test sequences used. During these tests, it is assumed that some mechanism is implemented that informs the streaming server of the required transmission bit rate.

Results

Three scenarios were tested: First, the quality is adapted down from the best to worst; second, the quality is upgraded from worst to best; and third, the quality varies in an additive increase/multiplicative decrease fashion. The first two tests are complementary and are designed to assess symmetrical perception, that is, whether subjects perceive quality increases and quality decreases uniformly. The third test is designed to test quality perception in a typical adaptive network environment. Of particular interest are the MOS scores when the quality is decreased.

Copyright © 2006, Idea Group Inc. Copying or distributing in print or electronic forms without written permission of Idea Group Inc. is prohibited.

Additive Decrease Quality Adaptation

In this test, the quality of the clip degrades from the best quality to the worst quality. Figure 8(a) shows the bit rate decreasing as the quality degrades. Figure 8(b) shows the encoding configuration of frame rate and resolution for each segment as the quality is adapting down in either the frame rate dimension only, or the resolution dimension only, or using the OAT adapting down in both the frame rate and resolution dimensions. Through time interval 0-45 seconds, the resolution and frame rate dimensions are perceived the same (Figure 8(c). In time interval 45-60 seconds, there appears to be an imperceptible difference between a decrease in resolution from 80_R to 70_R. Using the OAT, there is a smooth decrease in the MOS scores, which outperforms both one-dimensional

Figure 8. Time series during additive decrease (AD) in quality; (a) Segment average bit rate variations over time; (b) Video encoding parameter variations over time; (c) MOS Scores over time; (d) MOS Scores during period of lowest quality

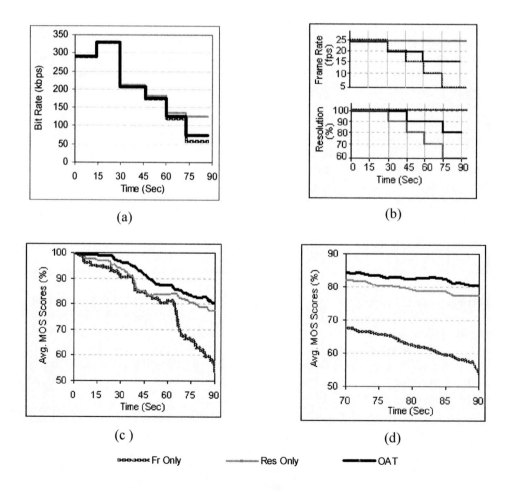

Copyright © 2006, Idea Group Inc. Copying or distributing in print or electronic forms without written permission of Idea Group Inc. is prohibited.

adaptation of frame rate and resolution. During time interval 45-60 seconds, there is high action in the content which may explain the sharp decrease in the MOS scores for adapting the frame rate only. When there is high action, subjects prefer smooth continuous motion. Further, when there is high action content, reductions in spatial resolution cannot be perceived as clearly as there is too much happening in the video clip for the detail to be perceived properly. Figure 8(d) shows a close up view of MOS scores during the lowest quality level in time interval 70-90 seconds, the frame rate is perceived worst of all while the resolution performs very well. This may be due to the fact that the bit rate for the resolution is significantly greater than the two other methods. It was undesirable to achieve a lower bit rate for the resolution at 60%, as this would require a target bit rate to be set in the encoder.

Figure 9. Time series during additive increase (AI) in quality; (a) Segment average bit rate variations over time; (b) Video encoding parameter variations over time; (c) MOS Scores over time; (d) MOS Scores during period of lowest quality

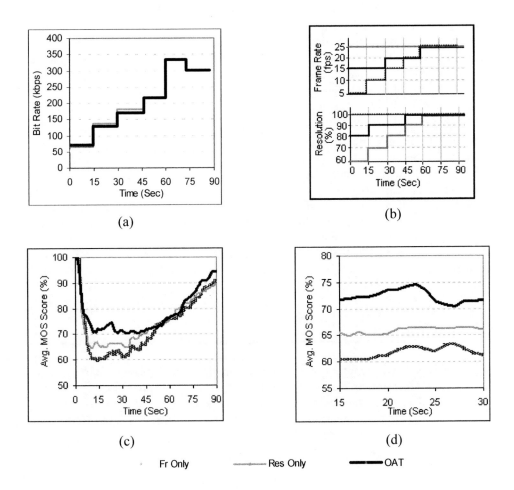

Copyright © 2006, Idea Group Inc. Copying or distributing in print or electronic forms without written permission of Idea Group Inc. is prohibited.

Additive Increase Quality Adaptation

In this test, the quality of the clip upgrades from the worst quality to the best quality. Figure 9(b) shows the encoding configuration of frame rate and resolution as the quality is adapting up in either the frame rate dimension only or the resolution dimension only or using the OAT adapting down in both the frame rate and resolution dimensions. During this experiment, the slider is placed at the highest quality value on the rating scale when the clip begins. It can be seen that it took subjects several seconds to react to the quality level and adjust the slider to the appropriate value (Figure 9(c)). At low quality, subjects perceive adaptation using the OAT better than one-dimensional adaptation. The quality is slowly increasing, however subjects do not seem to notice the quality increasing nor do they perceive it significantly differently – indicating that subjects are more aware of quality when it is low (Figure 9(d)).

AIMD Adaptation

This section presents the results for AIMD adaptation, as might be expected from a TCP-friendly rate control mechanism. The same bit rate variation patterns were obtained in these three sequences by adapting quality in the frame rate dimension only, the spatial resolution dimension only, or both frame rate and spatial resolution dimensions, as shown in Figure 10(a). The traces in Figures 10(b) show the encoding configuration of frame rate or resolution for each segment as the quality was adapted in either the frame rate dimension only, or the resolution dimension only, or using the OAT adapting in both the frame rate and resolution dimensions.

In Figure 10(a), it can be seen that although the first bit-rate reduction occurs at time 15 seconds, it is not fully perceived until time 28 seconds because there is a time delay for subjects to react to the quality adaptation. At time interval 70-90 seconds, a larger drop in bit rate occurs resulting in the lowest quality level that might reflect a mobile user entering a building. The MOS scores for adapting only the frame rate and spatial resolution are quick to reflect this drop. However, using the OAT, it takes subjects much longer to perceive this drop in quality. This is a high action part of the sequence and so the reduced frame rate is perceived more severely. The standard deviation of MOS scores using the OAT was much less than that for adapting frame rate only or spatial resolution only.

Discussion

From the experiments reported here, it appears that if a user's average bit rate changes from being quite near their maximum to near the minimum that they can tolerate, then a one-dimensional adaptation policy will cause the perceived quality to degrade quite severely. Using the two-dimensional adaptation strategy given by the OAT allows the bit rate to be dropped quite dramatically but maintain substantially better user-perceived quality.

Copyright © 2006, Idea Group Inc. Copying or distributing in print or electronic forms without written permission of Idea Group Inc. is prohibited.

Figure 10. Time series during additive increase multiplicative decrease (AIMD) in quality; (a) Segment average bit rate variations over time; (b) Video encoding parameter variations over time; (c) MOS Scores over time; (d) MOS Scores during period of lowest quality

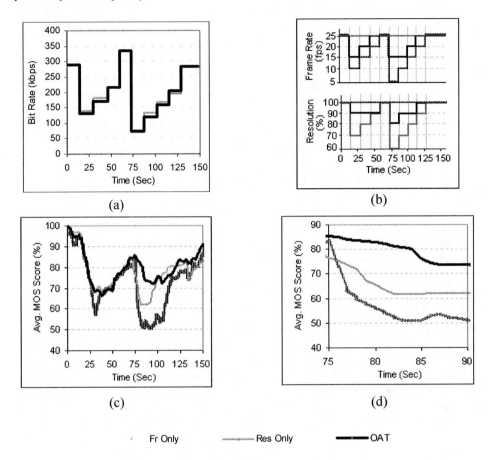

In addition to the greater bit rate adaptation range achieved using the OAT, adaptation using the two-dimensional OAT out-performs one-dimensional adaptation using frame rate or spatial resolution and reduces the variance of perception. From the various experiments conducted, subjects perceived adapting frame rate the worst, then resolution, and the OAT best of all. It was observed that there is a time delay of several seconds for subjects to react to quality adaptations. It was also observed that quality perception is asymmetrical when adapting the quality down and adapting quality up: Users are more critical of degradations in quality and less rewarding of increased quality. Similar observations were reported in Pinson and Wolf (2003).

Copyright © 2006, Idea Group Inc. Copying or distributing in print or electronic forms without written permission of Idea Group Inc. is prohibited.

Perception is strongly dependent on the spatio-temporal characteristics of the content. Given this understanding of user-perception, adaptation algorithms should consider the contents characteristics when making adaptation decisions. Also, frequent quality adaptation should be avoided to allow the users to become familiar with the video quality. In the experiments, the globally-averaged OAT was used, but the OAT can be dynamic if the contents' spatial and temporal characteristics are known at a given instant, thus making it more flexible to adapt according to the contents' characteristics and maximize user-perceived quality. It is expected that a dynamic OAT that adapted on the changing complexity of the content would yield even higher MOS scores.

Summary

This chapter provided a brief overview of adaptive streaming systems and identified key limitations of the techniques currently in use. Quite often, adaptation algorithms omit the user-perceived quality when making adaptation decisions. Recent work in multimedia adaptation has addressed this problem by incorporating objective video quality metrics into the adaptation algorithm, thereby making the adaptation process quality-aware. However, these objective metrics have limited efficacy in assessing the user-perceived quality. As a case study, we have focused on describing recent research that attempts to address both the limitations of objective video quality metrics and adaptation techniques.

This work proposed that there is an Optimal Adaptation Trajectory (OAT), which basically states that there is an optimal way video should be adapted that maximizes the user-perceived quality. More specifically, within the set of different ways to achieve a target bit rate given by an adaptation algorithm, there exists an encoding that maximizes the user-perceived quality. Furthermore, the OAT is dependent on the spatio-temporal characteristics of the content. We have described a subjective methodology to discover the OATs through subjective testing, and applied it to finding OATs for various MPEG-4 video clips. Further it was shown that using a two-dimensional adaptation strategy given by the OAT allows the bit rate to be dropped quite dramatically but maintain substantially better user-perceived quality over one-dimensional adaptation strategies. In addition to the greater bit rate adaptation range achieved using the OAT, adaptation using the two-dimensional OAT out-performs one-dimensional adaptation using frame rate or spatial resolution and reduces the variance of perception.

Future work will assess the possibility of using and/or modifying existing objective metrics in order to mimic the OATs found by subjective methods and enable the development of a dynamic OAT. This will involve a greater analysis of the relationship between content characteristics and the corresponding OAT to determine the sensitivity of an OAT to the particular video being transmitted.

Copyright © 2006, Idea Group Inc. Copying or distributing in print or electronic forms without written permission of Idea Group Inc. is prohibited.

Acknowledgment

The support of the Research Innovation Fund and Informatics Research Initiative of Enterprise Ireland is gratefully acknowledged.

References

Chiu, D. M., & Jain, R. (1989). Analysis of the increase and decrease algorithms for congestion avoidance in computer networks. *Elsevier Journal of Computer Networks and ISDN, 17*(1), 1-14.

Cranley, N., Murphy, L., & Perry, P. (2003, June). User-perceived quality aware adaptive delivery of MPEG-4 content. *Proceedings of the NOSSDAV '03*, Monterey, California (pp. 42-49).

Feamster, N., Bansal, D., & Balakrishnan, H. (2001). On the interactions between layered quality adaptation and congestion control for streamed video. *Proceedings of Packet Video.*

Ghinea, G., Thomas, J. P., & Fish, R. S. (1999). Multimedia, network protocols, and users - Bridging the gap. *Proceedings of ACM Multimedia '99*, Orlando, Florida (pp. 473-476).

Girod, B., (1993). What's wrong with mean-squared error. In A. B. Watson (Ed.), *Digital images and human vision* (pp. 207-220). Cambridge, MA: MIT Press.

Hekstra, A. P., Beerends, J. G., et al. (2002). PVQM - A perceptual video quality measure. *Signal Processing: Image Communication, 17*(10), 781-798.

ITU-R Recommendation BT.500-7 (1996). *Methodology for the subjective assessment of the quality of television pictures.* Geneva, Switzerland: International Telecommunication Union—Radiocommunications Sector.

ITU-T Recommendation J.143 (2000). *User requirements for objective perceptual video quality measurements in digital cable television.* Geneva, Switzerland: International Telecommunication Union—Radiocommunications Sector.

ITU-T Recommendation J.144 (2001). *Objective perceptual video quality measurement techniques for digital cable television in the presence of a full reference.* Geneva, Switzerland: International Telecommunication Union—Radiocommunications Sector.

ITU-T Recommendation J.148 (2003). *Requirements for an objective perceptual multimedia quality model.* Geneva, Switzerland: International Telecommunication Union—Radiocommunications Sector.

ITU-T Recommendation J.149 (2004). *Method for specifying accuracy and cross-calibration of Video Quality Metrics (VQM).* Geneva, Switzerland: International Telecommunication Union—Radiocommunications Sector.

Copyright © 2006, Idea Group Inc. Copying or distributing in print or electronic forms without written permission of Idea Group Inc. is prohibited.

ITU-T Recommendation P.910 (1999). *Subjective video quality assessment methods for multimedia applications.* Geneva, Switzerland: International Telecommunication Union—Radiocommunications Sector.

Masry, M., & Hemami, S. S. (2002, September). Models for the perceived quality of low bit rate video. *IEEE International Conference on Image Processing,* Rochester, NY.

Muntean, G. M., Perry, P., & Murphy, L. (2004, March). A new adaptive multimedia streaming system for all-IP multi-service networks. *IEEE Transactions on Broadcasting, 50*(1).

Pinson, M., & Wolf, S. (2003, July). Comparing subjective video quality testing methodologies. *SPIE Video Communications and Image Processing Conference,* Lugano, Switzerland.

Turletti, T., & Huitema, C. (1996). Videoconferencing on the Internet. *IEEE/ACM Transactions on Networking, 4*(3), 340-351.

van den Branden Lambrecht, C. J. (1996). Color moving pictures quality metric. *Proceedings of ICIP,* Lausanne, Switzerland (Vol. 1, pp. 885-888).

van den Branden Lambrecht, C.J., & Verscheure, O. (1996). Perceptual quality measure using a spatio-temporal model of the human visual system. *Proceedings of SPIE 96,* San Jose, CA.

Venkitaraman, N., Kim, T., Lee, K. W., Lu, S., & Bharghavan, V. (1999, May). Design and evaluation of congestion control algorithms in the future Internet. *Proceedings of ACM SIGMETRICS '99,* Atlanta, Georgia.

Video Quality Experts Group (VQEG) (2005). Retrieved from http://www.its.bldrdoc.gov/vqeg/

Wang, X., & Schulzrinne, H. (1999, June). Comparison of adaptive Internet applications. *Proceedings of IEICE Transactions on Communications, E82-B*(6), 806-818.

Wang, Y., Chang, S. F., & Loui, A. (2004, June). Subjective preference of spatio-temporal rate in video adaptation using multi-dimensional scalable coding. *IEEE International Conference On Multimedia and Expo (ICME),* Taipei, Taiwan.

Watson, A., & Sasse, M. A. (1998). Measuring perceived quality of speech and video in multimedia conferencing applications. *Proceedings of ACM Multimedia '98, 12-16 September 1998,* Bristol, UK (pp. 55-60).

Watson, A. B., Hu, J., & McGowan, J. F. (2001, January). DVQ: A digital video quality metric based on human vision. *Journal of Electronic Imaging, 10*(1), 20-29.

Winkler, S. (1999). A perceptual distortion metric for digital color video. *Proceedings of the SPIE,* San Jose, CA (Vol. 3644, pp. 175-184).

Wolf, S., & Pinson, M. (1999, September 11-22). Spatial-temporal distortion metrics for in-service quality monitoring on any digital video system. *SPIE International Symposium on Voice, Video, and Data Communications,* Boston.

Wu, D., Hou, T., Zhu, W., Lee, H. J., Chiang, T., Zhang, Y.Q., & Chao, H. J. (2002). MPEG-4 video transport over the Internet: A summary. *IEEE Circuits and Systems Magazine, 2*(1), 43-46.

Copyright © 2006, Idea Group Inc. Copying or distributing in print or electronic forms without written permission of Idea Group Inc. is prohibited.

Wu, D., Hou, Y. T., Zhu, W., Lee, H. J., Chiang, T., Zhang, Y. Q., & Chao, H. J. (2000, September). On end-to-end architecture for transporting MPEG-4 video over the Internet. *IEEE Transactions on Circuits and Systems for Video Technology, 10*(6), 923-941.

Yu, Z., & Wu, H. R. (2000, August). Human visual system based objective digital video quality metrics. *Proceedings of the International Conference on Signal Processing of IFIP World Computer Conference 2* (pp. 1088-1095).

Copyright © 2006, Idea Group Inc. Copying or distributing in print or electronic forms without written permission of Idea Group Inc. is prohibited.

Chapter XIII

A Three-Layer Framework for QoS-Aware Service Design

Greger Wikstrand, Umeå University, Sweden

Abstract

While increasing bandwidth might be an acceptable way to maximize user utility, it is not always sufficient. This chapter presents a three-layer model of networked multimedia application based on a division into network, application, and usage. Quality measures for each layer are presented. Mental workload, task performance, and enjoyment are proposed as the most important variables at the usage layer. Interactions between the layers are described. Examples are given of the model which has been used in actual service design work. Finally, it is suggested that the model could also be used for dynamic adaptations.

Introduction

Among the many things that a designer will have to consider when building a networked application or service, we will find things like:

Copyright © 2006, Idea Group Inc. Copying or distributing in print or electronic forms without written permission of Idea Group Inc. is prohibited.

- What is the purpose of the application?
- How will the users use it?
- Where will they use it (Hazas, Scott, & Krumm, 2004; Schmandt & Marmasse, 2004)?
- How should the application be built?
- What kind of network will the user have access to (Zhang, Zhu, & Zhang, 2005)?
- What kind of device will the user use when interacting with the service (Velez, Tremaine, Sarcevic, Dorohonceanu, Krebs, & Marsic, 2004)?

In this chapter, we shall consider a user of an application or service[1]. He or she will use it for a purpose and while connecting through a network. In reality, there might be any combination of a multitude of users, purposes, applications, and networks. Nevertheless, we shall focus on a single user of a single application. The objective of the designer of this application should be to maximize user *utility*, whatever that is, a most interesting question as we shall discover. The designer of the application is mainly able to maximize the utility by manipulations in the application or in the network.

In certain cases, for example, in the case of streaming video over a "free" network, the problem is simple to solve. The utility function simply becomes $Utility(t)=f(b(t))$ where $f()$ is an s-shaped, increasing function: no bandwidth, no utility; more bandwidth, more utility; threshold bandwidth, no further improvements in utility. In that case, all the application has to do is to ensure that $b(t)$ is maximized for a single user, or if there are several users to maximize $\sum f_i$ for all users $i \in 1..n$ and at the same time ensuring that

$\sum b_i \leq B$ where B is the total available bandwidth (Bocheck, Campbell, Chang, & Liao, 1999; Liao & Campbell, 2001).

When the factors mentioned in the very beginning of this chapter are considered, it becomes possible that one media channel might have a higher utility to the user even if it gets less bandwidth. A simple example illustrates this point. Suppose that a user is spectating a football match and that there are two audio commentator channels available: One supports the same team as the user while the other supports the opposing team. Spectators prefer listening to commentators that support their own team (Bryant & Raney, 2000) so even if the first commentator channel was allocated less bandwidth, the user would probably prefer it anyway. It would make the game more enjoyable and thus have higher utility for the user.

Structure of This Chapter

This chapter introduces a framework for understanding, developing, and controlling networked services and applications based on a division into three layers (Wikstrand, 2003). First the model is described. Then the question of utility as seen from the different layers. A number of examples of how the model was used in actual projects where the

Copyright © 2006, Idea Group Inc. Copying or distributing in print or electronic forms without written permission of Idea Group Inc. is prohibited.

author participated are described as well. Finally, future use of the model is discussed briefly.

The Three-Layer Model

Three–layer models similar to the one being introduced in this chapter have been used by other researchers. Yet other researchers have chosen to classify quality of service parameters in other ways. A few of these will be presented below. For a better understanding of how they are described below, the reader might want to have a look at how the layers are defined.

Other Models of Quality of Service

Fodor, Eriksson, and Tuoriniemi (2003) present many of the important issues involved in maintaining and controlling quality of service in an always best-connected network, that is, a network where the user (or his/hers device) roams between networks with different characteristics. They present an explicit model of how to negotiate quality of service at the application layer, for example, through SIP and at the network layer through IP QoS signaling. But as can be seen most clearly in their Figure 4, their view of the user/usage layer is rather crude: basically, a stylized human smiling to various degrees.

Lu, Kitagata, Suganuma, and Kinoshita (2003) present a model which might be considered a bit more developed on the user side. The main benefit of their model is that it explicitly involves the user in the decision-making process. A host of software agents are then employed to ensure that the human desires are met.

The model presented by Vogel, Kerhervé, Bochmann, and Gecsei (1995) is mainly concerned with application layer quality parameters as categorized into five broad categories: performance (e.g. delay and bit-rate), format (e.g. spatial and temporal resolution), synchronization, cost, and user. Naturally, many of these categories also have components from the network layer. The user-related category as presented by Vogel et al. (1995) remains on the application layer because it is not concerned with the utility or the usefulness of the usage, but with users' perception of the application layer quality of service.

Curcio (2002) presents only two categories: objective and subjective. In each of the categories, a number of measurements are presented: objective corruption versus "Degradation Category Rating". While two categories are presented, it is this author's view that both of them measure the same thing but with different instruments, that is, the quality of service at the application layer.

Jin and Nahrstedt (2004) divide QoS specifications into three layers. The user specifies his/her expectations at the *user* layer. These are then translated manually or automatically into an *application–layer* specification. This mapping is supposed to be independent from any knowledge of the underlying infrastructure. Finally, this specification is

Copyright © 2006, Idea Group Inc. Copying or distributing in print or electronic forms without written permission of Idea Group Inc. is prohibited.

translated into a concrete and specific *resource-layer* specification. The user-layer specification is specified in terms of perceptive media quality (excellent to bad), window size (big, medium, small), pricing model and range of price (high, medium, low). The application-layer specification is expressed in terms of quantitative issues (frame rate, resolution), qualitative issues (e.g. synchronization schemes), and adaptation rules. Finally, the resource-layer is concerned with the same general issues as the application level, but expressed in other terms such as throughput, delay, jitter, memory size, and more concrete adaptation rules.

The Three-Layer Model

A few other models have been proposed to serve as a framework for specifying quality of service in networked applications. Suffice it to say that they, like the models presented above, mostly strive to be content- and usage-independent. When the user is involved, it is mostly at the application layer, either by judging the quality at that layer or by giving some sort of instructions as to how it should be presented. Much thought seems to have been given to how to translate technical terms into user-understandable ones.

However, it was demonstrated above that it is quite possible that content quality might be more important than application layer quality, provided of course that the application layer quality meets some minimum requirements. What is needed is a model which explicitly takes the usage of the application into account. The following is a description of such a model. It is based on a division into the following three layers: usage, application, and network (also see Figure 1).

The **usage layer** is concerned with anything "above" the hardware/software layers such as, for instance, "watching football", "playing a game", or "talking to a friend". It is here that the meaning of the application resides, and it is here that the utility of the application must be ultimately judged. That can be done by asking the user, observing him/her, and/or by modeling the user's interaction with the application. The utility will be composed of several components, for example, task performance and emotional satisfaction.

The **application layer** is concerned with enabling the usage mentioned above. The analogue of the previous could be "streaming video", "a mobile game", and "conversational multi-media". The question of how to judge and measure quality at this and the network layer has been much discussed above and shall be more specifically addressed in the following.

The **network layer** is concerned with the transport of data and so on as well as connecting various pieces of the application with each other. Concepts on this layer include "broadband", "WCDMA", "WLAN".

The Layers in Action: An Example

Each of the three layers interacts with and puts constraints on the other layers. For instance, a person wishing to follow a football game on his/hers mobile phone might use a streaming video application over a mobile network. As the user roams, reception

Copyright © 2006, Idea Group Inc. Copying or distributing in print or electronic forms without written permission of Idea Group Inc. is prohibited.

Figure 1. Examples of concepts in each layer of the model; note that there is no one-to-one correspondence between layers, for example, streaming video can be used for other purposes than watching football

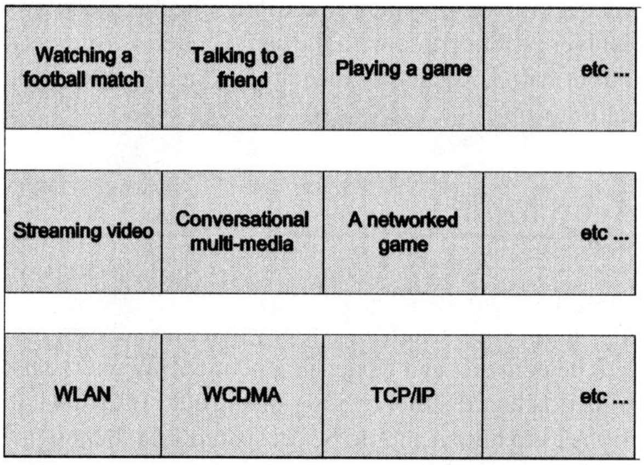

Watching a football match	Talking to a friend	Playing a game	etc ...
Streaming video	Conversational multi-media	A networked game	etc ...
WLAN	WCDMA	TCP/IP	etc ...

changes and the network service level becomes variable, and so does the streaming video quality. At this stage, adaptations can take place on all layers of the model. In the network layer, the user might change his mobility pattern to favor better reception or pay more for premium network services. In the application layer, the user might switch to a different viewing mode, for example, from streaming video to audio commentary only. In the usage layer, the user might even decide to do something completely different.

In this example, all the adaptations were made by the user. In principle, it is possible for agents in each of the layers to induce changes in the other layers or their own layer. These agents might be the user or another human. I think that in the sports example it would be beneficial to have a human producer who used his skill and judgment together with adapted tools to produce different versions of the event for different end user situations. The agents in question might also be software agents who negotiate among themselves in order to reach their objectives.

Delving further on how this run-time adaptation might take place goes beyond the scope of this chapter, but it is clearly an interesting area to consider in the future.

Quality Measures

Each of the layers provides a certain service level to the layer above it. For each of the layers used here, several measures of the service level have been presented and used in the literature (Wikstrand, 2003). The measures of the service level are often called the "Quality of Service".

Copyright © 2006, Idea Group Inc. Copying or distributing in print or electronic forms without written permission of Idea Group Inc. is prohibited.

Quality of Service (QoS) is a complex concept in many ways. On one hand, it refers to the quality of service as offered by a layer to the layer above it and as judged by the upper layer. On the other hand, it refers to the mechanism used to provide that quality of service and is measured in terms relevant to the layer offering the service (Pulakka, 2003).

Because of this dichotomy, it is important to know what we talk about when we talk about quality of service. Who is the judge of the quality and on which scale is it measured? How can measures used in one layer be compared to measures used in another layer? What is necessary to do in each layer in a given situation to achieve the intended goals?

In this section, a brief review of quality of service measures which have been proposed or used by other researchers is presented. Placing the quality measures in the model is important because it allows us to decide which measures are more important in a given application.[2]

Network Layer

In the network layer mostly "objective" performance measures are used. The most basic concepts are delay and bit–rate (sometimes called bandwidth) (Wang, Clayppol, & Zuo, 2001). Delay variability is called jitter and is sometimes more important than the delay itself (Wang et al., 2001). Especially in wireless networks, packet loss is an important measure which affects both of the previously-mentioned variables (Fong & Hui, 2001; Ghinea & Thomas, 2000; Hands & Wilkins, 1999; Koodli & Krishna, 1998; Patel & Turner, 2000; Verscheure, Frossard, & Hamdi, 1998). Obviously, if a packet is lost, the bit-rate goes down, and in certain protocols there will be a delay when the packet is retransmitted.

Application Layer

On the application layer, there is a rich flora of service level measurements. In streaming video and conversational multi-media, for instance, there are a host of measures, both objective and subjective, regarding the fidelity of the video compared to an imagined original which was not transported over a network.

There are many subjective fidelity measures. In the double stimuli continuous quality scale, a human judge expresses their level of preference between an original and a degraded version (Winkler, 2001). In many other approaches, the judge is asked to compare the degraded version with an imagined "broadcast quality" or the "live event" (Kies, Williges, & Rosson, 1997; Procter, Hartswood, McKinlay, & Gallacher, 1999). In cases where only a small level of degradation is to be expected, an error detection and annoyance level approach might be feasible (Moore, Foley, & Mitra, 2001; Steinmetz 1996).

There are also many "objective" measures of service levels here. For video, peak signal to noise ratio is often used, for example, when comparing video codecs (Verscheure et al., 1998; Winkler, 2001). Other common measures are bit-rate, frame rate, spatial and chromatic resolution, and so on (Wang et al., 2001).

Copyright © 2006, Idea Group Inc. Copying or distributing in print or electronic forms without written permission of Idea Group Inc. is prohibited.

Some measures try to predict, based on objectively measurable features of the video, and so forth, how a human judge would subjectively rate the service level. Such measures include the moving pictures quality metric (Verscheure et al., 1998) and the perceptual distortion metric (Winkler, 2001).

Usage Layer

It is under this heading that we should find the mythical "utility" mentioned above. However, no such single variable can be identified, and we shall have to be content with considering several which will be used together in various combinations to form the overall utility.

In traditional Human Factors work, three important objectives are usually considered. The "worker" should be as productive as possible while minimizing any harm done to him or her. He or she should also be able to find some kind of intrinsic motivation or enjoyment in performing the assigned task. The balance shifts between these three objectives, depending on the situation. For instance, a person who might not want to risk his or her health in the least way in the line of duty might happily go parachuting or swimming with sharks in his or her spare time.

In the usage layer, the most meaningful measure of the "harm" done to the user might be task-induced stress or mental workload (de Waard, 1996; Wickens, 1992; Wilson & Sasse 2000).

Task performance will obviously be measured in different ways depending on the actual task and usage situation. For video, examples include the ability to read lips, recognize faces and emotions, understanding of the presented material, and recall of things presented in the session (Barber & Laws, 1994; Procter et al., 1999; Kies et al., 1997). Specifically for conversational multimedia measures might include how well meaning is conveyed and agreed upon (Veinott, Olson, Olson, & Fu, 1999).

For games, usage layer quality measures can include the level of fun and enjoyment as well as the ability to "win the game" or get a high score.

In any case, it is crucial that the designer understands the usage area and what might be the relevant factors which influence the three main objectives in that particular context. Unlike in the other two layers where most of the same measures can be used regardless of what the actual context is, here it will be more specific. It is also important to consider how pre-existing differences between users will affect the actual measurements. For instance, it will be shown below how users' subject-matter expertise affected their preference for different application layer encodings.

Measures Used by the Author

As we shall see in the next section, the author has used this model in a variety of contexts. In all these cases, a basic model for measuring usage and, to some extent, application layer quality has been used. The model is largely based on self-report measures, as these are easier to administer and more likely to be possible to use in a semi–automated process.[3]

Copyright © 2006, Idea Group Inc. Copying or distributing in print or electronic forms without written permission of Idea Group Inc. is prohibited.

Mental workload was measured using the Rating Scale Mental Effort (RSME) (Ziljstra, 1993). The RSME has been validated and found to be accurate for workloads of around 60–90 seconds (Verwey & Veltman, 1996).

Task performance was measured objectively where possible, that is, through a quiz after each clip in the Bastard project (see below) and as the score in the networked games project (also below).

Enjoyment was measured through self-report. Certainly, a host of other measures have been proposed ranging from measuring hormone levels to observations of "laughs per hour". It was felt unnecessary to go to such lengths for this research.

Design Tool

The important issue is to be able to use the model as a design and analysis tool. Its justification lies in its ability to serve as a useful tool to those who want to build an optimal service.

Finding Out What is Important

First of all, just like in any human-computer related activity (and as explained above), it is necessary to find out what is important at the usage layer. This chapter is not intended as a full–fledged guide on design methodology, but I shall nevertheless mention some activities which I have found to be useful here.

- Reading scholarly works concerning what is important to persons involved in the activity: In the case of the Arena and the Bastard projects (see below), one of the main sources was Wann, Melnick, Russell, and Pease (2000).
- Interviewing current and potential users of the application: Prototypes at various levels of sophistication might be useful to focus the discussion.
- Involving a subject-matter expert in the design team: In the Bastard project, one of the team members was a player in the Swedish national football team.
- Looking at similar work as presented in scholarly or commercial contexts: This might show what is possible and desirable, and might also give new ideas.

Interactions Between Layers

One of the purposes of the model is to make us consider each of the three layers and the interactions between them. For instance, we can drop application layer quality if the "home" team is doing badly so that the user does not have to see them "crushed". Or we can implement a synchronization algorithm in the application layer to compensate for

Copyright © 2006, Idea Group Inc. Copying or distributing in print or electronic forms without written permission of Idea Group Inc. is prohibited.

problems in the network. These examples might be seen as far-fetched and maybe even as some kind of "cheating". Their main purpose is to get the reader started in thinking in new ways to take advantage of the interactions between the layers.[4]

Interactions with the Network Layer

Interactions between the network and the application layers have been described in much detail elsewhere. The reader should start by looking at the references provided earlier in this chapter. In general, the idea is that the application layer and the network layer will affect each other.

If, for instance, the bit-rate goes down at the network layer, the application layer might adapt in various ways. One way might be to demand of the network layer that it again increases the bit-rate, for example, by switching to a different network provider. A more realistic approach in most cases might be to make changes at the application layer. This could be done by changing, for example, the codec, the frame rate, or the spatial resolution.

Anyone who has used a mobile phone might have noticed another way to deal with the connectivity problem at the network layer through interactions with another layer, in this case the usage layer. An icon advises the user of reception conditions. The user might then use this information to first of all understand that connectivity is non-existent or poor and might then use this information to take action, for example, defer calling, finding another way to communicate, and so forth. See below for a similar solution for networked games.

Interactions with the Application Layer

The interactions between the application layer and the usage layer are perhaps the most important and interesting ones in the context of this chapter. Above, it was demonstrated that the content might be more important to the user than the application layer quality. That does not mean that if all else is equal, that the application layer quality is unimportant. On the contrary, as shown in the introduction, more is indeed generally better.

In Mueller and Agamanolis (2005), the information being transmitted between the two sites is exactly the same, how a ball travels through a room and hits a screen, but in one case the user interacts with the application through a traditional computer interface, and in the other case through a physical "exertion" interface. It turns out that not only did the users have more fun and get to know the other player better when using the exertion interface, they also thought that the transmitted audio and video quality was better. This is an example of how changes at the application layer (the interface) lead to changes in the usage layer (the interaction), which in turn leads to changes in how the application layer quality is perceived.

But there are simpler interactions between the two layers. In the Bastard project (see below) we were able to show a significant interaction between secondary task load, which

Copyright © 2006, Idea Group Inc. Copying or distributing in print or electronic forms without written permission of Idea Group Inc. is prohibited.

is arguably a usage layer characteristic, and perceived video quality. This shows that the designer must also consider the context of usage.

It might not be possible to provide an exhaustive list of all interesting interactions between the usage layer and the other two layers, and it shall not be attempted here. Suffice it to say that just as it was with objectives and measures as discussed above, so is it here. The specifics of the usage domain will have a vast influence on what interactions are possible and important. In the following, I shall give examples of how the three–layer model was used in projects in which I were involved.

Example: The Arena Project

I was working on a project[5] where we were investigating how to get an interactive, multimedia service distributed to a huge number of users in a sports arena. Each user would have a hand-held computer connected through a wireless local area network. We realized that the biggest bottleneck was the network layer. Performance in an IEEE 802.11 network will decrease dramatically with an increasing number of users, both in terms of system throughput and user throughput (Bianchi, 2000).

Proposed Solutions

On the usage layer, we wanted to give the user the impression that he or she had a fully interactive experience with instant access to live and recorded video from the game. The live video could be provided through multi-casting a video stream.

The recorded video was trickier. We decided that we would have to find a solution on the application layer, and we devised a distributed pre-caching scheme (Svanbro, Lindahl, Jonsson, Rosenqvist, & Wikstrand, 2003). This scheme also relied on multi-cast. The scheme was evaluated through simulations and by building a prototype, but because of the nature of how it would be used, with a large number of users, it was not deemed practical to perform a full–scale user test.

Through trials and simulations, we discovered that multi-cast in an IEEE 802.11 network suffers from a huge drawback. Packets will often collide with other traffic and be lost. We devised a media access algorithm which would solve the problem by changing the way that multi-cast worked in such networks (Nilsson, Wikstrand, & Eriksson, 2002). The solution was tested in simulations but not on real users.

So in summary, the usage situation with a lot of users in a limited area who wanted to use bandwidth-intensive applications which would overload the network demanded a solution where we worked with both the application and the network layer.

Example: The Bastard Project

An unsolved problem from the previous project was how to deal with a situation where a user wanted to access the same services but away from the arena where the access

Copyright © 2006, Idea Group Inc. Copying or distributing in print or electronic forms without written permission of Idea Group Inc. is prohibited.

would be through a different kind of network, for example, WCDMA or another 3G technology. What we saw as the biggest problems here were access cost and service level variability from the network. As the user moved around, and as other conditions changed in the network, different bandwidth would be available to the user at different costs. We saw a similar problem in the usage layer. The user would have different levels of interest during a game, and also would not be interested in paying for maximum bandwidth throughout a game, as that would be prohibitively expensive.

Proposed Solution

We devised a solution in the application layer which would meet the requirements posed by the network in the form of variable service levels and from the user in terms of different cost as a function of time. Our proposed solution was to code the game both as video (Figure 2(a)) and as positions (Figure 2(b)). The positions would then be rendered as simple graphics in the user terminal. Streaming the positions would require less than 1 kbps while the video might go as high as 350 kbps. Together the two different modes provided a wide range of requirements on the network. The system would then switch between different levels of video quality and the simple position-based graphics, based on input from the user as well as network conditions and the cost of available bandwidth.

We performed experiments based on short (about 90 seconds) sections of a game which we had encoded in three different ways: as animations, as low-quality video, and as high-quality video. The experiments were based on users watching and rating the clips as described above. There were a total of 89 participants in the experiment. We also tested the users' levels of sports fandom (Wann et al., 2000) and their level of knowledge about the sport (Knecht & Zenger, 1985). Both of these variables have been shown to affect how people react when spectating sports.

It turned out that the position-based animations were more popular than we had imagined, especially among users who were not as knowledgeable in the target sport in comparison to low bandwidth (20 kbps) video. The animations were easier to understand while the

Figure 2. (a) A video frame from the experiment and the corresponding animation; (b) frame

Copyright © 2006, Idea Group Inc. Copying or distributing in print or electronic forms without written permission of Idea Group Inc. is prohibited.

Figure 3. Significant interactions between coding and secondary task

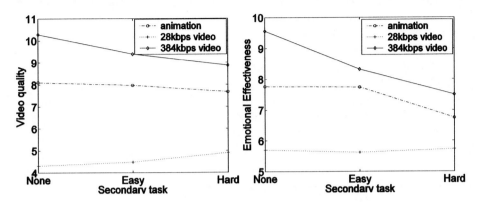

video was more emotional. Figure 3 illustrates how a secondary task affected the perceived emotional effectiveness and the video quality. Video quality was judged to increase with low bit-rate video when the secondary task was harder. (Wikstrand & Eriksson, 2002; Wikstrand, Eriksson, & Östberg, 2003; Wikstrand & Sun, 2004)

Example: Networked Games

A basic problem in networked games is that one or more users might be delayed in relationship to each other or to a central server if there is one. This will have an effect on the usage layer, as users will try to compensate for the resulting lack of responsiveness and inconsistent behavior which results (Vaghi, Greenhalgh, & Benford, 1999). Many researchers have looked at the gravity and impact of the problem and on how to compensate for it on the network and the application layer.

Possible Solutions

Solutions on the network layer are mostly concerned with or based on providing or obtaining differential quality of service or guaranteed quality of service.

On the application layer, many different solutions have been suggested. The most common might be so called dead reckoning (Aggarwal, Banavar, Khandelwal, Mukherjee, & Rangarajan, 2004). Other solutions are based on being able to go back to a previously-known safe state, for example, time warp and trailing state synchronization (Cronin, Filstrup, Kurc, & Jamin, 2002). The first solution is based on retracing the execution back to a safe point. The second solution is a simplified version of the first. In it, a delayed version of the game is run as a special process. When a problem occurs, its state is copied to all participating entities. Yet another type of solution is to avoid inconsistency

Copyright © 2006, Idea Group Inc. Copying or distributing in print or electronic forms without written permission of Idea Group Inc. is prohibited.

between different entities by using time buckets or breathing time buckets (Diot & Gautier, 1999; Steinman, 1993). In this type of solution, each message is delayed at the receiving end for a period of time, which is dependent on the known delay between the sending and receiving processes, and until it is safe to assume that all recipients have received all the messages for that time step.

Each of the three different types of solutions creates problems in the usage layer while attempting to solve problems in the network layer. Dead reckoning leads to inconsistencies when it turns out that the simulated track differs from the actual track. Time warp creates huge inconsistencies when the game suddenly moves back to a previous state. Time buckets introduce additional delay but mostly avoids inconsistencies.

Proposed Solution

We studied a solution which made a change in the application layer (Wikstrand, Schedin, & Elg, 2004). It was aimed at making the user in the usage layer more adapted to the delay problems in the network. In a simple two-player game, pong, we introduced a delay as well as some sort of feedback to the user about the level of the delay (Figure 4).

We performed an experiment where we let two users play pong against each other for about sixty seconds per round. We then varied the level of the delay and how the gravity of the delay was presented to the players.

It turns out that having the feedback does let the user adjust to the level of the delay in a better way, that is, the mental effort was more adapted to the delay and the user was more tolerant to the delay (Figure 5). Other researchers have performed a similar study regarding collaborative virtual environments (Gutwin, Benford, Dyck, Fraser, Vaghi, & Greenhalgh, 2004).

Figure 4. The pong game used in the project; a faint shadow on the left provides feedback to the user about the delay

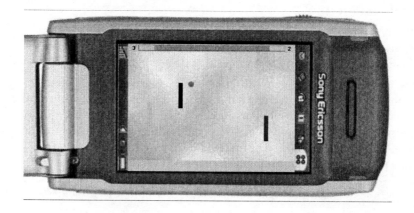

Copyright © 2006, Idea Group Inc. Copying or distributing in print or electronic forms without written permission of Idea Group Inc. is prohibited.

Figure 5. Effects on mental effort of getting feedback about the delay

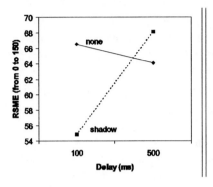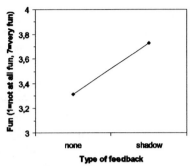

Summary

This paper has presented a three-layer model of networked services and applications. The model is based on three layers: usage, application, and network. By setting targets for and measuring "utility" in appropriate layers of the model, it becomes possible to consider new ways to solve problems. Changes can be made in or between each of the levels as shown through examples. The examples show that a problem in the network layer, which unaddressed would have to be dealt with in the usage layer, can be addressed in the network layer itself, in the application layer, or even in the usage layer.

The next step is to use the model for run-time adaptation. Adaptation can be performed in some measure in each layer of the model, and might also include negotiations between the layers.

The main contribution of this paper is the presentation of the three-layer model and its generalization to different applications and usages.

References

Aggarwal, S., Banavar, H., Khandelwal, A., Mukherjee, S., & Rangarajan, S. (2004). Accuracy in dead-reckoning based distributed multi-player games. *Proceedings of ACM SIGCOMM 2004 Workshops on Netgames '04* (pp. 161-165). New York: ACM Press.

Barber, P. J., & Laws, J. V. (1994). Image quality and video communication. In R. I. Damper, W. Hall, & J. W. Richards (Eds.), *Proceedings of the IEEE International Symposium on Multimedia Technologies and Future Applications* (pp. 163-178). London: Pentech Press.

Copyright © 2006, Idea Group Inc. Copying or distributing in print or electronic forms without written permission of Idea Group Inc. is prohibited.

Bianchi, G. (2000). Performance analysis of the IEEE 802.11 distributed coordination function. *IEEE Journal on Selected Areas in Communication, 18*(3), 535-547.

Bocheck, P., Campbell, A., Chang, S. -F., & Liao, R. -F. (1999). Utility-based network adaptation for MPEG-4 systems. *Proceedings of the 9th International Workshop on Network and Operating System Support for Digital Audio and Video* (pp. 279-288). Florham Park, NJ: AT&T Labs. Research.

Bryant, J., & Raney, A. A. (2000). Sports on the screen. In D. Zillman & P. Vorderer (Eds.), *Media entertainment - the psychology of its appeal* (pp. 153-174). Mahwah, NJ: Lawrence Erlbaum Associates.

Cronin, E., Filstrup, B., Kurc, A. R., & Jamin, S. (2002). An efficient synchronization mechanism for mirrored game architectures. *Netgames '02: Proceedings of the 1st Workshop on Network and System Support for Games* (pp. 67-73). New York: ACM Press.

Curcio, I. (2002). Mobile video QoS metrics. *International Journal of Computers & Applications, 24*(2), 41-51.

de Waard, D. (1996). *The measurement of drivers' mental workload.* Unpublished doctoral dissertation, University of Groningen.

Diot, C., & Gautier, L. (July-Aug. 1999). A distributed architecture for multiplayer interactive applications on the internet. *IEEE Network, 13*(4), 6-15.

Fodor, G., Eriksson, A., & Tuoriniemi, A. (2003, July). Providing quality of service in always best connected networks. *IEEE Communications Magazine, 41*(7), 154-163.

Fong, A. C. M., & Hui, S. C. (2001). Low-bandwidth internet streaming of multimedia lectures. *Engineering Science and Education Journal, 10,* 212-218.

Ghinea, G., & Thomas, J. P. (2000). Impact of protocol stacks on quality of perception. *Proceedings of the International Conference on Multimedia and Expo* (pp. 847-850). New York: IEEE.

Gutwin, C., Benford, S., Dyck, J., Fraser, M., Vaghi, I., & Greenhalgh, C. (2004). Revealing delay in collaborative environments. *CHI '04: Proceedings of the SIGCHI Conference on Human Factors in Computing Systems* (pp. 503-510). New York: ACM Press.

Hands, D., & Wilkins, M. (1999). A study of the impact of network loss and burst size on video streaming quality and acceptability. In M. Diaz, P. Owezarski, & P. Sénac (Eds.), *IDMS '99* (pp. 45-57). Berlin, DE: Springer Verlag.

Hazas, M., Scott, J., & Krumm, J. (2004). Location-aware computing comes of age. *Computer, 37*(2), 95-97.

Jin, J., & Nahrstedt, K. (2004). QoS specification languages for distributed multimedia applications: A survey and taxonomy. *IEEE Multimedia,* 74-87.

Kies, J. K., Williges, R. C., & Rosson, M. B. (1997). Evaluating desktop video conferencing for distance learning. *Computers & Education, 28,* 79-91.

Knecht, R. S., & Zenger, B. R. (1985). Sports spectator knowledge as a predictor of spectator behavior. *International Journal of Sport Psychology, 16,* 270-279.

Copyright © 2006, Idea Group Inc. Copying or distributing in print or electronic forms without written permission of Idea Group Inc. is prohibited.

Koodli, R., & Krishna, C. M. (1998). A loss model for supporting QoS in multimedia applications. *ISCA CATA-98* (pp. 234-237). Cary, NC: ISCA.

Liao, R. R. -F., & Campbell, A. T. (2001). A utility-based approach for quantitative adaptation in wireless packet networks. *Wireless Networks, 7*(5), 541-557.

Lu, L., Kitagata, G., Suganuma, T., & Kinoshita, T. (2003). Adaptive user interface for multimedia communication system based on multiagent. *Proceedings of the 17th International Conference on Advanced Information Networking and Applications (AINA '03)* (pp. 53-58). Piscataway, NJ: IEEE.

Moore, M. S., Foley, J. M., & Mitra, S. K. (2001). Comparison of the detectability and annoyance value of embedded MPEG-2 artifacts of different type, size, and duration. *Proceedings of the SPIE: Vol. 4299*, 90-101.

Mueller, F. F., & Agamanolis, S. (2005). Sports over a distance. *Comput. Entertainment, 3*(3), 4-4.

Nilsson, T., Wikstrand, G., & Eriksson, J. (2002). Early multicast collision detection in CSMA/CA networks. In M. Gerla & C. G. Omidyar (Eds.), *MWCN, 2002* (pp. 294-298). Piscataway, NJ: IEEE.

Patel, D., & Turner, L. F. (2000). Effects of ATM network impairments on audio-visual broadcast. *IEE Proceedings - Vision, Image, and Signal Processing, 147*, 436-444.

Procter, R., Hartswood, M., McKinlay, A., & Gallacher, S. (1999). An investigation of the influence of network quality of service on the effectiveness of multimedia communication. *Proceedings of the International ACM SIGGROUP Conference on Supporting Group Work* (pp. 160-168). New York: ACM Press.

Pulakka, K. (2003). A dynamic control system for adjusting prices and quality of service in DS enabled networks. *Network control and engineering for QoS and mobility* (pp. 241-252). Norwell, MA: Kluwer Academic Publishers.

Schmandt, C., & Marmasse, N. (2004). User-centered location awareness. *Computer, 37*(10), 110-111.

Steinman, J. S. (1993). Breathing time warp. *PADS '93: Proceedings of the Seventh Workshop on Parallel and Distributed Simulation* (pp. 109-118). New York: ACM Press.

Steinmetz, R. (1996). Human perception of jitter and media synchronization. *IEEE Journal on Selected Areas in Communications, 14*, 61-72.

Svanbro, K., Lindahl, G., Jonsson, E., Rosenqvist, E., & Wikstrand, G. (2003). *A system and a method relating to communication of data* (Patent application No. WO 03/069507 A1). Stockholm, Sweden: Telefonaktiebolaget L M Ericsson.

Vaghi, I., Greenhalgh, C., & Benford, S. (1999). Coping with inconsistency due to network delays in collaborative virtual environments. *Proceedings of the ACM Symposium on Virtual Reality Software and Technology* (pp. 42-49). New York: ACM Press.

Veinott, E. S., Olson, J., Olson, G. M., & Fu, X. (1999). Video helps remote work: Speakers who need to negotiate common ground benefit from seeing each other. *CHI '99: Proceedings of the SIGCHI Conference on Human Factors in Computing Systems* (pp. 302-309). New York: ACM Press.

Copyright © 2006, Idea Group Inc. Copying or distributing in print or electronic forms without written permission of Idea Group Inc. is prohibited.

Velez, M., Tremaine, M. M., Sarcevic, A., Dorohonceanu, B., Krebs, A., & Marsic, I. (2004). Who's in charge here? Communicating across unequal computer platforms. *ACM Trans. Computer-Human Interactions*, *11*(4), 407-444.

Verscheure, O., Frossard, P., & Hamdi, M. (1998). Joint impact of MPEG-2 encoding rate and ATM cell losses on video quality. *IEEE Globecom* (pp. 71-76). Piscataway, NJ: IEEE.

Verwey, W. B., & Veltman, H. A. (1996). Detecting short periods of elevated workload: A comparison of nine workload assessment techniques. *Journal of Experimental Psychology: Applied*, *2*, 270-285.

Vogel, A., Kerhervé, B., von Bochmann, G., & Gecsei, J. (1995). Distributed multimedia and QOS: A survey. *IEEE Multimedia*, (Summer), 10-19.

Wang, Y., Clayppol, M., & Zuo, Z. (2001). An empirical study of RealVideo performance across the Internet. *Internet Measurement Workshop* (pp. 295-309). New York: ACM Press.

Wann, D. L., Melnick, M. J., Russell, G. W., & Pease, D. G. (2000). *Sport fans: The psychology and social impact of spectators*. New York: Routledge.

Wickens, C. D. (1992). *Engineering psychology and human performance* (2nd ed.). New York: HarperCollins.

Wikstrand, G. (2003). *Improving user comprehension and entertainment in wireless streaming media*. Licentiate thesis, Umeå University, Umeå, Sweden.

Wikstrand, G., & Eriksson, S. (2002). Football animations for mobile phones. In O. W. Bertelsen, S. Bødker, & K. Kuutti (Eds.), *Proceedings of NordiCHI '03* (pp. 255-258). New York: ACM Press.

Wikstrand, G., Eriksson, S., & Östberg, F. (2003). Designing a football experience for a mobile device. In M. Rauterberg, M. Menozzi, & J. Wesson (Eds.), *Proceedings of Interact '03* (pp. 940-943). Amsterdam, The Netherlands: IOS Press.

Wikstrand, G., Schedin, L., & Elg, F. (in press). *High and low ping and the game of pong - Effects of delay and feedback*. (Accepted for publication at Network and System Support for Games).

Wikstrand, G., & Sun, J. (in press). *Determining utility functions for streaming low bit rate football video*. (Accepted for publication at IASTED International Conference on Internet and Multimedia Systems and Applications, IMSA 2004).

Wilson, G., & Sasse, M. (2000). Listen to your heart rate: Counting the cost of media quality. *Affective interactions: Towards a new generation of computer interfaces* (p. 9-20).

Winkler, S. (2001). Visual fidelity and perceived quality: Toward comprehensive metrics. In B. E. Rogowitz & T. N. Pappas (Eds.), *Human Vision and Electronic Imaging VI - Proceedings of SPIE* (pp. 114-125). Bellingham, WA: SPIE.

Zhang, Q., Zhu, W., & Zhang, Y. (2005). End-to-end QoS for video delivery over wireless internet. *Proceedings of the IEEE*, *93*, 123-134.

Copyright © 2006, Idea Group Inc. Copying or distributing in print or electronic forms without written permission of Idea Group Inc. is prohibited.

Ziljstra, F. R. H. (1993). *Efficiency in work behavior: A design approach for modern tools.* Unpublished doctoral dissertation, Delft University, Delft, The Netherlands.

Endnotes

[1]   Although it might be argued that an application and a service are not the same thing, the boundary is fuzzy. In this chapter, the terms are used interchangeably to refer to a situation where a user is engaged in computer-mediated communication with a remote entity.

[2]   Using these measures requires knowing how they are defined and operationalized. Giving a full description of each of these measures goes beyond the scope of this chapter. The interested reader should look up the references given here and earlier.

[3]   The interested reader might want to turn to de Waard (1996) for an excellent discussion on the appropriateness of objective versus subjective, or as he calls them self-report, measures (of mental workload).

[4]   This Website, http://www.mycoted.com/creativity/techniques/, has a thorough listing of creativity techniques which might be useful.

[5]   http://www.cdt.luth.se/projects/arena/.

Copyright © 2006, Idea Group Inc. Copying or distributing in print or electronic forms without written permission of Idea Group Inc. is prohibited.

Chapter XIV

Adaptation and Personalization of Web-Based Multimedia Content

Panagiotis Germanakos, National & Kapodistrian University of Athens, Greece

Constantinos Mourlas, National & Kapodistrian University of Athens, Greece

Abstract

A traditional multimedia system presents the same static content and suggests the same next page to all users, even though they might have widely differing knowledge of the subject. Such a system suffers from an inability to be all things to all people, especially when the user population is relatively diverse. The rapid growth of mobile and wireless communication allowed service providers to develop new ways of interactions, enabling users to become accustomed to new means of multimedia-based service consumption in an anytime, anywhere, and anyhow manner. This chapter investigates the new multi-channel constraints and opportunities emerged by these technologies, as well as the new user-demanding requirements that arise. It further examines the relationship between the adaptation and personalization research considerations, and proposes a three-layer architecture for adaptation and personalization of Web-based multimedia content based on the "new" user profile, with visual, emotional, and cognitive processing parameters incorporated.

Copyright © 2006, Idea Group Inc. Copying or distributing in print or electronic forms without written permission of Idea Group Inc. is prohibited.

Introduction

Since 1994, the Internet has emerged as a fundamental information and communication medium that has generated extensive enthusiasm. The Internet has been adopted by the mass market more quickly than any other technology over the past century, and is currently providing an electronic connection between progressive entities and millions of users whose age, education, occupation, interest, and income demographics are excellent for sales or multimedia-based service provision.

The explosive growth in the size and use of the World Wide Web, as well as the complicated nature of most Web structures, may lead in orientation difficulties, as users often lose sight of the goal of their inquiry, look for stimulating rather than informative material, or even use the navigational features unwisely. To alleviate such navigational difficulties, researchers have put huge amounts of effort to identify the peculiarities of each user group, and design methodologies and systems that could deliver an adapted and personalized Web-content. To this date, there has not been a concrete definition of personalization. However, the many solutions offering personalization features meet an abstract common goal: to provide users with what they want or need without expecting them to ask for it explicitly (Mulvenna, Anand, & Buchner, 2000). A complete definition of personalization should include parameters and contexts such as user intellectuality, mental capabilities, socio-psychological factors, emotional states, and attention- grabbing strategies, since these could affect the apt collection of users' customization requirements, offering in return the best adaptive environments to the user preferences and demands.

With the emergence of wireless and mobile technologies, new communication platforms and devices, apart from PC-based Internet access, are now emerging, making the delivery of content available through a variety of media. Inevitably, this increases user requirements which are now focused upon an *"anytime, anywhere, and anyhow"* basis. Nowadays, researchers and practitioners not only have to deal with the challenges of adapting to the heterogeneous user needs and user environment issues such as current location and time (Panayiotou & Samaras, 2004), but they also have to face numerous considerations with respect to multi-channel delivery of the applications concerning multimedia, services, entertainment, commerce, and so forth. To this end, personalization techniques exploit Artificial Intelligence, agent-based, and real-time paradigms to give presentation and navigation solutions to the growing user demands and preferences.

This chapter places emphasis on the adaptation of the Web-based multimedia content delivery, starting with an extensive reference to the mobility and wireless emergence that sub-serves the rapid development of the multi-channel multimedia content delivery, and the peculiarities of the user profiling that significantly vary from the desktop to the mobile user. Furthermore, it approaches the existing adaptation (adaptive hypermedia) and personalization (Web personalization) techniques and paradigms that could work together in a coherent and cohesive way, since they are sharing the same goal, to provide the most apt result to the user. Lastly, having analyzed the aforementioned concepts, it defines a three-layer adaptation and personalization Web-based multimedia content architecture that is based on the introduction of a "new" user profile that incorporates user characteristics such as user perceptual preferences, on top of the "traditional" ones,

Copyright © 2006, Idea Group Inc. Copying or distributing in print or electronic forms without written permission of Idea Group Inc. is prohibited.

and the semantic multimedia content that includes, amongst others, the perceptual provider characteristics.

Mobility Emergence

The rapid development of the wireless and mobile advancements and infrastructures has evidently given "birth" to Mobile Internet. It is considered fundamental to place emphasis on its imperative existence, since statistics show that in the future the related channels will take over as the most sustainable mediums of Web-based (multimedia) content provision. Mobile Internet could be considered as a new kind of front-end access to Web-based content with specific capabilities of delivering on-demand real-time information. Nowadays, many sectors (governmental, private, educational, etc.) start to offer multimedia-based services and information via a variety of service delivery channels apart from the Web (Germanakos, Samaras, & Christodoulou, 2005). Two of these mobile multimedia-based service delivery channels are mobile telephony and PDAs. These channels become more important considering the much faster growth of the mobile penetration rate compared to desktop-based Internet access. The most significant future development will be the growth of mobile broadband multimedia-based services, once the potential of third generation mobile (3G) and its enhancements, as well other wireless technologies, including W4, RLAN, satellite, and others, is realized. The dissemination of these technologies represents a paradigm shift that enables the emergence of new data multimedia-based services, combining the benefits of broadband with mobility, delivered over high-speed mobile networks and platforms.

Multi-Channel Web-Based Content Delivery Characteristics

"To struggle against the amplification of the digital divide and therefore to think 'user interaction' whatever the age, income, education, experience, and the social condition of the citizen" (Europe's Information Society, 2004).

The specific theme above reveals exactly the need for user-centered multimedia-based service development and personalized content delivery. In many ways, the new technology provides greater opportunities for access. However, there are important problems in determining precisely what users want and need, and how to provide Web-based content in a user-friendly and effective way. User needs are always conditioned by what they already get, or imagine they can get. A channel can change the user perception of a multimedia application; when users have a free choice between different channels to access an application, they will choose the channel that realizes the highest relative value for them. However, separate development of different channels for a single multimedia content (multi-channel delivery) can lead to inconsistencies such as different data formats or interfaces. To overcome the drawbacks of multiple-channel content delivery, the different channels should be integrated and coordinated.

Copyright © 2006, Idea Group Inc. Copying or distributing in print or electronic forms without written permission of Idea Group Inc. is prohibited.

Since successful multimedia-based service delivery depends on a vast range of parameters, there is not a single formula to fit all situations. However, there have been reported particular steps (IDA, 2004) that could guide a provider throughout the channel selection process. Moreover, it should be mentioned that the suitability and usefulness of channels depends on a range of factors, out of which technology is only one element. Additional features that could affect the service channels assessment could be: directness, accessibility and inclusion, speed, security and privacy, and availability. To realize though their potential value, channels need also to be properly implemented and operated.

The design and implementation complexity is rising significantly with the many channels and their varying capabilities and limitations. Network issues include low bandwidth, unreliable connectivity, lack of processing power, limited interface of wireless devices, and user mobility. On the other hand, mobile devices issues include small size, limited processing power, limited memory and storage space, small screens, high latency, and restricted data entry.

Initial Personalization Challenges and Constraints

The needs of mobile users differ significantly from those of desktop users. Getting personalized information *"anytime, anywhere, and anyhow"* is not an easy task. Researchers and practitioners have to take into account new adaptivity axes, along which the personalized design of mobile Web-based content would be built. Such applications should be characterized by flexibility, accessibility, quality, and security in a ubiquitous interoperable manner. User interfaces must be friendlier enabling active involvement (information acquisition), giving the control to the user (system controllability), providing easy means of navigation and orientation (navigation), tolerating users' errors, supporting system-based and context-oriented correction of users' errors, and finally enabling customization of multi-media and multi-modal user interfaces to particular user needs (De Bra, Aroyo, & Chepegin, 2004; De Bra & Nejdl, 2004). Intelligent techniques have to be implemented that will enable the development of an open Adaptive Mobile Web (De Bra & Nejdl, 2004), having as fundamental characteristics the directness, high connectivity speed, reliability, availability, context-awareness, broadband connection, interoperability, transparency and scalability, expandability, effectiveness, efficiency, personalization, security, and privacy (Lankhorst, Kranenburg, Salden, & Peddemors, 2002; Volokh, 2000).

Personalization Considerations in the Context of Desktop and Mobile User

The science behind personalization has undergone tremendous changes in recent years while the basic goal of personalization systems was kept the same, to provide users with what they want or need without requiring them to ask for it explicitly. Personalization is

Copyright © 2006, Idea Group Inc. Copying or distributing in print or electronic forms without written permission of Idea Group Inc. is prohibited.

the provision of tailored products, multimedia-based services, Web-based multimedia content, information, or information relating to products or services. Since it is a multi-dimensional and complicated area (covering also recommendation systems, customization, adaptive Web sites, Artificial Intelligence), a universal definition that would cover all its theoretical areas has not been given so far. Nevertheless, most of the definitions that have been given to personalization (Kim, 2002; Wang & Lin, 2002) are converging to the objective that is expressed on the basis of delivering to a group of individuals relevant information that is retrieved, transformed, and/or deduced from information sources in the format and layout as well as specified time intervals.

Comprehensive User Requirements and the Personalization Problem

The user population is not homogeneous, nor should be treated as such. To be able to deliver quality knowledge, systems should be tailored to the needs of individual users providing them personalized and adapted information based on their perceptions, reactions, and demands. Therefore, a serious analysis of user requirements has to be undertaken, documented, and examined, taking into consideration their multi-application to the various delivery channels and devices. Some of the user (customer) requirements and arguments anticipated could be clearly distinguished into Top of the Web (2003) and CAP Gemini Ernst & Young (2004): (a) General User Service Requirements (flexibility: anyhow, anytime, anywhere; accessibility; quality; and security), and (b) Requirements for a Friendly and Effective User Interaction (information acquisition; system controllability; navigation; versatility; errors handling; and personalization).

Although one-to-one multimedia-based service provision may be a functionality of the distant future, user segmentation is a very valuable step in the right direction. User segmentation means that the user population is subdivided, into more or less homogeneous, mutually-exclusive subsets of users who share common user profile characteristics. The subdivisions could be based on: demographic characteristics (i.e. age, gender, urban- or rural-based, region); socio-economic characteristics (i.e. income, class, sector, channel access); psychographic characteristics (i.e. life style, values, sensitivity to new trends); individual physical and psychological characteristics (i.e. disabilities, attitude, loyalty).

The issue of personalization is a complex one with many aspects and viewpoints that need to be analyzed and resolved. Some of these issues become even more complicated once viewed from a moving user's perspective, in other words when constraints of mobile channels and devices are involved. Such issues include, but are not limited to: what content to present to the user, how to show the content to the user, how to ensure the user's privacy, how to create a global personalization scheme. As clearly viewed, user characteristics and needs, determining user segmentation, and thus provision of the adjustable information delivery, differ according to the circumstances and change over time (Panayiotou and Samaras, 2004).

There are many approaches to address these issues of personalization, but usually, each one is focused upon a specific area, that is, whether this is profile creation, machine learning and pattern matching, data and Web mining, or personalized navigation.

Copyright © 2006, Idea Group Inc. Copying or distributing in print or electronic forms without written permission of Idea Group Inc. is prohibited.

Beyond the "Traditional" User Profiling

One of the key technical issues in developing personalization applications is the problem of how to construct accurate and comprehensive profiles of individual users and how these can be used to identify a user and describe the user behavior, especially if they are moving (Adomavicious & Tuzhilin, 1999). According to Merriam-Webster dictionary, the term profile means "a representation of something in outline". User profile can be thought of as being a set of data representing the significant features of the user. Its objective is the creation of an information base that contains the preferences, characteristics, and activities of the user. A user profile can be built from a set of keywords that describe the user-preferred interest areas compared against information items.

User profiling is becoming more and more important with the introduction of the heterogeneous devices used, especially when published contents provide customized views on information. User profiling can either be static, when it contains information that rarely or never changes (e.g. demographic information), or dynamic, when the data change frequently. Such information is obtained either explicitly, using online registration forms and questionnaires resulting in static user profiles, or implicitly, by recording the navigational behavior and/or the preferences of each user. In the case of implicit acquisition of user data, each user can either be regarded as a member of a group and take up an aggregate user profile or be addressed individually and take up an individual user profile. The data used for constructing a user profile could be distinguished into: (a) the Data Model which could be classified into the demographic model (which describes who the user is), and the transactional model (which describes what the user does); and (b) the Profile Model which could be further classified into the factual profile (containing specific facts about the user derived from transactional data, including the demographic data, such as "the favorite beer of customer X is Beer A"), and the behavioral profile (modeling the behavior of the user using conjunctive rules, such as association or classification rules. The use of rules in profiles provides an intuitive, declarative, and modular way to describe user behavior (Adomavicious & Tuzhilin, 1999)). Additionally, in the case of a mobile user, by user needs is implied both the thematic preferences (i.e., the traditional notion of profile) as well as the characteristics of their personal device called "device profile". Therefore, here, adaptive personalization is concerned with the negotiation of user requirements and device abilities.

But, could the user profiling be considered complete incorporating only these dimensions? Do the designers and developers of multimedia-based services take into consideration the real user preferences in order to provide them a really personalized Web-based multimedia content? Many times this is not the case. How can user profiling be considered complete, and the preferences derived optimized, if it does not contain parameters related to the user perceptual preference characteristics? We could define User Perceptual Preference Characteristics as all the critical factors that influence the visual, mental, and emotional processes liable of manipulating the newly received information and building upon prior knowledge, that is different for each user or user group. These characteristics determine the visual attention, cognitive, and emotional processing taking place throughout the whole process of accepting an object of perception (stimulus) until the comprehensive response to it. It has to be noted at this

Copyright © 2006, Idea Group Inc. Copying or distributing in print or electronic forms without written permission of Idea Group Inc. is prohibited.

point that the user perceptual preference characteristics are directly related to the "traditional" user characteristics since they are affecting the way a user approaches an object of perception.

It is true that nowadays, there are not so many researches that move towards the consideration of user profiling to incorporate optimized parameters taken from the research areas of visual attention processing and cognitive psychology. Some serious attempts have been made on approaching e-learning systems providing adapted content to the students, but most of them are lying to restricted analysis and design methodologies considering particular cognitive learning styles, including Field Independence vs. Field Dependence, Holistic-Analytic, Sensory Preference, Hemispheric Preferences, and Kolb's Learning Style Model (Yuliang & Dean, 1999), applied to identified mental models, such as concept maps, semantic networks, frames, and schemata (Ayersman & Read, 1999). In order to deal with the diversified students' preferences, they are matching the instructional materials and teaching styles with the cognitive styles, and consequently they are satisfying the whole spectrum of the students' cognitive learning styles by offering a personalized Web-based educational content.

A Comprehensive Overview of Adaptation and Personalization Techniques and Paradigms: Similarities and Differences

When we are considering adaptation and personalization categories and technologies, we refer to Adaptive Hypermedia and Web Personalization respectively, due to the fact that together these can offer the most optimized adapted content result to the user.

Adaptive Hypermedia Overview

Adaptive Hypermedia is a relatively old and well-established area of research counting three generations: The first "pre-Web" generation of adaptive hypermedia systems explored mainly adaptive presentation and adaptive navigation support and concentrated on modeling user knowledge and goals. The second "Web" generation extended the scope of adaptive hypermedia by exploring adaptive content selection and adaptive recommendation based on modeling user interests. The third "New Adaptive Web" generation moves adaptive hypermedia beyond traditional borders of desktop hypermedia systems embracing such modern Web trends as "mobile Web", "open Web", and "Semantic Web" (Brusilovsky & Maybury, 2002).

Adaptivity is a particular functionality that alleviates navigational difficulties by distinguishing between interactions of different users within the information space (De Bra & Nejdl, 2004; Eklund & Sinclair, 2000). Adaptive Hypermedia Systems employ

Copyright © 2006, Idea Group Inc. Copying or distributing in print or electronic forms without written permission of Idea Group Inc. is prohibited.

adaptivity by manipulating the link structure or by altering the presentation of information, based on a basis of a dynamic understanding of the individual user, represented in an explicit user model (Brusilovsky, 1996; De Bra et al., 1999; Eklund, & Sinclair, 2000). In 1996, Brusilovsky identified four user characteristics to which an Adaptive Hypermedia System should adapt. These were: user's knowledge, goals, background and hypertext experience, and user's preferences. In 2001, further two sources of adaptation were added to this list, user's interests and individual traits, while a third source of different nature having to deal with the user's environment had also been identified.

Generally, Adaptive Hypermedia Systems can be useful in application areas where the hyperspace is reasonably large and the user population is relatively diverse in terms of the above user characteristics. A review by Brusilovsky has identified six specific application areas for adaptive hypermedia systems since 1996 (Brusilovsky, 2001). These are: educational hypermedia, on-line information systems, information retrieval systems, institutional hypermedia, and systems for managing personalized view in information spaces. Educational hypermedia and on-line information systems are the most popular, accounting for about two- thirds of the research efforts in adaptive hypermedia. Adaptation effects vary from one system to another. These effects are grouped into three major adaptation technologies: adaptive content selection (De Bra & Nejdl, 2004), adaptive presentation (or content-level adaptation), and adaptive navigation support (or link-level adaptation) (Brusilovsky, 2001; De Bra et al., 1999; Eklund & Sinclair, 2000) and are summarized in Figure 1.

Figure 1. Adaptive hypermedia techniques

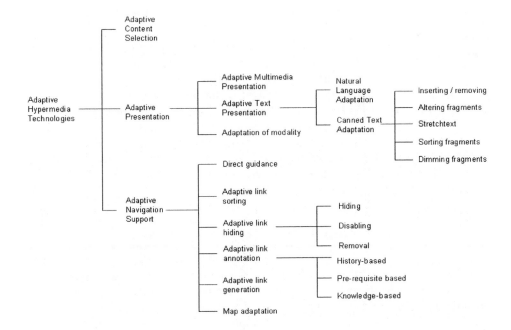

Copyright © 2006, Idea Group Inc. Copying or distributing in print or electronic forms without written permission of Idea Group Inc. is prohibited.

Figure 2. INSPIRE's components and the interaction with the learner

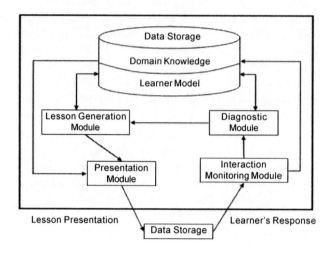

As mentioned earlier, successful adaptation attempts have been made in the e-learning research field to provide the students with adapted content according to their different learning styles or knowledge level and goals. A typical case of such a system could be considered the INSPIRE (Intelligent System for Personalized Instruction in a Remote Environment) architecture, see Figure 2, where throughout its interaction with the learner, the system dynamically generates lessons that gradually lead to the accomplishment of the learning goals selected by the learner (Papanikolaou, Grigoriadou, Kornilakis, & Magoulas, 2002). INSPSIRE architecture has been designed so as to facilitate knowledge communication between the learner and the system and to support its adaptive functionality.

INSPIRE comprises of five different modules: (a) the *Interaction Monitoring Module* that monitors and handles learner's responses during his/her interaction with the system, (b) the *Learner's Diagnostic Module* that processes data recorded about the learner and decides on how to classify the learner's knowledge, (c) the *Lesson Generation Module* that generates the lesson contents according to learner's knowledge goals and knowledge level, (d) the *Presentation Module* which functions to generate the educational material pages sent to the learner, and (e) the *Data Storage*, which holds the *Domain knowledge* and the *Learner's Model*.

Web Personalization Overview

Web Personalization refers to the whole process of collecting, classifying, and analyzing Web data, and determining based on these the actions that should be performed so that the information is presented in a personalized manner to the user. As inferred from its name, Web Personalization refers to Web applications solely (with popular use in e-

Copyright © 2006, Idea Group Inc. Copying or distributing in print or electronic forms without written permission of Idea Group Inc. is prohibited.

business multimedia-based services), and generally is a relatively new area of research. Web personalization is the process of customizing the content and structure of a Web site to the specific needs of each user by taking advantage of the user's navigational behavior. Being a multi-dimensional and complicated area, a universal definition has not been agreed to date. Nevertheless, most of the definitions given to Web personalization (Cingil, Dogac, & Azgin, 2000; Kim, 2002) agree that the steps of the Web personalization process include: (1) the collection of Web data, (2) the modeling and categorization of these data (pre-processing phase), (3) the analysis of the collected data, and the determination of the actions that should be performed. Moreover, many argue that emotional or mental needs, caused by external influences, should also be taken into account.

Web Personalization could be realized in one of two ways: (a) Web sites that require users to register and provide information about their interests, and (b) Web sites that only require the registration of users so that they can be identified (De Bra et al., 2004). The main motivation points for personalization can be divided into those that are primarily to facilitate the work, and those that are primarily to accommodate social requirements. The former motivational subcategory contains the categories of enabling access to information content, accommodating work goals, and accommodating individual differences, while the latter contains the categories of eliciting an emotional response and expressing identity.

Personalization levels have been classified into: Link Personalization (involves selecting the links that are more relevant to the user, changing the original navigation space by reducing or improving the relationships between nodes), Content Personalization (user interface can present different information for different users providing substantive information in a node, other than link anchors), Context Personalization (the same information (node) can be reached in different situations), Authorized Personalization (different users have different roles and therefore they might have different access authorizations) and Humanized Personalization (involves human computer interaction) (Lankhorst et al., 2002; Rossi, Schwade, & Guimaraes, 2001). The technologies that are employed in order to implement the processing phases mentioned above as well as the Web personalization categories are distinguished into: Content-Based Filtering, Rule-Based Filtering, Collaborative Filtering, Web Usage Mining, Demographic-Based Filtering, Agent Technologies, and Cluster Models (Mobasher, 2002; Pazzani, 1999; Perkowitz & Etzioni, 2003).

The use of the user model is the most evident technical similarity of Adaptive Hypermedia and Web Personalization to achieve their goal. However, the way they maintain the user profile is different; Adaptive Hypermedia requires a continuous interaction with the user, while Web Personalization employs algorithms that continuously follow the users' navigational behavior without any explicit interaction with the user. Technically, two of the adaptation/personalization techniques used are the same. These are adaptive-navigation support (of Adaptive Hypermedia and else referred to as link-level adaptation) and Link Personalization (of Web Personalization) and adaptive presentation (of Adaptive Hypermedia and else referred to as content-level adaptation) and Content Personalization (of Web Personalization).

An example of a Web personalization application for the wireless user is the mPERSONA system, depicted in Figure 3. The mPERSONA system architecture combines existing

Copyright © 2006, Idea Group Inc. Copying or distributing in print or electronic forms without written permission of Idea Group Inc. is prohibited.

Figure 3. Detailed view of the mPERSONA architecture

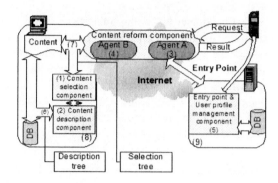

techniques in a component-based fashion in order to provide a global personalization scheme for the wireless user. The mPERSONA is a flexible and scalable system that focuses towards the new era of wireless Internet and the moving user. The particular architecture uses autonomous and independent components avoiding this way tying up to specific wireless protocols (e.g., WAP). To achieve a high degree of independence and autonomy, mPERSONA is based on mobile agents and mobile computing models such as the "client intercept model" (Panayiotou & Samaras, 2004).

The architectural components are distinguished based on their location and functionality: a) the *Content description* component (Figure 3: 2 & 6), creates and maintains the content's provider metadata structure that describes the actual content, (b) the *Content selection* component (Figure 3: 1 & 7), selects the content that will be presented to the user when "applying" his profile, (c) the *Content reform* component (Figure 3: 3 & 4), reforms and delivers the desired content in the needed (by the user's device) form, and (d) the *User profile management* component (Figure 3: 5), registers and manages user profiles. The user's profile is split into two parts: the device profile (covers the user's devices) and the theme profile (preferences).

A Three-Layer Architecture for Adaptation and Personalization of Web-Based Multimedia Content

Based on the above considerations, a three-layer architecture for adaptation and personalization of Web-based multimedia content will now be presented, trying to convey the essence and the peculiarities encapsulated, and further answering the question why adaptation and personalization of Web-based content is considered vital for the sustainable provision of quality multi-channel Web-based multimedia content/ multimedia-based services.

Copyright © 2006, Idea Group Inc. Copying or distributing in print or electronic forms without written permission of Idea Group Inc. is prohibited.

Figure 4. Adaptation and personalization of Web-based multimedia content architecture

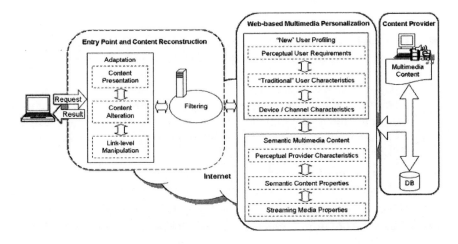

The current architecture, depicted in Figure 4, Adaptation and Personalization of Web-based Multimedia Content Architecture, is composed of three interrelated parts/layers. Each *layer* for the purpose of the infrastructure functionality may be composed of *components*, and each component may be broken down into *elements*, as detailed below:

Front-End Layer (Entry Point and Content Reconstruction)

The front-end layer is the primary layer, and it is the user access interface of the system. It is called "Entry Point and Content Reconstruction", and it accepts multi-device requests. It enables the attachment of various devices on the infrastructure (such as mobile phones, PDAs, desktop devices, tablet PC, satellite handset, etc.) identifying their characteristics and preferences as well as the location of the user currently active (personalization/locationbased). It also handles multi-channel requests. Dut to the variety of multi-channel delivery (i.e., over the Web, telephone, interactive kiosks, WAP, MMS, SMS, satellite, and so on), this layer identifies the different characteristics of the channels. It directly communicates with the middle layer exchanging multi-purpose data. It consists of two components, each one assigned for a different scope:

- **Adaptation:** This component comprises of all the access-control data (for security reasons) and all the information regarding the user profile. These might include user preferences, geographical data, device model, age, business type, native language, context, and so forth. It is the entry point for the user, enabling the login to the architecture. This component is directly communicating with the middle layer where the actual verification and profiling for the user is taking place. Once the

Copyright © 2006, Idea Group Inc. Copying or distributing in print or electronic forms without written permission of Idea Group Inc. is prohibited.

whole processing has been completed, it returns the adapted results to the user. It is comprised of three elements:

- **Content Presentation (or Adaptive Presentation):** It adapts the content of a page to the characteristics of the user according to the user profile and personalization processing. The content is individually generated or assembled from pieces for each user, to contain additional information, pre-requisite information, or comparative explanations by conditionally showing, hiding, highlighting, or dimming fragments on a page. The granularity may vary from word replacement to the substitution of pages to the application of different media.

- **Content Alteration (or Adaptive Content Selection):** When the user searches for a particular content, that is, related information to his/her profile, the system can adaptively select and prioritize the most relevant items.

- **Link-Level Manipulation (or Adaptive Navigation Support):** It provides methods that restrict the user's interactions with the content or techniques that aid the user in their understanding of the information space, aiming to provide either orientation or guidance (i.e. adaptive link, adaptive link hiding/annotation). Orientation informs the user about his/her place in the information space, while guidance is related to a user's goal.

- **Filtering:** This component is considered the main link of the front-layer with the middle layer of the architecture. It actually transmits the data accumulated both directions. It is responsible for making the the low-level reconstruction and filtering of the content, based on the personalization rules created, and to deliver the content for adaptation.

Middle Layer (Web-Based Multimedia Personalization)

The middle layer is the main layer of the architecture and it is called "Web-Based Multimedia Personalization". At this level all the requests are processed. This layer is responsible for the custom tailoring of information to be delivered to the users, taking into consideration their habits and preferences, as well as, for mobile users mostly, their location ("location-based") and time ("time-based") of access. The whole processing varies from security, authentication, user segmentation, multimedia content identification, to provider perceptual characteristics, user perceptions (visual, mental and emotional), and so forth. This layer accepts requests from the front-end and, after the necessary processing, either sends information back or communicates with the next layer (back-end) accordingly. The middle layer is comprised of the following two components:

- **"New" User Profiling:** It contains all the information related to the user, necessary for the Web Personalization processing. It is directly related to the Semantic Multimedia Content component and is composed of three elements:

Copyright © 2006, Idea Group Inc. Copying or distributing in print or electronic forms without written permission of Idea Group Inc. is prohibited.

- **Perceptual User Requirements:** This is the new element/dimension of the user profile. It contains all the visual attention and cognitive processes (cognitive and emotional processing parameters) that completes the user perception and fulfills the user profile. It is considered a vital element of the user profile since it identifies the aspects of the user that is very difficult to be revealed and measured but, however, might determine his/her exact preferences and lead to a more concrete, accurate, and optimized user segmentation.

- **"Traditional" User Characteristics:** This element is directly related to the Perceptual User Requirements element and provides the so-called "t r a d i-tional" characteristics of a user: knowledge, goals, background, experience, preferences, activities, demographic information (age, gender), socio-economic information (income, class, sector, etc.), and so forth. Both elements are completing the user profiling from the user's point of view.

- **Device/Channel Characteristics:** This element is referring to all the characteristics that referred to the device or channel that the user is using and contains information like: bandwidth, displays, text-writing, connectivity, size, power processing, interface and data entry, memory and storage space, latency (high/low), and battery lifetime. These characteristics are mostly referred to mobile users and are considered important for the formulation of a more integrated user profile, since it determines the technical aspects of it.

- **Semantic Multimedia Content:** This component is based on metadata describing the content (data) available from the Content Provider (back-end layer). In this way, a common understanding of the data, that is, semantic interoperability and openness is achieved. The data manipulated by the system/architecture is described using metadata that comprises of all needed information to unambiguously describe each piece of data and collections of data. This provides semantic interoperability and a human-friendly description of data. This component is directly related to the "New" User Profile component, providing together the most optimized personalized Web-based multimedia content result. It is consisted of three elements:

  - **Perceptual Provider Characteristics:** It identifies the provider characteristics assigned to the Web-based multimedia content or multimedia based service. They are involving all these perceptual elements that the provider has been based upon for the design of the content (i.e., actual content/data of the service, layout preferences, content presentation, etc.)

  - **Semantic Content Properties:** This element performs the identification and metadata description of Web-based multimedia content or multi-media-based service based on predetermined ontologies. It is implemented in a transparent manner, removing data duplication and the problem of data consistency.

  - **Streaming Media Properties:** It contains data transition mechanisms and the databases. These databases contain the Web-based multimedia content or

Copyright © 2006, Idea Group Inc. Copying or distributing in print or electronic forms without written permission of Idea Group Inc. is prohibited.

multimedia-based services as supplied by the provider (without at this point being further manipulated or altered).

Back-End Layer (Content Provider)

This is the last layer of the architecture and is directly connected to the middle layer. It contains transition mechanisms and the databases of Web-based multimedia content or multimedia-based services as supplied by the provider without been through any further manipulation or alteration.

The proposed three-layer architecture for adaptation and personalization of Web-based multimedia content will allow users to receive the Web-based multimedia content or multimedia-based service which they access in an adapted style according to their preferences, increasing in that way efficiency and effectiveness of use.

Implementation Considerations

So far, the functionality and interrelation of three-layer achitecture components that provide adapted and personalized Web-based content have been extensively investigated This section will focus on the concepts and parameters that take part in the construction of a comprehensive user profile, and how these could be used in order to collect all the relevant information. As it has already been mentioned, a lot of research has been done for the implementation of the "traditional" user profiling. Many adaptation and personalization techniques have been developed, and common semantic libraries have been set up that give basically specific and ad-hoc solutions.. However, to our knowledge, implementations that incorporate visual attention, cognitive, and emotional processing parameters to the user profile have not been reported as yet, and such parameters would definitely lead to a comprehensive accumulation of user perceptual preference characteristics and hence, provide users with more sustainable personalized content. Therefore, main emphasis in the following section is given to the construction of the "new" comprehensive user profiling, incorporating these user perceptual preference characteristics mentioned above.

Further examining the middle layer of the proposed architecture and the Perceptual User Requirements element of the "New" User Profiling component, we can see that the User Perceptual Preference Characteristics could be described as a continuous mental processing starting with the perception of an object in the users' attentional visual field (stimulus) and going through a number of cognitive, learning, and emotional processes giving the actual response to that stimulus. This is depicted in Figure 5.

These processes formulate a three-dimensional approach to the problem, as depicted in Figure 6. The three dimensions created are the Learning Styles, the Visual and Cognitive Processing, and the Emotional Processing dimensions.

The *User Learning Processing* dimension is a selection of the most appropriate and technologically feasible learning styles, such as Witkin's Field-Dependent and Field-Independent and Kolb's Learning Styles, being in a position to identify how users

Copyright © 2006, Idea Group Inc. Copying or distributing in print or electronic forms without written permission of Idea Group Inc. is prohibited.

Figure 5. User perceptual preference characteristics

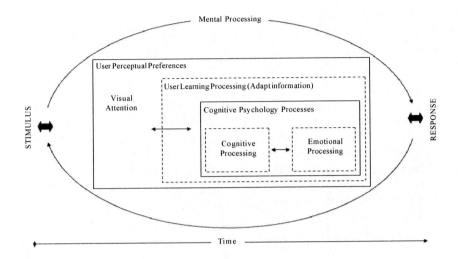

Figure 6. Three-dimensional approach

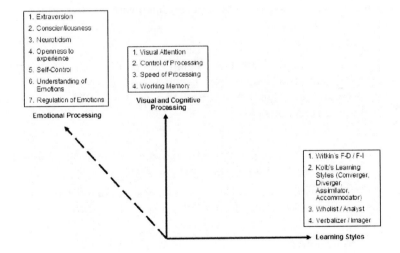

Copyright © 2006, Idea Group Inc. Copying or distributing in print or electronic forms without written permission of Idea Group Inc. is prohibited.

transform information into knowledge (constructing new cognitive frames) and if they could be characterized as a converger, diverger, assimilator, accommodator, wholist, analyst, verbalizer, or imager.

The *Visual and Cognitive Processing* dimension is being distinguished from:

- **Visual Attention Processing:** It is composed from the pre-attentive and the limited-capacity stage; the pre-attentive stage of vision subconsciously defines objects from visual primitives, such as lines, curvature, orientation, color, and motion, and allows definition of objects in the visual field. When items pass from the pre-attentive stage to the limited-capacity stage, these items are considered as selected. Interpretation of eye movement data is based on the empirically-validated assumption that when a person is performing a cognitive task, while watching a display, the location of his/her gaze corresponds to the symbol currently being processed in working memory and, moreover, that the eye naturally focuses on areas that are most likely to be informative.

- **Control of Processing:** It refers to the processes that identify and register goal-relevant information and block out dominant or appealing but actually irrelevant information.

- **Speed of Processing:** It refers to the maximum speed at which a given mental act may be efficiently executed (cognitive processing efficiency).

- **Working Memory:** It refers to the processes that enable a person to hold information in an active state while integrating it with other information until the current problem is solved.

The *Emotional Processing* dimension is composed of these parameters that could determine a user's emotional state during the whole response process. This is vital so as to determine the level of adaptation (user needs per time interval) during the interaction process. These parameters include:

- **Extroversion:** Extraverts are sociable, active, self-confident, and uninhibited; while introverts, are withdrawn, shy, and inhibited.

- **Conscientiousness:** Conscientious individuals are organized, ambitious, determined, reliable, and responsible; while individuals low in conscientiousness are distractible, lazy, careless, and impulsive.

- **Neuroticism:** Individuals high in neuroticism are confident, clear-thinking, alert, and content.

- **Open to experience:** Individuals who are open to experience are curious and with wide interests, inventive, original, and artistic; individuals who are not open to experience are conservative, cautious, and mild.

- **Understanding of emotions:** It is the cognitive processing of the emotions; it is the ability of understanding and analysis of the complex emotions and the chain reactions of the emotions, that is, how one emotion generates another.

Copyright © 2006, Idea Group Inc. Copying or distributing in print or electronic forms without written permission of Idea Group Inc. is prohibited.

- **Regulation of emotions:** It is the control and regulation of personal and other people's emotions for the emotional and intellectual development; it is the human's ability to realize what is hidden behind an emotion, like fear, anxiety, anger, or sadness, and to find each time the most suitable ways to confront them.

- **Self control:** It includes processes referring to the control of attention, the provision of intellectual resources, and the selection of the specialized procedures and skills liable for the evaluation of a problem's results or a decision's uptake; it is a superior control system that coordinates the functioning of other, more specialized control systems.

These parameters must be filtered even more so that the final optimized model is achieved. Once this is established, a series of tests (some in the form of questionnaires and others with real-time interaction metrics) will be constructed which will attempt to reveal users' perceptual preference characteristics. These features, along with the *"Traditional" User Characteristics*, could complete the "New" User Profile, and therefore adaptation and personalization schemes could be adjusted to deliver even more personalized Web-based content accordingly. The next step is to identify what is the correlation between the various users and/or user groups (i.e., to investigate similarities and differences between them) and if it would be feasible to refer to the term "users' segmentation" (i.e., users sharing similar "new" user profiling characteristics). In case the latter is true, personalization mechanisms will be based upon these parameters and considering users' device/channel characteristics, and the semantic content will provide them with the corresponding adapted result. Eventually, this methodology will be implemented with personalization algorithms and paradigms so to automatically gather all the related information and construct the "new" user profiling, giving the users the adapted and personalized result without their actual intervention.

Summary and Future Trends

When referring to adapted multimedia-based services or Web-based multimedia content provision, it is implied that the content adaptation and personalization is based not only on the "traditional" user characteristics, but on a concrete and comprehensive user profiling that covers all the dimensions and parameters of the users preferences. However, knowing the user traditional characteristics and channel/device capabilities, providers can design and offer an apt personalized result. Most of the times, though the providers tend to design multimedia applications based on their own preferences and what they think should be offered. However, the concept of adaptation and personalization is much more complicated than that. This is the reason why until today there is not any sustainable related definition of personalization. A profile can be considered complete when it incorporates the users' perceptual preference characteristics that mostly deal with intrinsic parameters and are very difficult to be technologically measured. Visual attention (that can be thought of as the gateway to conscious perception and memory) and cognitive psychology processes (cognitive and emotional

Copyright © 2006, Idea Group Inc. Copying or distributing in print or electronic forms without written permission of Idea Group Inc. is prohibited.

processing parameters) should be in combination investigated and analyzed in a further attempt to complete the desktop and mobile users' preferences.

This chapter made an extensive reference to the mobility emergence and the extensive use of the new channels that tend to satisfy the new user requirements (desktop and mobile) for *"anytime, anyhow, and anywhere"* multimedia-based services and Web-based multimedia content provision in general. The problem of personalization as well as challenges created has been investigated supporting the view of why the provision of adapted content, based on a comprehensive user profile, is considered critical nowadays. Moreover, an Adaptation (Adaptive Hypermedia) and Personalization (Web Personalization) categories and paradigms review has been presented identifying common grounds and objectives of these two areas. Eventually, a three-layer architecture for the adaptation and personalization of Web-based multimedia content was reviewed, making use of the aforementioned adaptation and personalization concepts and technologies, the new user profiling (that incorporates the user perceptual preference characteristics), as well as the semantic multimedia content..

The basic objective of this chapter was to introduce a combination of concepts coming from different research areas, all of which focus upon the user. It has been attempted to approach the theoretical considerations and technological parameters that can provide the most comprehensive user profiling, supporting the provision of the most apt and optimized adapted and personalized multimedia result.

References

Adomavicious, G., & Tuzhilin, A. (1999). User profiling in personalization applications through rule discovery and validation. *Proceedings of the ACM Fifth International Conference on Data Mining and Knowledge Discovery (KDD '99)* (pp. 377-381).

Ayersman, D. J., & Reed, W. M. (1998). Relationships among hypermedia-based mental models and hypermedia knowledge. *Journal of Research on Computing in Education, 30*(3), 222-238.

Brusilovsky, P. (1996). Adaptive hypermedia: An attempt to analyze and generalize. In P. Brusilovsky, P. Kommers, & Streitz (Eds.), *Multimedia, hypermedia, and virtual reality* (pp. 288-304). Berlin: Springer-Verlag.

Brusilovsky, P. (1996). Methods and techniques of adaptive hypermedia. *User Modeling and User Adapted Interaction, 6*(2-3), 87-129.

Brusilovsky, P. (2001). Adaptive hypermedia. *User Modeling and User-Adapted Interaction, 11*, 87-110.

Brusilovsky, P., & Maybury, M. T. (2002). From adaptive hypermedia to the adaptive Web. In P. Brusilovsky & M. T. Maybury (Eds.), *Communications of the ACM, 45*(5), *Special Issue on the Adaptive Web*, 31-33.

CAP Gemini Ernst & Young. (2004). Online availability of public services: How is Europe progressing? *European Commission DG Information Society.*

Copyright © 2006, Idea Group Inc. Copying or distributing in print or electronic forms without written permission of Idea Group Inc. is prohibited.

Cingil, I., Dogac, A., & Azgin, A. (2000). A broader approach to personalization. *Communications of the ACM, 43*(8), 136-141.

De Bra, P., Aroyo, L., & Chepegin, V. (2004). The next big thing: Adaptive Web-based systems. *Journal of Digital Information, 5*(1), Article no. 247.

De Bra, P., Brusilovsky, P., & Houben, G. (1999). Adaptive hypermedia: From systems to framework. *ACM Computing Surveys, 31*(4es), 12.

De Bra, P., & Nejdl, W. (2004). Adaptive hypermedia and adaptive Web-based systems. *Proceedings of the Third International Conference (AH 2004)*, Springer Lecture Notes in Computer Science, 3137.

Eklund, J., & Sinclair, K. (2000). An empirical appraisal of the effectiveness of adaptive interfaces of instructional systems. *Educational Technology and Society, 3*(4), 165-177.

Europe's Information Society. (2004). *User interaction.* Retrieved from http://europa.eu.int/information_society/activities/egovernment_research/focus/user_interaction/index_en.htm

Germanakos, P., Samaras, G., & Christodoulou, E. (2005 10-12). Multi-channel delivery of services—the road from e-government to m-government: Further technological challenges and implications. *Proceedings of the 1st European Conference on Mobile Government (Euro mGov 2005)*, Brighton (pp. 210-220).

Interchange of Data between Administrations. (2004). *Multi-channel delivery of e-government services.* Retrieved from http://europa.eu.int/idabc/

Kim, W. (2002). Personalization: Definition, status, and challenges ahead. *JOT, 1*(1), 29-40.

Lankhorst, M. M., Kranenburg, Salden, A., & Peddemors A. J. H. (2002). Enabling technology for personalizing mobile services. *Proceedings of the 35th Annual Hawaii International Conference on System Sciences (HICSS-35'02): Vol. 3*(3) (p. 87).

Mobasher, B., Dai, H., Luo, T., Nakagawa, M., & Wiltshire, J. (2002). Discovery of aggregate usage profiles for Web personalization. *Data Mining and Knowledge Discovery, 6*(1), 61-82.

Mulvenna, M. D., Anand, S. S., & Buchner, A. G. (2000). Personalization on the net using Web mining. *Communications of the ACM, 43*(8), 123-125.

Panayiotou, C., & Samaras, G. (2004). mPersona: Personalized portals for the wireless user: An agent approach. *Journal of ACM/ Baltzer Mobile Networking and Applications (MONET), Special Issue on Mobile and Pervasive Commerce, 9*(6), 663-677.

Papanikolaou, K.A., Grigoriadou, M., Kornilakis, H., & Magoulas, G.D. (2002). INSPIRE: An intelligent system for personalized instruction in a remote environment. In S. Reich, M. M. Tzagarakis, & P. M. E. De Bra (Eds.), *OHS/SC/AH 2001, LNCS 2266* (pp. 215-225). Springer-Verlag.

Pazzani, J. M. (1999). A framework for collaborative, content-based, and demographic filtering. *Artificial Intelligence Review, 13*(5-6), 393-408.

Copyright © 2006, Idea Group Inc. Copying or distributing in print or electronic forms without written permission of Idea Group Inc. is prohibited.

Rossi, G., Schwade, D., & Guimaraes, M. R. (2001). Designing personalized Web applications. *ACM Proceedings of the 10th International Conference on World Wide Web* (pp. 275-284).

Top of the Web (2003). Survey on quality and usage of public e-services. Top of the Web. Retrieved from http://www.idt.unisg.ch/org/idt/ceegov.nsf/0/1ae4025175a16a90 c1256df6002a0fef/$FILE/Final_report_2003_quality_and_ usage.pdf

Volokh, E. (2000). Personalization and privacy. *The Communications of the Association for Computing Machinery, 43*(8), 84.

Wang, J., & Lin, J. (2002). Are personalization systems really personal? Effects of conformity in reducing information overload. *Proceedings of the 36th Hawaii International Conference on Systems Sciences (HICSS '03)*. 0-7695-1874-5/03.

Yuliang, L., & Dean, G. (1999). Cognitive styles and distance education. *Online Journal of Distance Learning Administration, 2*(3), Article 005.

Copyright © 2006, Idea Group Inc. Copying or distributing in print or electronic forms without written permission of Idea Group Inc. is prohibited.

About the Authors

Gheorghita Ghinea is a senior lecturer in the School of Information Systems, Computing, and Mathematics at Brunel University. He holds a BSc (Honors) in computer science and mathematics, an MSc in computer science (with Distinction, 1996), and a PhD in computer science (2000). He has published more than 60 research papers in leading international journals and peer-reviewed conferences, and has consulted for both public and private organizations in his areas of research interest, which are: distributed multimedia (especially perceptual aspects), Web-based systems, ubiquitous computing, and telemedicine. Dr. Ghinea is a member of the IEEE and the British Computer Society.

Sherry Y. Chen is a senior lecturer in the School of Information Systems, Computing, and Mathematics at Brunel University. She obtained her PhD from the University of Sheffield in 2000. Her current research interests include human-computer interaction, data mining, digital libraries, and educational technology. She has published widely in these areas. Dr. Chen was the co-editor of the books, *Adaptive and Adaptable Hypermedia Systems* and *Advances in Web-based Education: Personalized Learning Environments*. She is a member of the editorial boards of five computing journals. She has been invited to give several talks, including the 9th International Conference on User Modelling and the EPSRC Network of Women in Computer Science colloquium.

***

Copyright © 2006, Idea Group Inc. Copying or distributing in print or electronic forms without written permission of Idea Group Inc. is prohibited.

Lesley Axelrod, MSc, studied human communication and worked as a registered speech and language therapist before turning to research. She worked on human communication-related projects as a senior research fellow at University College, London, and at Surrey University. She studied human-computer interaction with ergonomics at University College, London, and her research interests now center on how people communicate with technology, and the design of usable user-centered systems. She is currently a PhD student at Brunel University, School of Information Systems, Computing, and Mathematics, where she is a member of the VIVID Centre, working on a project to investigate how giving computers the ability to recognize human emotions might improve the quality of human-computer interaction.

Linda D. Bussell is the director of research and development for Kinder Magic Software and a consultant to industry. She received her EdD degree in educational technology from San Diego State University and the University of San Diego. Previously, she earned a Bachelor of Fine Arts degree in painting from the University of Hawaii at Manoa, and a Master's degree in educational technology from San Diego State University. She has been designing and developing educational multimedia materials and games for children and adults since 1992, covering content ranging from science and mathematics to language arts and workplace skills for home and school markets, corporate clients, and funding agencies. Her research interests lie in the design of effective interactive environments for teaching and learning.

Andrea Cavallaro received his MSc (Summa cum Laude) from the University of Trieste, Italy, in 1996, and a PhD from the Swiss Federal Institute of Technology, Lausanne, Switzerland, in 2002, both in electrical engineering. Since 2003, he has been a lecturer at Queen Mary, University of London. His research interests include multimedia signal processing, perceptual semantics, and interactive media computing. Dr. Cavallaro was awarded a research fellowship with British Telecommunications (BT) and a Drapers' Prize for the development of learning and teaching in 2004. He is a member of the program committee of various IEEE and ACM conferences, and he is the author of more than 40 papers, including four book chapters.

Edward Y. Chang received his MS in computer science and PhD in electrical engineering at Stanford University in 1994 and 1999, respectively. Since 2003, he is an associate professor of electrical and computer engineering at the University of California, Santa Barbara. His recent research activities are in the areas of machine learning, data mining, high-dimensional data indexing, and their applications to image databases and video surveillance. Recent research contributions of his group include methods for learning image/video query concepts via active learning with kernel methods, formulating distance functions via dynamic associations and kernel alignment, managing and fusing distributed video-sensor data, and categorizing and indexing high-dimensional image/video information. Professor Chang has served on several ACM, IEEE, and SIAM conference program committees. He co-founded the annual ACM Video Sensor Network Workshop and has co-chaired it since 2003. He will be co-chairing major multimedia

Copyright © 2006, Idea Group Inc. Copying or distributing in print or electronic forms without written permission of Idea Group Inc. is prohibited.

conferences such as MMM 2006 and ACM Multimedia 2006. He serves as an associate editor for *IEEE Transactions on Knowledge and Data Engineering* and *ACM Multimedia Systems Journal*. Professor Chang is a recipient of the IBM Faculty Partnership Award and the NSF Career Award. He is a co-founder of VIMA Technologies, which provides image searching and filtering solutions.

Nicola Cranley is a post-doctoral research fellow at the Dublin Institute of Technology. Her research interests include video streaming, video adaptation, and wireless networks. She has a BSc in applied physics with French from Dublin City University, an MSc in computing for commerce and industry from the Open University, and a PhD in computer science from University College, Dublin.

Martha E. Crosby is a professor and chair in the Department of Information and Computer Sciences at the University of Hawaii. She conducts research in the human use of computing systems, individual differences of users, and the evaluation of innovative educational environments. She has conducted several empirical studies of reading comprehension and problem solving and is interested in developing user models and in the evaluation of human use of computer interfaces for educational applications. The long-range goal of her research is to gain insights and knowledge that will be useful in the design of human-computer interfaces in educational software systems.

Panagiotis Germanakos is a PhD candidate of the Faculty of Communication and Media Studies of the National and Kapodistrian University of Athens with research interests in Web adaptation and personalization environments and systems based on user profiling encompassing, among others, visual attention and cognitive psychology (mental and emotional) processes, implemented on desktop and mobile/wireless platforms. He obtained a master's degree in international marketing management, a bachelor's degree in computer science and an HND Diploma of Technician Engineer in the field of computer studies. He is also a research associate of the Department of Computer Science of the University of Cyprus, collaborating on nationally- and internationally-funded projects in the areas of e-government and e-health. Additionally, he has over four years of experience in the provision of consultancy of large-scaled IT solutions and implementations in the business sector.

Kostas Giannakis is an IT researcher at the Foundation of the Hellenic World (Athens, Greece). He received a PhD in computing science from Middlesex University (London, UK) and also holds an MSc in cognitive science and intelligent computing from the University of Westminster (London, UK). His research interests include, but are not limited to, novel user interfaces, interaction design for the disabled, information visualization, computer music, and multimedia. At present, he is actively involved in various projects incorporating state-of-the-art computing for the promotion and dissemination of cultural and historical content. More information is available at http://www.mixedupsenses.com.

Copyright © 2006, Idea Group Inc. Copying or distributing in print or electronic forms without written permission of Idea Group Inc. is prohibited.

Stephen R. Gulliver is a lecturer in the School of Information Systems, Computing, and Mathematics at Brunel University, West London. He received a BEng (Honors) degree in micro-electronics, an MSc degree (distributed information systems), and a PhD in 1999, 2001, and 2004, respectively. His research interests cover human factors and include the perceptual, usability, accessibility, and information acquisition aspects of computer and multimedia systems. Current studies incorporate eye-tracking, 3D model realization, attention analysis, display adaptation, and C4I systems (with particular focus on situational awareness). His personal interests include: classic VW beetles, films, reading, drumming, wine, and gardening.

Kate Hone is a senior lecturer at Brunel University, School of Information Systems, Computing, and Mathematics, where she is a member of the VIVID Centre. After studying psychology, work design, ergonomics, and interactive dialogue systems at Birmingham University, she went on to lecture on human factors at the University of Nottingham. Her research centers on the human aspects of computing. She is interested in the design of usable interfaces for computer systems, particularly those using recognition technology. She has done most of her research on the design and use of systems using speech recognition technology and also on teleworking. She also heads a project investigating the human factors issues of "affective" or emotional computing.

Curtis S. Ikehara is an assistant researcher in the Information and Computer Sciences Department of the University of Hawaii, Manoa. Currently, he is working on several projects related to user modeling in an emergency environment. The recent DARPA project (2002-2003) focused on using physiological sensors with respect to applications in augmented cognition and the identification of cognitive processes. Dr. Ikehara and Dr. Crosby have developed a novel sensor that measures the hand and finger pressures applied to a computer mouse, and have applied for a patent for its use as a biometric device.

Slava Kalyuga holds a PhD in cognition and instruction from the University of New South Wales (1998). He specializes in evidence-based instructional design principles for multimedia learning, diagnostic assessment of organized knowledge structures in complex domains, and adaptive e-learning applications. His research papers on cognitive load effects relative to levels of learner expertise (the expertise reversal effect), the redundancy effect in multimedia learning, and rapid diagnostic assessment techniques were published by leading international journals in the field. He contributed to the *Cambridge Handbook of Multimedia Learning*. His current research interests are in optimizing cognitive load in adaptive learning environments using embedded dynamic monitoring of levels of learner performance.

Michael May has a background in psychology and cultural sociology from the University of Copenhagen (1985), and a PhD from 1994 on the relation of the logic and semiotics of C.S. Peirce to modern cognitive semantics. Dr. May has been employed by RISOE National Lab at the Department for Systems Analysis (in the Cognitive Systems

Copyright © 2006, Idea Group Inc. Copying or distributing in print or electronic forms without written permission of Idea Group Inc. is prohibited.

Engineering group) from 1991 to 1994, where he worked with semiotic aspects of graphics and multimedia communication. In 1998-2003, he worked within the field of maritime human factors for the former Danish Maritime Institute (now a part of Force Technology) and within the field of human-machine interaction in the former Centre for Human-Machine Interaction. Main parts of his research deal with semiotic and conceptual foundations of human-machine interaction. In 1994-98 and again from 2003-present, he has also been working with problems of conceptual understanding in engineering education and with instructional design. He is presently employed by the Technical University of Denmark, where he works as an educational consultant at the LearningLab DTU. He is also designing and teaching a course on human-machine interaction in safety-critical domains, together with colleagues from Ørsted-DTU.

Constantinos Mourlas is an assistant professor in the National and Kapodistrian University of Athens (Greece), Department of Communication and Media Studies since 2002. He was born in Athens (Greece) in 1966. He graduated from the University of Crete in 1988 with a diploma in computer science and obtained his PhD from the Department of Informatics, University of Athens in 1995. In 1998, he was an ERCIM fellow for post-doctoral studies through research. He was employed as lecturer at the University of Cyprus, Department of Computer Science, from 1999-2002. His initial research work has been focused on distributed multimedia systems, quality of service issues, streaming media, and the Internet. His current main research interest is the development of environments that provide adaptive and personalized context to the Internet users according to their needs, preferences, and cognitive characteristics. He currently coordinates a national funded research project on Web personalization.

Liam Murphy is a senior lecturer in computer science at University College Dublin, where he is director of the Performance Engineering Laboratory. His current research projects involve mobile and wireless systems, computer network convergence issues, and Web services performance issues. Dr. Murphy has a BE in electrical engineering from University College Dublin, and an MSc and a PhD in electrical engineering and computer sciences from the University of California, Berkeley.

Cecilia Sik-Lányi is an associate professor at the University of Veszprém, Department of Image Processing and Neurocomputing. She received an MSc in mathematics and computer science at the József Attila University, Szeged. She obtained She obtained a doctoral degree in physical chemistry and a PhD in computer science. She reads and leads laboratory studies on multimedia and virtual reality. Her research area is in multimedia and virtual reality in education and rehabilitation.

Klara Nahrstedt is a full professor at the University of Illinois at Urbana-Champaign, Computer Science Department. Dr. Nahrstedt received her BA in mathematics from Humboldt University, Berlin, in 1984, and MSc degree in numerical analysis from the same university in 1985. She was a research scientist in the Institute for Informatik in Berlin until 1990. In 1995, she received her PhD from the University of Pennsylvania in the

Copyright © 2006, Idea Group Inc. Copying or distributing in print or electronic forms without written permission of Idea Group Inc. is prohibited.

Department of Computer and Information Science. She is the member of IEEE, ACM, and SIG Multimedia.

Adam Tilinger is a PhD student at the University of Veszprém, Department of Image Processing and Neurocomputing. His research area is virtual reality in rehabilitation and virtual reality ergonomics.

Greger Wikstrand is a PhD student at Umeå University from which he obtained the Licenciate of Engineering degree in 2003. He was born in Uppsala, Sweden, in 1972. He obtained the Master of Science in engineering physics degree from Uppsala University in 1998. He was a doctoral student at the University of Linköping during 1998 and 1999. In 2000, he was with Procter & Gamble in Stockholm, Sweden. During 2001-2002, he worked as a research project manager at Ericsson AB where, among other things, he was responsible for the development of Testplats Botnia — a platform for testing mobile services and applications. He is a student member of the ACM and the IEEE.

Stefan Winkler received an MSc in electrical engineering from the University of Technology in Vienna, Austria, in 1996, and a PhD in electrical engineering from the Swiss Federal Institute of Technology (EPFL) in Lausanne in 2000 for research on video quality measurement. He also spent one year at the University of Illinois at Urbana-Champaign as a Fulbright student. Dr. Winkler has worked as a post-doctoral fellow at EPFL and as an assistant professor at the University of Lausanne. In 2001, he co-founded Genimedia (now Genista Corporation), a company developing perceptual quality metrics for multimedia applications, where he is currently chief scientist. He has published more than 30 papers on vision modeling and quality assessment and is the author of a book on digital video quality.

Copyright © 2006, Idea Group Inc. Copying or distributing in print or electronic forms without written permission of Idea Group Inc. is prohibited.

Index

Copyright © 2006, Idea Group Inc. Copying or distributing in print or electronic forms without written permission of Idea Group Inc. is prohibited.

Copyright © 2006, Idea Group Inc. Copying or distributing in print or electronic forms without written permission of Idea Group Inc. is prohibited.

E

e-motional 110
early vision 207
Educational Modelling Language (EML) 71
educational software 133
electrocutaneous stimulators 84
electrorheological devices 84
embodied conversational agents (ECAs) 112
embodied knowledge 133, 134
emergent properties 56
emergent semantics 61
emotion in recognition in computing (ERIC) project 117
emotional multimedia 111
emotional processing 300
emotional recognition 110
entry point 295
equal average bit rate (EABR) 251
expert-novice difference 206
expertise reversal effect 209
exploit 31
explore 30
extroversion 300

F

face detection 7
facial action coding scheme (FACS) 115
facial actions 116
feature space 27
feature structures 60
feature-based multimedia semantics 47, 60
feature-based taxonomy 66
feature-structures 60
FEELitPaddle.exe 139
field dependence/independence 190
field-independent learners 190
filtering 296
fire-alarm simulation 235
first-step diagnostic method 213
force feedback 133, 137
force feedback mouse 139
forced choice methodology 256
formal concept analysis (FCA) 62
formal context 61

frame-dropping filter 247
frequency filter 247
front-end layer 295
full-reference quality metric 12

G

galvanic skin resistance (GSR) 98
Game 118
gaze 116
generative 27
geographical information systems (GIS) 56
gestic media 54
granularity 67, 156
graph comprehension 73
graphical content 47
graphical design 49
graphical media 53

H

hall monitor 10
hand preference 225
haptic 54, 83, 116, 132
haptic feedback 132
haptic interfaces 135
haptic media 54
Hasse-diagram 64
highway 10
hit-rate 39, 40
holistic features 36
holistic perceptual feature 26
human factors 272
human visual system (HVS) 2, 249
human-computer interaction (HCI) 111
hybrid rate control 247

I

iconic 59
illustrations 48
image annotation 21
image schemata 55
implementation considerations 298
implicit 55
independent component analysis (ICA) 32
indexical 59
individual differences 188

Copyright © 2006, Idea Group Inc. Copying or distributing in print or electronic forms without written permission of Idea Group Inc. is prohibited.

Copyright © 2006, Idea Group Inc. Copying or distributing in print or electronic forms without written permission of Idea Group Inc. is prohibited.

Copyright © 2006, Idea Group Inc. Copying or distributing in print or electronic forms without written permission of Idea Group Inc. is prohibited.

Copyright © 2006, Idea Group Inc. Copying or distributing in print or electronic forms without written permission of Idea Group Inc. is prohibited.

Copyright © 2006, Idea Group Inc. Copying or distributing in print or electronic forms without written permission of Idea Group Inc. is prohibited.

Copyright © 2006, Idea Group Inc. Copying or distributing in print or electronic forms without written permission of Idea Group Inc. is prohibited.

Experience the latest full-text research in the fields
of Information Science, Technology & Management

InfoSci-Online

InfoSci-Online is available to libraries to help keep students,
faculty and researchers up-to-date with the latest research in
the ever-growing field of information science, technology, and
management.

The InfoSci-Online collection includes:
- Scholarly and scientific book chapters
- Peer-reviewed journal articles
- Comprehensive teaching cases
- Conference proceeding papers
- All entries have abstracts and citation information
- The full text of every entry is downloadable in .pdf format

Some topics covered:
- Business Management
- Computer Science
- Education Technologies
- Electronic Commerce
- Environmental IS
- Healthcare Information Systems
- Information Systems
- Library Science
- Multimedia Information Systems
- Public Information Systems
- Social Science and Technologies

**InfoSci-Online
features:**
- Easy-to-use
- 6,000+ full-text
  entries
- Aggregated
- Multi-user access

"...The theoretical bent
of many of the titles
covered, and the ease
of adding chapters to
reading lists, makes it
particularly good for
institutions with strong
information science
curricula."
— Issues in Science and
Technology Librarianship

To receive your free 30-day trial access subscription contact:
Andrew Bundy
Email: abundy@idea-group.com • Phone: 717/533-8845 x29
Web Address: www.infosci-online.com

InfoSci⊞Online
Full Text · Cutting Edge · Easy Access

A PRODUCT OF IDEA GROUP INC.
Publishers of Idea Group Publishing, Information Science Publishing, CyberTech Publishing, and IRM Press

infosci-online.com

Introducing

IGI Teaching Case Collection

The new **IGI Teaching Case Collection** is a full-text database containing hundreds of teaching cases related to the fields of information science, technology, and management.

Key Features

- Project background information
- Searches by keywords and categories
- Abstracts and citation information
- Full-text copies available for each case
- All cases are available in PDF format
- Cases are written by IT educators, researchers, and professionals worldwide

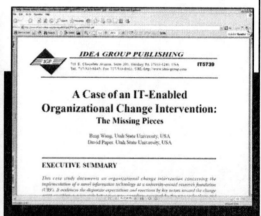

IDEA GROUP PUBLISHING

IT5739

A Case of an IT-Enabled Organizational Change Intervention: The Missing Pieces

Bing Wang, Utah State University, USA
David Paper, Utah State University, USA

EXECUTIVE SUMMARY

View each case in full-text, PDF form. **Hundreds of cases** provide a **real-world edge** in information technology classes or research!

The Benefits of the IGI Teaching Case Collection

- Frequent updates as new cases are available
- Instant access to all full-text articles saves research time
- No longer necessary to purchase individual cases

For More Information Visit
www.igi-online.com

Recommend to your librarian today!

A Product Of

IDEA GROUP INC.